T0314192

Wireless Semantic Communications

Wireless Semantic Communications

Concepts, Principles, and Challenges

Edited by

Yao Sun
University of Glasgow
Glasgow, UK

Lan Zhang
Michigan Technological University
Houghton, USA

Dusit Niyato
Nanyang Technological University
Singapore

Muhammad Ali Imran
University of Glasgow
Glasgow, UK

This edition first published 2025

Registered Offices
John Wiley & Sons, Inc., 111 River Street, Hoboken, NJ 07030, USA
John Wiley & Sons Ltd, The Atrium, Southern Gate, Chichester, West Sussex, PO19 8SQ, UK

For details of our global editorial offices, customer services, and more information about Wiley products visit us at www.wiley.com.

Wiley also publishes its books in a variety of electronic formats and by print-on-demand. Some content that appears in standard print versions of this book may not be available in other formats.

Library of Congress Cataloging-in-Publication Data applied for:

Hardback: 9781394223305

Cover Design: Wiley
Cover Image: © Yuichiro Chino/Getty Images

Set in 9.5/12.5pt STIXTwoText by Straive, Chennai, India
Printed and bound by CPI Group (UK) Ltd, Croydon, CR0 4YY
C9781119909378_190924

Contents

List of Contributions

Xuyang Chen
College of Electronics and Information Engineering
Shenzhen University
Shenzhen
China

Runze Cheng
James Watt School of Engineering
University of Glasgow
Glasgow
Lanarkshire
UK

Xiangyi Deng
Glasgow College
University of Electronic Science and Technology of China
Chengdu
Sichuan
China

Hongyang Du
School of Computer Science and Engineering
Nanyang Technological University
Singapore

Daquan Feng
College of Electronics and Information Engineering
Shenzhen University
Shenzhen
China

Lei Feng
The State Key Laboratory of Networking and Switching Technology
Beijing University of Posts and Telecommunications
Beijing
China

Biqian Feng
Department of Electronic Engineering
Shanghai Jiao Tong University
China

Chenyuan Feng
Eurecom
France

Zhipeng Gao
State Key Laboratory of Networking and Switching Technology
Beijing University of Posts and Telecommunications
Beijing
China

Shuaishuai Guo
School of Control Science and Engineering
Shandong University
Jinan
China

and

Shandong Key Laboratory of Wireless Communication Technologies
Shandong University
Jinan
China

Qi He
National Key Laboratory of Science and Technology on Communications
University of Electronic Science and Technology of China
Chengdu
China

Chongwen Huang
College of Information Science and Electronic Engineering
Zhejiang University
Hangzhou
Xihu Region
China

Muhammad Ali Imran
James Watt School of Engineering
University of Glasgow
Glasgow
Lanarkshire
UK

Rongpeng Li
College of Information Science and Electronic Engineering
Zhejiang University
Hangzhou
China

Wenjing Li
The State Key Laboratory of Networking and Switching Technology
Beijing University of Posts and Telecommunications
Beijing
China

Chengsi Liang
James Watt School of Engineering
University of Glasgow
Glasgow
Lanarkshire
UK

Yijing Lin
State Key Laboratory of Networking and Switching Technology
Beijing University of Posts and Telecommunications
Beijing
China

Yijie Mao
School of Information Science and Technology
ShanghaiTech University
Shanghai
China

Fei Ni
College of Information Science and Electronic Engineering
Zhejiang University
Hangzhou
China

Dusit Niyato
School of Computer Science and Engineering
Nanyang Technological University
Singapore

Yao Sun
James Watt School of Engineering
University of Glasgow
Glasgow
Lanarkshire
UK

Bingyan Wang
College of Information Science and Electronic Engineering
Zhejiang University
Hangzhou
China

Bohao Wang
College of Information Science and Electronic Engineering
Zhejiang University
Hangzhou
Xihu Region
China

Jiacheng Wang
School of Computer Science and Engineering
Nanyang Technological University
Singapore

Yanhu Wang
School of Control Science and Engineering
Shandong University
Jinan
China

and

Shandong Key Laboratory of Wireless Communication Technologies
Shandong University
Jinan
China

Yiwen Wang
School of Information Science and Technology
ShanghaiTech University
Shanghai
China

Le Xia
James Watt School of Engineering
University of Glasgow
Glasgow
Lanarkshire
UK

Xiang-Gen Xia
Department of Electrical and Computer Engineering
University of Delaware
Newark, DE
USA

Ruopeng Xu
College of Information Science and Electronic Engineering
Zhejiang University
Hangzhou
Xihu Region
China

Zhaohui Yang
College of Information Science & Electronic Engineering
Zhejiang University
Hangzhou
Xihu Region
China

Honggang Zhang
College of Information Science and Electronic Engineering
Zhejiang University
Hangzhou
China

and

Zhejiang Lab
Hangzhou
China

Xuefei Zhang
School of Information and Communication Engineering
Beijing University of Posts and Telecommunications
Beijing
China

Zhifeng Zhao
College of Information Science and Electronic Engineering
Zhejiang University
Hangzhou
China

and

Zhejiang Lab
Hangzhou
China

Jiakang Zheng
School of Electronic and Information Engineering
Beijing Jiaotong University
Beijing
China

Yu Zhou
The State Key Laboratory of Networking and Switching Technology
Beijing University of Posts and Telecommunications
Beijing
China

Preface

Tremendous traffic demands have increased in current wireless networks to accommodate for the upcoming pervasive network intelligence with a variety of advanced smart applications. In response to the ever-increasing data rates along with stringent requirements for low latency and high reliability, it is foreseeable that available communication resources like spectrum or energy will gradually become scarce in the upcoming years. Combined with the almost insurmountable Shannon limit, these destined bottlenecks are, therefore, motivating us to hunt for bold changes in the new design of future networks, i.e., making a paradigm shift from bit-based traditional communication to context-based semantic communication.

The concept of semantic communication was first introduced by Weaver in his landmark paper, which explicitly categorizes communication problems into three levels, including the technical problem at the bit level, the semantic problem at the semantic level, and the effectiveness problem at the information exchange level. Nowadays, the technical problem has been thoroughly investigated in the light of classical Shannon information theory, while the evolution toward semantic communication is just beginning to take shape, with the core focus on meaning delivery rather than traditional bit transmission.

Concretely, semantic communication first refines semantic features and filters out irrelevant content by encoding the semantic information (i.e., semantic encoding) at the source, which can greatly reduce the number of required bits while preserving the original meaning. Then, the powerful semantic decoders are deployed at the destination to accurately recover the source meaning from received bits (i.e., semantic decoding), even if there are intolerable bit errors at the syntactic level. Most importantly, through further leveraging matched background knowledge with respect to the observable messages between source and destination, users can acquire efficient exchanges for the desired information with ultralow semantic ambiguity by transmitting fewer bits.

While semantic communication offers these attractive and valuable benefits, it also faces many challenges. For example, when considering the computing limitation of terminal devices, personalized background knowledge, as well as unstable wireless channel conditions, how to design semantic encoder and decoder should be a challenging issue. Moreover, in the networking layer, it is nontrivial to seek the optimal wireless resource management strategy to optimize its overall network performance in a semantics-aware manner. Therefore, before fully enjoying the superiorities of semantic communications, this book would like to explore the following fundamental issues: (i) How many benefits can be achieved by using semantic communication? (ii) How much cost (mainly consumed resources) is incurred to guarantee the required performance, such as semantic ambiguity? and (iii) How can we maximize the benefits of semantic communications applied to different wireless networks with constraints of cost?

This book will explore recent advances in the theory and practice of semantic communication. In detail, the book covers the following aspects:

(1) Principles and fundamentals of semantic communication.
(2) Transceiver design of semantic communications.
(3) Resource management in semantic communication networks.
(4) Semantic communication applications to vertical industries and some typical communication scenarios.

Chapter 1 delves into the transceiver design for semantic communications. Specifically, we first summarize established designs for key components in semantic communications, with a key focus on the knowledge base and semantic encoders/decoders crucial for single-user semantic communications. Our discussion then extends to multiuser SC, specifically emphasizing the synergy between various multiple-access schemes and semantic communications, including orthogonal multiple access (OMA), space-division multiple access (SDMA), non-orthogonal multiple access (NOMA), rate-splitting multiple access (RSMA), and model-division multiple access (MDMA). In the end, we explore various applications of semantic communications and analyze the potential alterations in transceiver design required for these applications.

Chapter 2 studies semantic communications from a networking perspective, particularly focusing on the upper layer. Our primary objective is to investigate optimal wireless resource management strategies within the semantic communications-enabled network (SC-Net) to enhance overall network performance in a semantics-aware manner. This entails addressing the unique challenge of ensuring background knowledge alignment between multiple mobile users (MUs) and multitier base stations (BSs). Efficient resource management remains paramount within the SC-Net, offering numerous benefits such as guaranteeing high-quality SemCom services and enhancing spectrum utilization. By devising effective resource allocation strategies, we aim to optimize network performance and facilitate seamless communication within the SC-Net ecosystem.

Chapter 3 innovatively proposes the transformation theory of the semantic domain–spatial domain, projecting the knowledge graph onto a three-dimensional tensor in the spatial domain. By mapping the entity's semantic ambiguity with the intensity of discrete points, the knowledge graph is reconstructed from background knowledge libraries and three-dimensional tensors at the receiving end. Additionally, this chapter proposes a graph-to-graph semantic similarity (GGSS) metric based on graph optimal transport theory to evaluate the similarity of semantic information before and after transmission, as well as a semantic-level image-to-image semantic similarity (IISS) metric that aligns with human perception. Finally, we demonstrate the effectiveness and rationality of the framework through simulations.

Chapter 4 first introduces the problem that neural networks in semantic communication are very vulnerable to adversarial attacks, then proposes robust semantic communication systems for image and speech transmission. Meanwhile, this chapter discusses the privacy issue caused by the difference of knowledge bases between transmitter and receiver in semantic communication and proposes a knowledge discrepancy-oriented privacy protection (KDPP) method for semantic communication to reduce the risk of privacy leakage while retaining high data utility.

Chapter 5 investigates the means of knowledge learning in semantic communication with a particular focus on the utilization of Knowledge Graphs (KGs). Specifically, we first review existing efforts that combine semantic communication with knowledge learning. Subsequently, we introduce a KG-enhanced semantic communication system, wherein the receiver is carefully calibrated to leverage knowledge from its static knowledge base for ameliorating the decoding performance. Contingent upon this framework, we further explore potential approaches that can empower the system to operate in evolving an knowledge base more effectively. Furthermore, we investigate the possibility of integration with large language models (LLMs) for data augmentation, offering additional perspective into the potential implementation means of semantic communication. Extensive numerical results demonstrate that the proposed framework yields superior performance on top of the KG-enhanced decoding and manifests its versatility under different scenarios.

Chapter 6 presents a novel framework called VISTA (VIdeo transmission over Semantic communicaTion Approach) for video transmission by exploiting semantic communications. VISTA comprises three key modules: the semantic segmentation module and the frame interpolation module, responsible for semantic encoding and decoding, respectively, and the joint source-channel coding (JSCC) module, designed for SNR-adaptive wireless transmission.

Chapter 7 proposes a content-aware robust semantic communication framework for image transmission based on generative adversarial networks (GANs). Specifically, the accurate semantics of the image are extracted by the semantic encoder and divided into two parts for different downstream tasks: regions of interest (ROI) and regions of non-interest (RONI). By reducing the quantization accuracy of RONI, the amount of transmitted data volume is reduced significantly. During the transmission process of semantics, a signal-to-noise ratio (SNR) is randomly initialized, enabling the model to learn the average noise distribution. The experimental results demonstrate that by reducing the quantization level of RONI, transmitted data volume is reduced up to 60.53% compared to using globally consistent quantization while maintaining comparable performance to existing methods in downstream semantic segmentation tasks. Moreover, our model exhibits increased robustness with variable SNRs.

Chapter 8 first proposes an integrated framework for bridging meanings of semantic information in the Metaverse to achieve efficient interaction between physical and virtual worlds. This chapter then presents a Zero Knowledge Proof-based verification mechanism to secure the authenticity of the extracted information. This chapter also introduces a diffusion model-based resource allocation mechanism to maximize the utility of resources. Simulation results are presented to validate the authenticity and efficiency of the proposed mechanisms. Additionally, this chapter discusses future directions to further advance SemCom in the metaverse.

Chapter 9 discusses the method of leveraging large language model (LLM) to assist semantic communication systems. Specifically, this chapter first discusses leveraging LLM to define the semantic loss of communications, based on which a signal-shaping method is proposed to minimize the semantic loss for semantic communications with a few message candidates. Then, this proposed a more generalized method to quantify the semantic importance of a word/frame using LLM and investigate semantic importance-aware communications (SIAC) to reliably convey the semantics with limited communication and network resources. Finally, this chapter points out the future direction of using LLM for semantic correction. Experiments are conducted to verify the effectiveness of leveraging LLM to assist semantic communications.

Chapter 10 explores cutting-edge advancements in RIS for semantic communication. It delves into three pivotal areas: optimizing beamforming in RIS-aided systems for enhanced communication in complex digital environments like the Metaverse; employing physical layer strategies for robust privacy protection in semantic communication systems; and leveraging deep learning for advanced interpreting and prioritizing data and encoding and decoding semantic transmission in wireless communications. These topics collectively highlight the potential of RIS in transforming communication paradigms, emphasizing efficiency, security, and intelligent data processing in semantic communication.

Yao Sun
Lan Zhang
Dusit Niyato
Muhammad Ali Imran
April 2024

1

Intelligent Transceiver Design for Semantic Communication

Yiwen Wang[1], *Yijie Mao*[1], *and Zhaohui Yang*[2]

[1] *School of Information Science and Technology, ShanghaiTech University, Shanghai, China*
[2] *College of Information Science & Electronic Engineering, Zhejiang University, HangZhou, Xihu Region, China*

1.1 Knowledge Base

In this section, we first introduce the basic structure of a knowledge base (KB) and then describe the development of KB-assisted semantic communication (SC) systems. After that, we introduce multimodal SC systems that are assisted by KB.

1.1.1 Basic Structures of the Knowledge Base

The KB plays a vital role in supporting the process of SC. It involves real-world information such as facts, relationships, and possible reasoning methods that can be understood, recognized, and learned by all participants in the communication [Shi et al., 2021].

One of the typical KBs is the knowledge graph (KG), which contains entities as nodes and relationships as edges [Strinati and Barbarossa, 2020]. The knowledge hidden in the KG is embedded into a continuous vector space [Wang et al., 2017]. One popular KG design is to describe the relationship between two entities using triples. These triples are typically structured as (head entity, relation, and tail entity) or (entity, relation, and attribute). Figure 1.1 shows two different famous KG-based KBs, namely DBpedia and Oxford Art. Both Figure 1.1a and b show detailed information about the same artist, Eugenio Bonivento. Different entities connected with edges are tagged by their relationship. In the DBpedia-based KB shown in Figure 1.1a, information about Eugenio Bonivento is stored using text triples. For example, the triple (Eugenio Bonivento, Date of birth, 1880-06-08) in Figure 1.1a represents the meaning that Eugenio Bonivento's date of birth is 1880-06-08. Moreover, an entity can be connected with more than one entity or attribute. In the Oxford Art KG shown in Figure 1.1b, the artist is represented by an id "OOUNzqQ2." Unlike DBpedia, which utilizes text triples to store information about Eugenio Bonivento, the Oxford Art KG incorporates entities with multimodal data, including images and texts. Such multimodal data-based KB facilitates multimodal SC and provides ways to understand information from different modal information.

1.1.2 KB-Assisted Single-Modal SC

KB can help to assist SC systems in many ways. However, it is worth mentioning that some simple transmission frameworks in SC do not involve KB. Instead, they may take advantage of the

Wireless Semantic Communications: Concepts, Principles, and Challenges, First Edition.
Edited by Yao Sun, Lan Zhang, Dusit Niyato, and Muhammad Ali Imran.

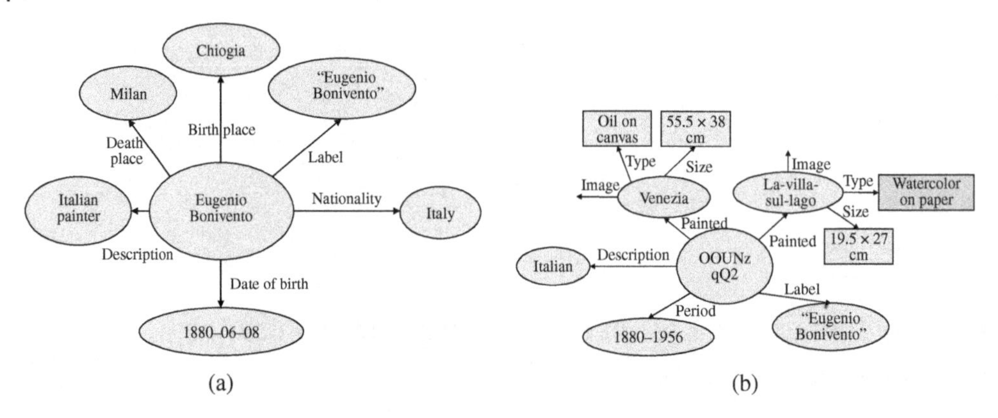

Figure 1.1 Two examples of KG systems based on KB. (a) DBpedia KB and (b) Oxford Art KB. Source: Adapted from [Collarana et al., 2017].

background knowledge during the training process, but they do not construct a KB for assistance during the encoding and decoding process when applying the trained SC [Xie et al., 2021b; Hu et al., 2022; Jiang et al., 2022a; Zhang et al., 2023a]. The structures of both the training and inference phases for an SC system without a KB are shown in Figure 1.2. In this example, neural networks are deployed in the semantic encoding and semantic decoding blocks, and the background knowledge is used in the training phase only to get a fine-tuned neural network. Once the training phase is complete, the background knowledge is unused, and only the pretrained neural networks work in the inference phase.

As mentioned previously, a widely recognized KB definition is an underlying set of facts, assumptions, and rules to help a computer system solve a problem. In general, both the transmitter and receiver have their own KBs, called "source KB" and "destination KB." Those two KBs typically share a part of common knowledge while keeping a part of private knowledge. With the help of the common knowledge in KBs, the reasoning ability of the semantic encoder and decoder is enhanced. However, the difference between the two KBs would inevitably cause semantic noise since the same message can be understood differently with different background knowledge. This semantic noise can be eliminated by sharing the knowledge between the source and destination KBs [Wang et al., 2017].

As shown in Figure 1.3, a structured KB assists an SC system during both the training and inference phases. At the transmitter, the information is transformed into triples, and the triples are subsequently transmitted through conventional communication approaches and reconstructed using the KB at the decoder.

1.1.3 KB-Assisted Multimodal SC

Typically, different modalities of information require different types of KB. As mentioned earlier, various types of modalities exist, such as text, image, speech, or video [Luo et al., 2022a]. With the development of SC, it becomes imperative to design KBs capable of bridging different modalities.

Li et al. [2022] designed and introduced a novel KB that is adaptable across various modalities. To extract semantic knowledge and build a cross-modal KG (CKG), different knowledge extraction models are designed for different modalities after clustering multimodal samples. Then, knowledge entities, relations, and attributes are identified through similarity measurement from the pairwise similarity scores obtained from the previous step to construct a cross-modal KG. Through semantic extraction, the original message is mapped into multiple triples, which are then transmitted using traditional communication ways and restored at the receiver's side [Zhou et al., 2022]. It is noted

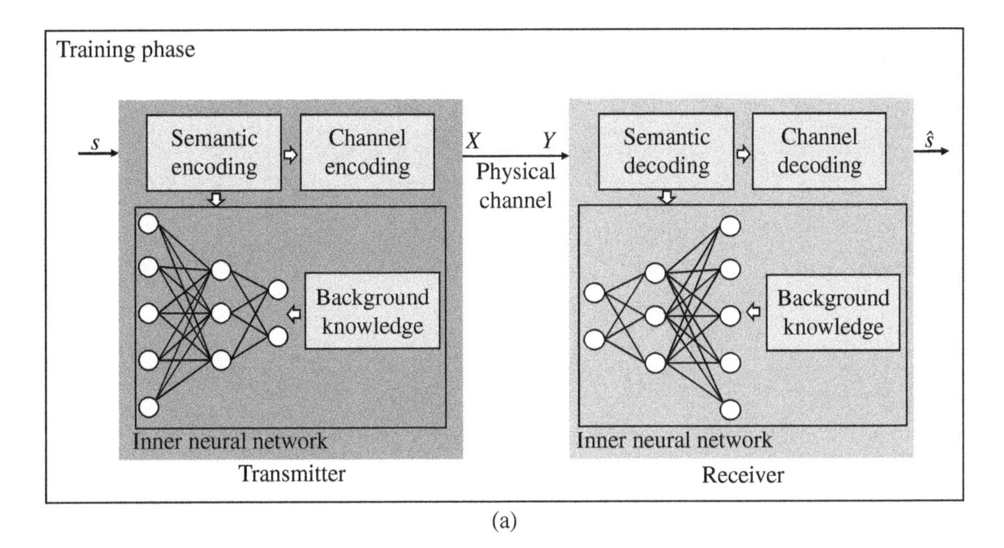

(a)

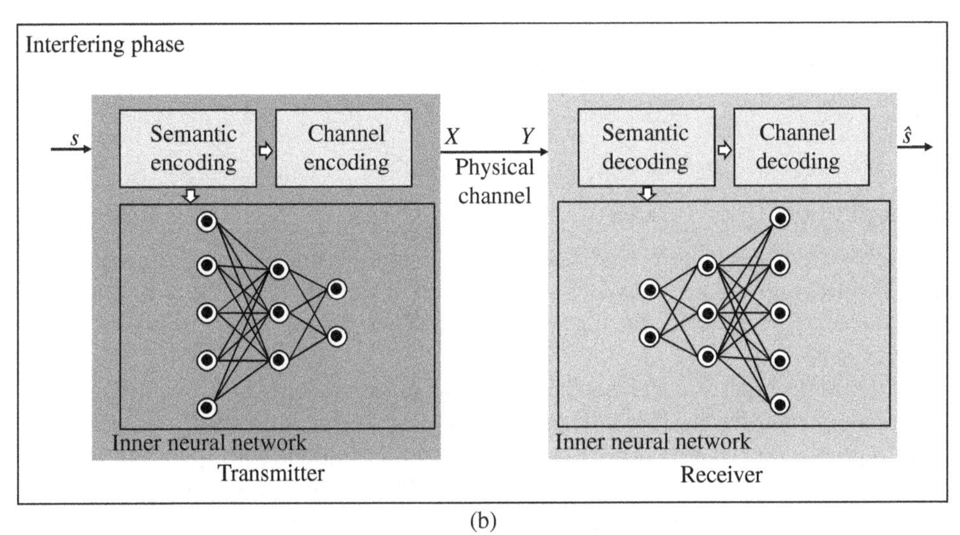

(b)

Figure 1.2 An SC system without a KB. (a) Training phase of an SC system without a KB and (b) interfering phase of an SC system without a KB.

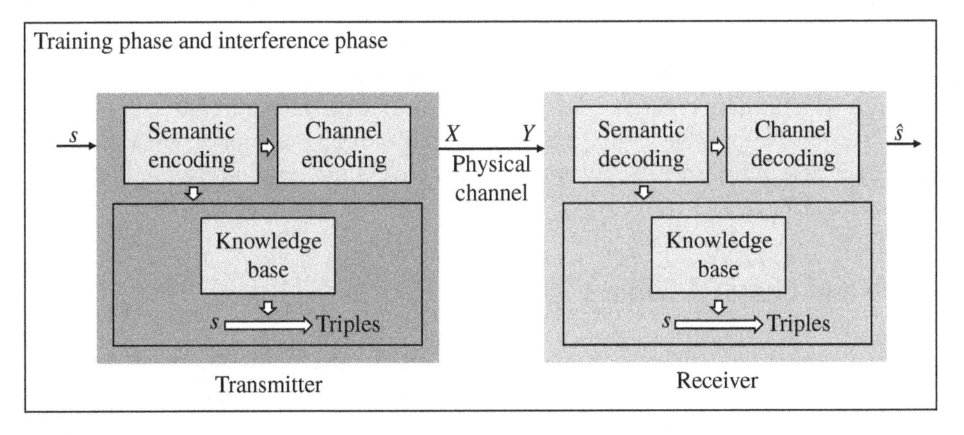

Figure 1.3 An SC system with KBs at both transmitter and receiver.

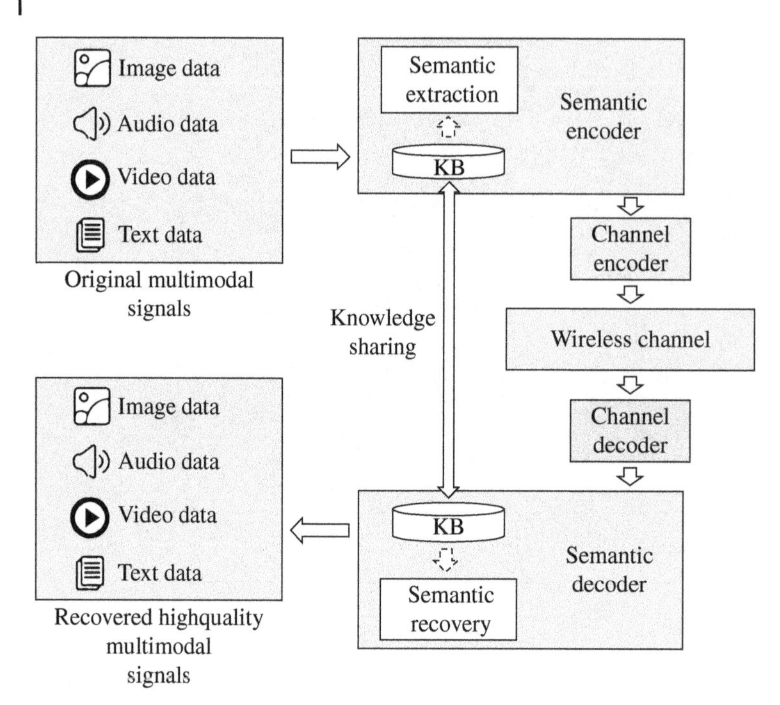

Figure 1.4 A multi-modal SC system assisted by KBs.

that the restored message is not exactly the same as the original message, but they are semantically equivalent. Moreover, this work proposed a CKG fusion to aggregate multi-source, heterogeneous, and diverse knowledge obtained by new knowledge extraction and existing CKGs. Based on those CKGs, the semantic encoder can acquire essential signal patches, thereby improving the quality of the recovered signals.

Jiang et al. [2023a] proposed a personalized large language model (LLM)–based KB (LKB) to extract semantics from the data and reconstruct the data. It should be noted that this LKB is dedicated to processing text data. The proposed large artificial intelligence (AI) model-based multimodal SC (LAM-MSC) framework is capable of handling multimodal data such as image, audio, video, and text. The key point to achieve multimodal data processing is to transform different modal data into text data before semantic extraction based on LKB and data transmission. After data transmission, the data are first recovered based on LKB, and then further recovered based on multimodal alignment (MMA).

Both Li et al. [2022] and Jiang et al. [2023a] proposed KB-assisted multimodal SC systems following the process shown in Figure 1.4, which handle the multimodal data by extracting the hidden semantics. The primary difference between them lies in the semantic extraction part of the semantic encoder, where Li et al. [2022] employed the CKG to get the implicit semantics, and Jiang et al. [2023a] first transformed other non-text modal data into text data before extracting semantics using the LKB.

1.2 Source and Channel Coding

In this section, we first introduce the components of the source and channel encoders and decoders in SC. Then, two different transceiver architectures with different levels of integration between source and channel coding are delineated.

1.2.1 Preliminaries of Source and Channel Encoders/Decoders

In the traditional communication process, there are two essential components at the transceiver: source coding and channel coding. Source coding, typically implemented with a source encoder at the transmitter, takes the source message and applies encoding techniques to compress the data in order to minimize the number of bits to represent the information while maintaining the effectiveness of communication. The channel coding implemented with a channel encoder introduces redundancy to data before transmission so that the system has the ability for error correction and anti-interference. It can greatly avoid the occurrence of errors during transmission. Corresponding to the source and channel encoders at the transmitter, there are source and channel decoders at the receiver to perform the reverse decoding processes, as illustrated in Figure 1.5.

In the SC system, as Figure 1.6, the source encoder in the traditional communication system is typically replaced by a semantic encoder, while the semantic decoder is used to replace the source decoder. A semantic encoder is able to infer the meaning of the source message and make sure the meaning can be successfully delivered. This contrasts with a conventional source encoder, which typically ignores the meaning of the message and encodes every bit of the source message in order to make sure all message bits can be recovered at the receiver [Shi et al., 2021]. With this alteration, semantic encoding and decoding in the SC system not only extract the semantic information but also enhance the effectiveness of communication by reducing the number of transmission bits.

Empowered by the emerging deep-learning (DL) technologies, semantic encoder, semantic decoder, channel encoder, and channel decoder in the SC system can be implemented by deep neural networks (DNNs). Compared to traditional communication systems, SC enables manipulation of messages at the semantic level. The design of semantic and channel encoders/decoders exploits the semantics of the messages, leading to a deeper and more accurate understanding of the intended message.

1.2.2 Different Transceiver Design in SC

Based on the design of semantic and channel encoders as well as semantic and channel decoders, the SC system can be divided into two categories, as illustrated in Figures 1.7 and 1.8. The first

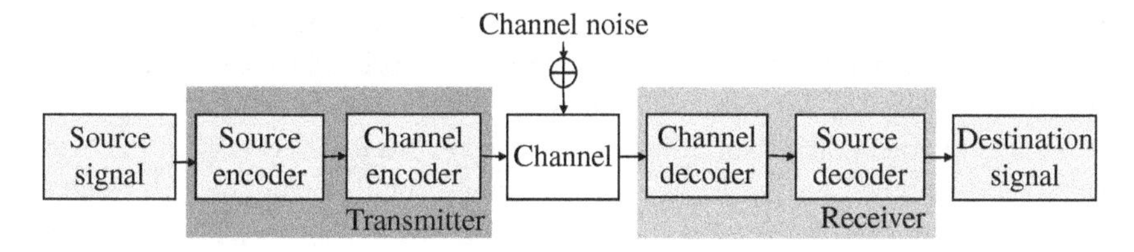

Figure 1.5 The transceiver architecture of a conventional communication system. Source: Adapted from Lan et al. [2021].

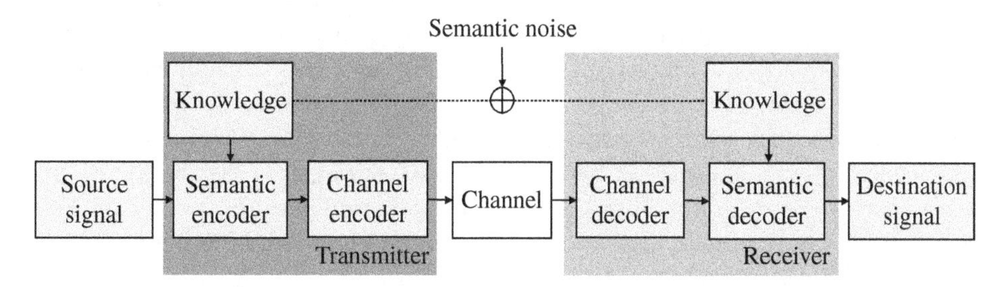

Figure 1.6 The transceiver architecture of an SC system. Source: Adapted from Lan et al. [2021].

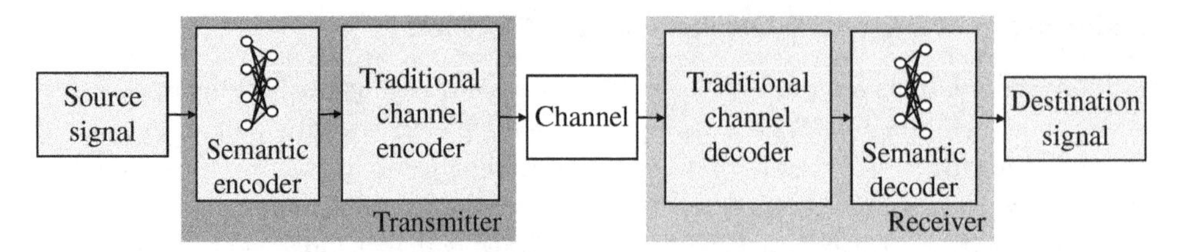

Figure 1.7 The transceiver architecture of an SC with independent source and channel coding.

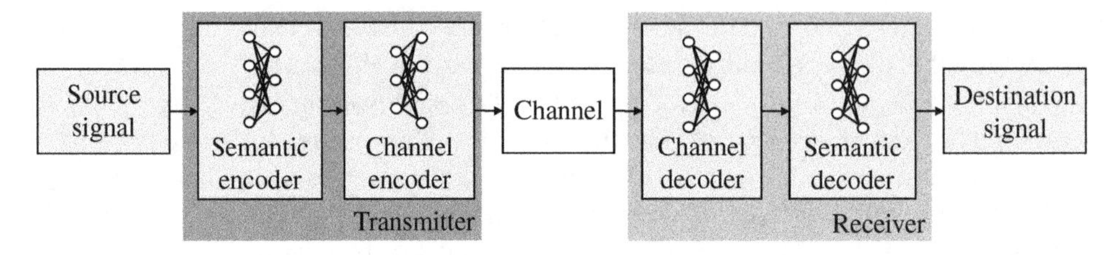

Figure 1.8 The transceiver architecture of an SC with intelligent joint source and channel coding. Source: Adapted from Lan et al. [2021].

type of SC considers the semantic encoder/decoder and channel encoder/decoder as two separate modules, which are independently designed. Typically, the design of semantic encoders and decoders relies on DNNs, and the channel encoders and decoders are implemented in a traditional way. Extracted knowledge from the semantic extraction is processed and transmitted in a more traditional manner. The second type of SC enables joint source–channel coding by designing semantic and channel encoders and decoders using DNNs. These modules are jointly trained in an end-to-end fashion so as to achieve a unified design purpose.

In the first type of SC, as shown in Figure 1.7, the semantic information is first extracted from the original messages and stored as the directional possibility graph (DPG) or an other KG. It should be noted that the KG mentioned here keeps the same structure as the KG mentioned in Section 1.1 but with different content. The KG for KB in Section 1.1 stores all background information that may help to improve the quality of SC, while the KG mentioned in this section stores the semantic information extracted from the messages that need to be transmitted. The transformation from the message to the KG or the counterpart is always based on a neural network, which can be assumed as the semantic encoder and decoder. After that, the semantic information stored in the DPG or other KG is encoded into a bitstream and then transmitted. Many existing SC systems of this type have been proposed, i.e., the probability graph-based semantic information compression system proposed by Zhao et al. [2023], an SC system based on the KG proposed by Jiang et al. [2022b], and another SC assisted by rate-splitting multiple access (RSMA), which uses the KG proposed by Yang et al. [2023b].

In the second type of SC, as shown in Figure 1.8, the transmitter and receiver are typically designed by two separate DNNs. The DNN at the transmitter carries out the roles of semantic and channel encoders, while the DNN at the receiver carries out the roles of both semantic and channel decoders. Some existing works also consider separate DNNs for the channel encoder/decoder and semantic encoder/decoder, which are connected to each other. Typically, all DNNs in the same SC system are trained jointly with the separated but connected KBs. Again, many existing SC systems of this type have been proposed, i.e., the DeepSC system for text transmission proposed by Xie et al. [2021b], the MR DeepSC system for one-to-many communication proposed by Hu et al. [2022], and the DeepWive system, which is an end-to-end video transmission scheme proposed by Tung and Gündüz [2022]. Different neural network designs are adopted in these works; however, the basic architectures of these SC systems remain the same.

1.3 Multiuser SC

Single-user communication refers to a point-to-point transmission mode, where a single transmitter establishes direct communication with a single receiver. Conversely, multiuser communication involves a scenario where multiple users or devices engage in communication within a shared system or network. The majority of studies on SC focus on resolving issues within a single-user SC system, where there is no interference from other users [Bourtsoulatze et al., 2018; Farsad et al., 2018; Xie et al., 2021b; Jiang et al., 2022a; Tung and Gündüz, 2022; Weng and Qin, 2021]. In these works, the transceiver design is centered on communication between a single transmitter and receiver. However, communication systems nowadays usually involve multiple users, demanding an efficient way to distinguish messages from different users. In this section, we focus on multiuser SC, introducing the state-of-the-art approaches to address the interference issue.

1.3.1 Interference Management Approaches in SC

In contrast to the traditional communication systems, SC introduces the concepts of semantics and takes advantage of them to better manage inter-user interference.

In general, there are several approaches to managing inter-user interference in SC. The first approach is employing personalized KBs for individual users. This ensures the absence of shared knowledge among users, allowing each individual to understand only their own messages and thereby facilitating interference-free communication. However, in real multiuser SC systems, the overlap of the KBs among users is inevitable, leading to interference from other users [Zhou et al., 2023].

To manage inter-user interference, some recent works introduced novel models at the transceiver in order to better utilize the semantic information for different users. Hu et al. [2022] introduced DistilBERT, a semantic recognizer built upon the pretrained model. It is designed to distinguish messages from different users based on the emotions of those sentences. Concurrently, Luo et al. [2022b] proposed a method for channel-level information fusion, where signals from different users are merged in wireless channels. Therefore, the signal received at the receiver is a fused single-modal information, eliminating multiuser interference.

Besides the aforementioned two approaches, various multiple access (MA) techniques have been introduced in multiuser SC systems to manage inter-user interference. In Section 1.3.2, we will introduce different MA-assisted SCs.

1.3.2 Different MA-Assisted SC

Various MA schemes have been proposed to manage inter-user interference and enhance SC performance, such as orthogonal frequency-division multiple access (OFDMA), space-division multiple access (SDMA), non-orthogonal multiple access (NOMA), RSMA, model division multiple access (MDMA), and deep multiple access (DeepMA). The transceiver design for these MA-assisted SCs is delineated in this subsection.

1.3.2.1 OFDMA-Assisted SC

OFDMA divides the spectrum into multiple subchannels/subcarriers, allowing multiple users to communicate simultaneously on different subcarriers, thus improving spectral efficiency. Shao and Gunduz [2022] designed a passband transceiver for the OFDMA SC system, in which a differentiable clipping operation is incorporated into the training process.

The OFDMA-assisted SC proposed by Shao and Gunduz [2022] is shown in Figure 1.9. After joint source channel coding (JSCC) encoding, the source signal is encoded into symbols, which are then normalized and transformed into a power-normalized real vector. Then, for each block, the signal vector is mapped onto subcarriers. The inverse discrete Fourier transform (IDFT) is applied to each orthogonal frequency-division multiplexing (OFDM) symbol, and then the cyclic prefix (CP) is appended. After pulse shaping, we have the baseband continuous-time signal, with which we can construct a passband signal. Moreover, the receiver converts the received signal to the baseband, matches filters, and samples the baseband signal. After that, the signal is recovered by OFDM demodulation, in-phase and quadrature (IQ) remapping, and JSCC decoding.

1.3.2.2 SDMA-Assisted SC

SDMA is capable of utilizing the spatial domain to serve multiple users at the same time–frequency resource. It introduces the concept of precoding, which adjusts the phase, amplitude, and spatial properties of the signal to improve the overall signal quality. Guo and Yang [2022] proposed an SC system that designed the precoding strategy with the help of semantic information. Convolutional neural networks (CNNs) are used to learn the semantic/channel encoder/decoder and modulator/demodulator, while a graph neural network (GNN) is adopted to learn the precoding policy managing the interference. All those CNNs and a GNN are jointly trained to achieve the expected classification accuracy in a system that delivers images to multiple users. The SDMA-assisted SC system proposed by Guo and Yang [2022] is illustrated in Figure 1.10.

Simulation results provided by Guo and Yang [2022] show that the learned precoding policy requires much less bandwidth than the fixed low-complexity precoding schemes for achieving the same expected classification accuracy.

1.3.2.3 NOMA-Assisted SC

A semi-NOMA-assisted SC system, designed by Mu and Liu [2022], is shown in Figure 1.11. This semi-NOMA-assisted SC considers two types of users, namely semantic-based users (S-users) and bit-based users (B-users). S-users are capable of decoding both the semantic signal and the bit signal, while the B-users are only capable of decoding the bit signal. For traditional NOMA

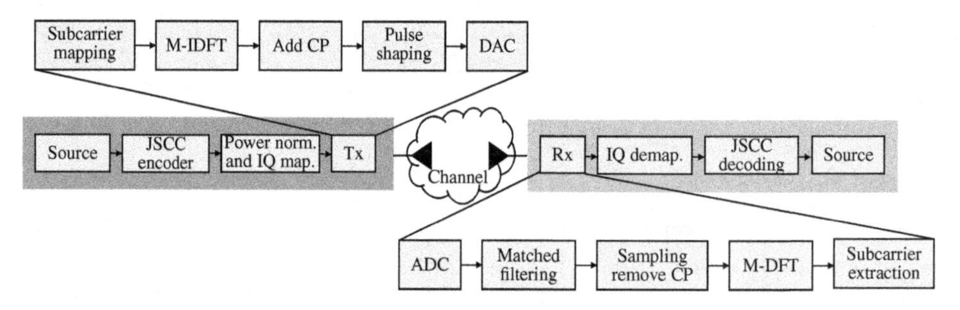

Figure 1.9 The architecture of an OFDMA-assisted SC system. Source: Adapted from Shao and Gunduz [2022].

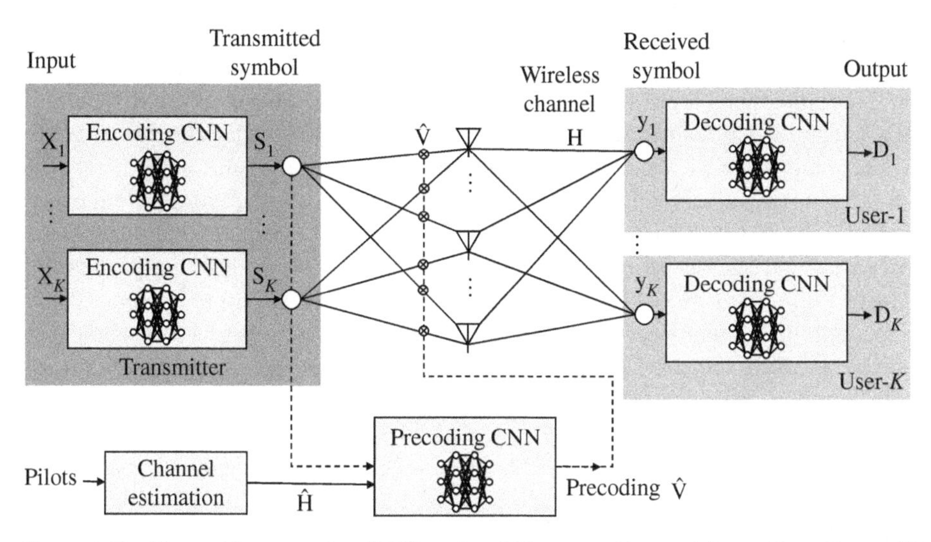

Figure 1.10 The architecture of an SDMA-assisted SC system. Source: Adapted from Guo and Yang [2022].

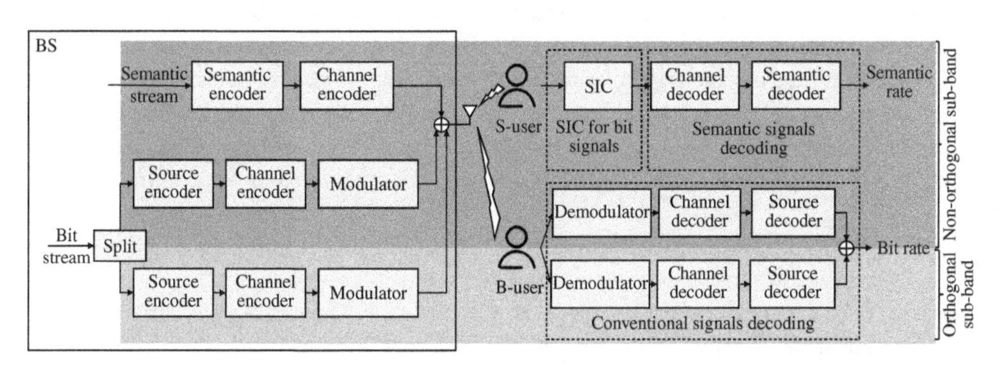

Figure 1.11 The architecture of a NOMA-assisted SC system. Source: Adapted form Mu and Liu [2022].

schemes, users who have higher channel conditions are supposed to decode the signal of users having lower channel conditions. However, in the SC system here, the transmission of semantic information between S-users and the base station (BS) requires a dedicated semantic decoder. Therefore, it is impossible for B-users to decode the signal from the S-users. Conversely, if S-users are required to decode all of the signals from B-users, it leads to substantial interference from the S-users to the B-users. To address this issue, the system splits the bitstream of the B-user and respectively allocates them to non-orthogonal and orthogonal sub-bands. The signal from the non-orthogonal sub-band is decoded using NOMA schemes at the S-users, while the signal from the orthogonal sub-band is decoded according to the orthogonal multiple access (OMA) scheme. As illustrated in Figure 1.11, in the non-orthogonal sub-band, each S-user first decodes and removes the signal of the B-user based on successive interference cancellation (SIC); it then decodes the intended semantic signals. The proposed semi-NOMA achieves the largest semantic-versus-bit rate region compared with those achieved by OMA and NOMA. Moreover, numerical results show that NOMA with SC provides a significant performance gain compared to the traditional NOMA.

1.3.2.4 RSMA-Assisted SC

RSMA is a newly emerged MA scheme, which shows powerful capability in interference management [Mao and Clerckx, 2020]. Compared with traditional MAs such as SDMA and NOMA, RSMA can partially decode the interference and partially regard the interference as noise, thereby providing a more flexible interference management ability [Mao et al., 2017]. In conventional multiuser communication systems, RSMA has been shown to enhance spectral efficiency (SE), energy efficiency (EE), user fairness, and quality of service (QoS) for both underloaded and overloaded scenarios and various user deployments [Mao et al., 2022]. In order to process the interference in two different ways simultaneously, user messages are split into common parts, which are combined into a common message and then decoded by all users, and private parts, which are only decoded by the corresponding users. At the receiver side, the common stream encoded by the common message is first decoded and then removed by successive interference cancellation (SIC) at each user. After that, the intended private stream is decoded with interference from other private streams [Mao et al., 2019]. The flexible interference management ability of RSMA makes it a great potential for SC.

When designing the SC system assisted by RSMA, to better take advantage of the features of SC, a specific strategy to find the common stream is proposed by Yang et al. [2023b] and Zeng et al. [2023]. As mentioned in Section 1.1, the SC systems assisted by KBs always contain two separate KBs respectively at the receiver and transmitter. However, since those two separate KBs collaborate with each other during the transmission, there must exist knowledge-sharing between the KB at the transmitter and the KB at the receiver. The common knowledge shared between the transmitter and receiver should be equally shared with all users. Therefore, Yang et al. [2023b] and Zeng et al. [2023] make sure that the common knowledge is transmitted by the common stream of RSMA.

In other SC systems proposed by Cheng et al. [2023a,2023b], with the application of the RSMA, the common semantic information is transmitted as the common message while the private semantic information of each user is transmitted as the private message. The messages of different users often contain part of the common semantic information, especially users in the same scenario. For the downlink scenario, all users' messages are processed by the same transmitter. Therefore, the common semantic information is accessible to each user. After encoding via the semantic encoder, the semantic information of each user's message is extracted and the common semantic information is obtained.

However, the key ideas of those three works are the same as mentioned earlier, except that the common stream transmits different messages, which is further explained in Figure 1.12. The one-layer layer RSMA, which is the simplest rate-splitting (RS) model, is applied in all three works.

As illustrated in Figure 1.12, there are K users and one transmitter in this RSMA-assisted SC system. At the transmitter, K user messages $W_1, W_2, \dots, W_K$ are first encoded into semantic information and then split according to their semantic meaning. It should be noted that the RSMA-assisted SC system is different from the traditional RSMA communication system since the common message is not combined with all common parts from all users. As all common parts are the same, only a single common part is transmitted via the common message. After semantic encoding, we separate the original messages into the common message W_c and private messages $W_{p,1}, W_{p,2}, \dots, W_{p,K}$, which are then encoded as common stream s_c and private streams

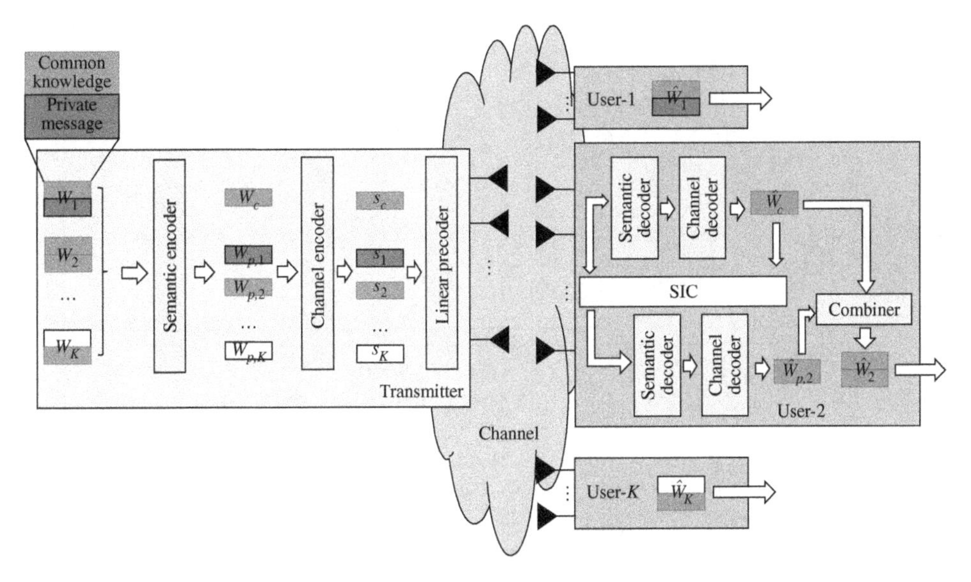

Figure 1.12 The architecture of an RSMA-assisted SC system.

$s_{p,1}, s_{p,2}, \dots, s_{p,K}$. As user-$k$, the common message is first decoded into $\hat{W}_c$ and then removed by SIC. After that, the private message is decoded as $\hat{W}_{p,k}$. The messages $\hat{W}_{p,k}$ and $\hat{W}_c$ are combined as the user-k's recovered message $\hat{W}_k$.

Due to the inevitability of the semantic overlap, RSMA can be used to enhance the downlink performance of SC. Transmission efficiency, multiplexing gains, and latency reduction are improved by integrating RSMA and SC [Cheng et al., 2023b; Zeng et al., 2023].

Yang et al. [2023b] is the first to combine the RSMA with the SC systems. It focuses on the wireless resource allocation problem and semantic information extraction problem to achieve better energy-efficient SC systems. Zeng et al. [2023] study the RSMA-assisted SC adopted at the control center of the downlink ultra-reliability and low-latency communication (URLLC) in the wireless control system (WCS). Both Yang et al. [2023b] and Zeng et al. [2023] choose to store semantic information and represent the knowledge with the DPG. Cheng et al. [2023a, 2023b] propose a task-oriented SC system assisted by RSMA for image transmission. In these works, common information is recognized by segmenting the image and then classifying the sub-images via object recognition.

The common message of RSMA, which is broadcast to all users, has the inherent similarity of the common knowledge and common semantic information. Therefore, there is great potential that combining RSMA and SC can enhance each other's performance. With the numerical results provided by Cheng et al. [2023b], RSMA-assisted SC outperforms SDMA-assisted SC. Zeng et al. [2023] confirmed that RSMA-assisted SC achieves overloaded connections with large multiplexing gains and reduced latency.

1.3.2.5 MDMA-Assisted SC

Zhang et al. [2023b] proposed a novel MA named "MDMA" in SC based on the exploitation of the semantic resources. Specifically, MDMA extracts the semantic information and maps it into a high-dimensional space, which is named "the semantic model space." By exploiting

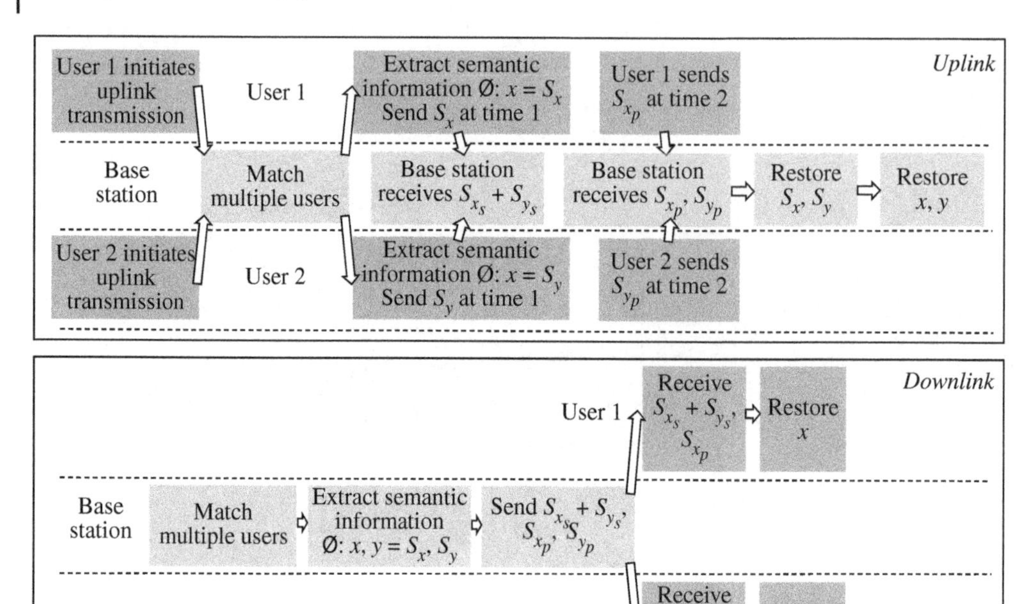

Figure 1.13 The architecture of an MDMA-assisted SC system. Source: Zhang et al. [2023b]/Springer Nature.

the personalized semantic information for each individual user in this high-dimensional semantic space, multiuser interference can be dramatically reduced. The uplink and downlink MDMA-assisted SC system is shown in Figure 1.13.

In the uplink, once the BS receives the signals from the users, it matches multiple users with the shared and personalized information, which facilitates the recovery of the original semantic information and the source signal. In the downlink system, the BS uses a series of model-based semantic-channel encoders to obtain the semantic information of each user. Then, the shared information is broadcast to all users, and the personalized information is only transmitted to the corresponding user. After each user receives the shared information and personalized information, the original signal can be restored accordingly. Zhang et al. [2023b] show that the MDMA schemes have greater feasible regions than the traditional MA methods, which means the service capability of MDMA is better with a two-user channel. It is shown that multiuser transmission with MDMA can save 25% bandwidth when compared to image-based SC.

1.3.2.6 DeepMA-Assisted SC

The DeepMA proposed by Zhang et al. [2023c], shown in Figure 1.14, is a kind of MA scheme that uses DL methods to achieve JSCC and orthogonal signal modulation. The input data are encoded as mutually orthogonal semantic symbol vectors (SSVs) Therefore, the receivers can decode the received mixed SSV (MSSV) and recover their own target data. DeepMA supports sending multiple orthogonal SSV, which helps to achieve channel multiplexing.

Compared to OFDMA and time division multiple access (TDMA), the DeepMA can achieve comparable bandwidth efficiency with better noise resilience.

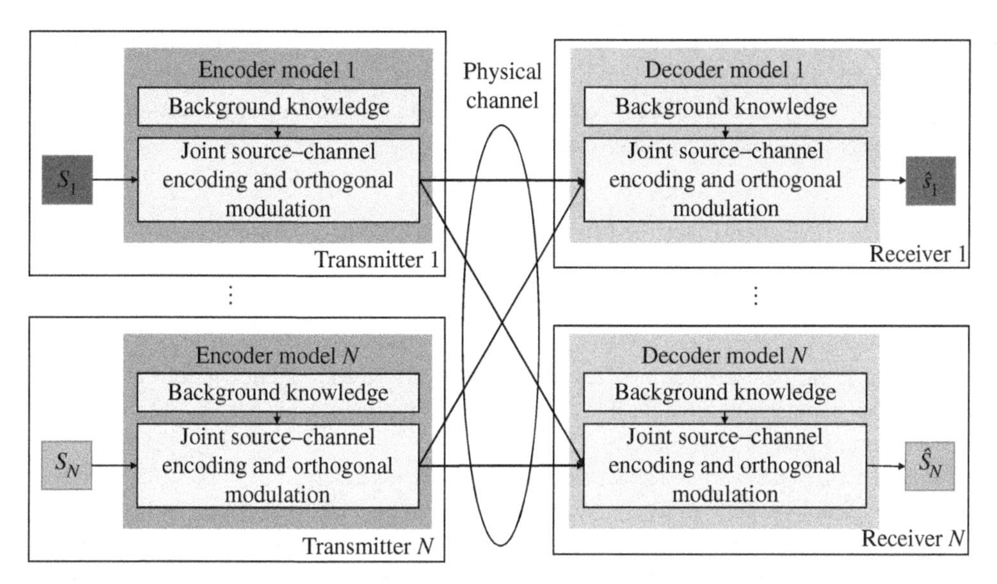

Figure 1.14 The architecture of a DeepMA-assisted SC system. Source: Adapted from Zhang et al. [2023b].

1.4 Transceiver Design for Single-Modal and Multimodal Data

The transceiver design of SC is highly dependent on the modal of transmitting data. In practical applications, there are four types of modal data, including text, image, speech, and video. The unique characteristics of each modal data and the challenges posed by multimodal data call for different transceiver designs. In this section, we introduce the existing transceiver designs tailored to the aforementioned four types of modal data and explore their extension for multimodal scenarios.

1.4.1 Text

The SC systems, which aim at text transmission, leverage the advancements in natural language processing (NLP). Sequential in nature, text data typically hold a high dimensionality, features sparsity, and maintains a strong connection with its context. All those features of text data inspire the usage of recurrent neural networks (RNNs) and transformers. RNNs specialize in handling sequential data, with the capability to retain past information so as to understand subsequent text. Transformers also excel in processing text sequences by capturing global and local relationships.

Inspired by the promising performance of transformers in the research field of NLP, Xie et al. [2020] proposed an SC system named "DeepSC." In this work, texts are embedded as vectors before being fed into a stacked bidirectional long short-term memory (BLSTM) network [Farsad et al., 2018]. After transmission through an erasure channel, the received signal is decoded by a stack of long short-term memory (LSTM) networks, which are a variant of the RNNs.

Building upon DeepSC, several variants have been introduced, including those proposed by Xie et al. [2021b, 2022], Xie and Qin [2021], Qin et al. [2023], Park et al. [2022], among others. Xie et al. [2021b] applied DeepSC across diverse communication scenarios by using deep transfer learning. Xie and Qin [2021] proposed L-DeepSC, a lite-distributed SC system tailored to address the constrained computing capabilities of internet-of-things (IoT) devices by introducing a novel semantic constellation. Qin et al. [2023] proposed Mem-DeepSC, which introduces the memory module to enhance the semantic decoding capability.

1.4.2 Image

Different from the text data, images are composed of a pixel grid, where each pixel contains spatial and visual information. To effectively capture these features, convolutional operations are essential in images. Therefore, CNNs find extensive use in image-related tasks and demonstrate remarkable performance across various applications. Moreover, generative adversarial networks (GANs) have been proposed for generating new data that resemble a given dataset. It is beneficial to image SC systems for generating decoded results.

Bourtsoulatze et al. [2018] adopted an autoencoder for wireless image transmission. The encoder and decoder are implemented through two CNNs with a non-trainable layer in the middle, representing the presence of a noisy communication channel. Huang et al. [2020] jointly matched the images and sentences produced by LSTM based on the semantic concepts extracted from the image using the multiregional multi-label CNN. Meanwhile, Huang et al. [2021] incorporated GANs in image semantic coding, exploring elements such as convolutional encoder, quantizer, conditional spatially-adaptive (DE) normalization (SPADE) generator, residual coding, and perceptual losses in order to propose a coarse-to-fine image SC system. Yang et al. [2021] adopted a CNN to directly map the source image into complex-valued baseband samples for OFDM transmission, demonstrating efficacy in handling multipath fading channels.

1.4.3 Speech

Speech data share many similar features with the text data such as sequentiality and high dimensionality. However, speech data are more susceptible to influences such as noise, accent, and speech variations compared to text data. This has motivated the SC system to develop novel methods to address these challenges.

Weng et al. [2021] proposed DeepSC-S, a variant tailored for speech signals based on the original DeepSC designed for text data. The transceiver relies on an attention mechanism with mean-square error (MSE) as the loss function. The signal-to-distortion ratio (SDR) is used to evaluate this system, which measures the quality of recovered speech information. The lower the MSE, the higher the SDR. Perceptual evaluation of speech quality (PESQ) is also employed for evaluation. Weng et al. [2023] proposed another variant of DeepSC named "DeepSC-ST." It uses text as an intermediary modal, requiring fewer resources for transmission compared to directly transmitting the speech signal. The text-related semantic features are generated by a semantic encoder, constructed using CNN and gated recurrent unit (GRU)–based bidirectional RNN (BRNN) modules. The Tacotron 2, proposed by Shen et al. [2018], is used to reconstruct the speech signal after decoding in the receiver. Additionally, Tong et al. [2021] proposed an autoencoder that consists of CNNs and implements it by federated learning (FL) over multiple devices and a server.

1.4.4 Video

Video data contain both speech and image features. It constitutes a sequential set of images with a certain time thread. Effectively processing video data involves tackling temporal sequences and spatial correlations, necessitating the use of more complex models and techniques to comprehend and process this information.

Tung and Gündüz [2022] proposed DeepWiVe, an SC system that directly maps video signals to channel symbols using DNNs. To maximize the overall visual quality under limited available channel bandwidth, reinforcement learning (RL) is used to allocate bandwidth. Jiang et al. [2023b] proposed a semantic video conferencing (SVC) network based on keypoint transmission since the background of conferencing is almost static and speakers are often quite still. Wang et al. [2023] proposed a framework named "deep video semantic transmission" (DVST), featuring a novel adaptive transmission mechanism that customizes deep joint source–channel coding for video sources.

1.4.5 Multimodal

The majority of existing works on SC focus on solving single-modal tasks; however, multimodal is more common in the real world. For example, users may have different requirements to process different types of modal data. Therefore, there is a growing need to develop an SC system that can handle tasks with multiple modalities. Figure 1.15 illustrates the transceiver architecture of most multimodal SC systems, where different encoder and decoder pairs process distinct modal signals from the transmitter and receiver.

Zhang et al. [2022] proposed a unified DL-enabled SC system named "U-DeepSC." During the training phase, domain adaptation is used to specify the task-specific features for processing different modalities of data. Encoded features are selectively transmitted, focusing only on those relevant to the specific task at hand. Luo et al. [2022b] proposed channel-level information fusion, which considers the received signal as the fused information. DeepSC-VQA, proposed by Xie et al. [2022], and MU-DeepSC, proposed by Xie et al. [2021a], are both the variants of DeepSC. DeepSC-VQA, designed for visual question answering (VQA), extracts the semantic information from different modalities and fuses it at the receiver. In contrast, the transceiver for MU-DeepSC aims to capture the features from the correlated multimodal data based on the use of memory, attention, and composition neural networks.

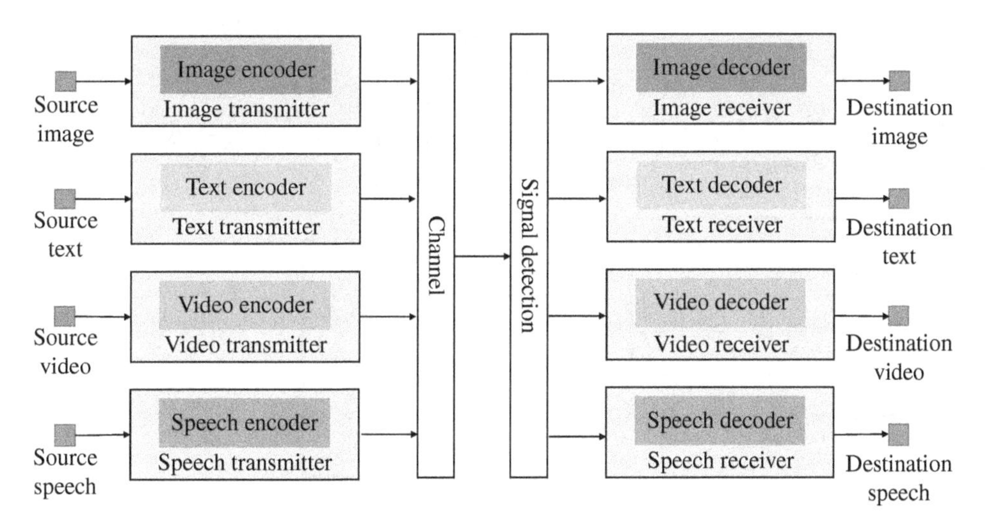

Figure 1.15 An example of a multimodal SC system.

1.5 Challenges and Future Directions

SC is a promising technique for advancing communication, offering higher data rates and lower latency to support increasingly intelligent applications in the future. In this section, we introduce the future directions and the corresponding challenges with the SC system. The five directions to be discussed in this subsection, including the IoT system, unmanned aerial vehicles (UAVs) system, holographic telepresence (HT), intelligent transport system, and smart cities, are shown in Figure 1.16.

1.5.1 Internet of Things System

The IoT is a vast network of interconnected devices that can collect and exchange data with each other. An IoT system includes devices, networks, and platforms or cloud systems for collecting and analyzing user data. The efficiency and real-time performance of the IoT system are profoundly influenced by the network connecting these devices. Given the time-sensitive nature of data transmission in IoT, the introduction of SC offers a novel approach for accurate and real-time data transmissions while requiring minimal radio resources. Additionally, SC exhibits enhanced robustness as it is less sensitive to channel noise.

However, the classical design of the semantic/channel encoders and decoders is not viable for IoT systems. This limitation arises from the limited computation and storage capabilities of devices within IoT systems, making it challenging to implement simplified neural networks in SC or introduce FL to facilitate network training in IoT networks.

1.5.2 Unmanned Aerial Vehicles System

UAVs show great potential in serving as aerial BSs or as relays [Hayat et al., 2016; Zhou et al., 2021]. UAV systems, which consist of a group of collaborative UAVs, provide a more adaptable

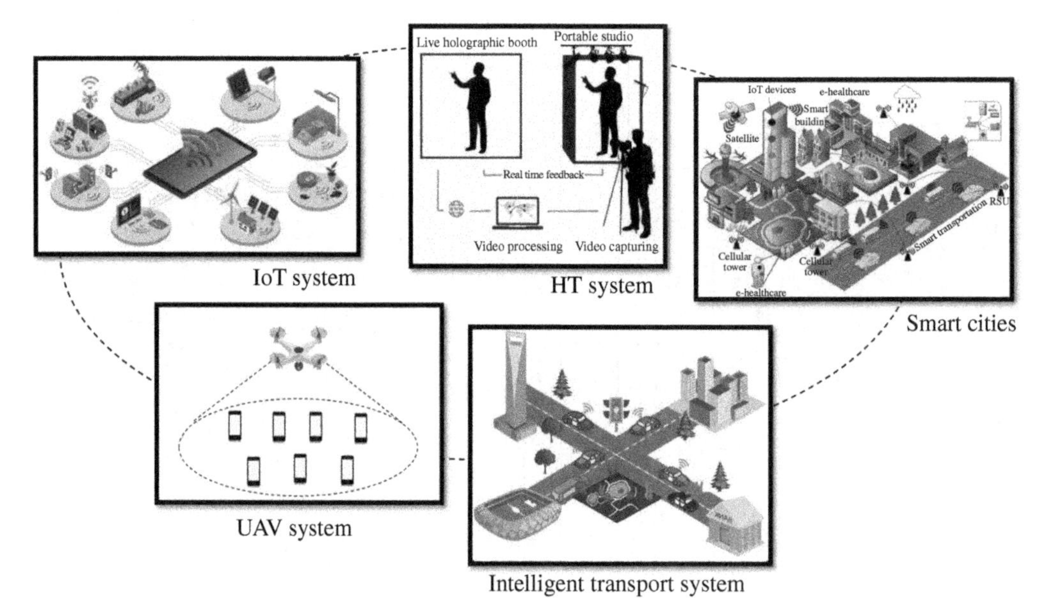

Figure 1.16 The challenges and future directions of SC system.

deployment than static ground BSs. However, the communication among those UAVs poses a challenge since the UAVs aim to establish long-term communication with each other with limited energy.

The energy constraints of the UAVs call for a more efficient communication strategy. This has led to the integration of SC, which is able to reduce the amount of information that needs to be transmitted. When UAVs serve as the relay, a semantic process-and-forward method is proposed by Yang et al. [2023a] to meet the requirements of the UAV–SC system. Beyond serving as a relay, UAVs can also be deployed as a semantic encoder and/or decoder. Moreover, SC is beneficial in assisting navigation within a UAV swarm, as demonstrated by Yun et al. [2021]. By using a graph attention exchange network for UAV swarm navigation, such UAV–SC systems significantly lower latency compared to the current state-of-the-art methods.

1.5.3 Holographic Telepresence

HT is a cutting-edge technology that can project three-dimensional, lifelike representations of individuals in remote locations, allowing individuals to interact as if they were in the same room [Movassaghi et al., 2014]. It can be applied in virtual conferencing, virtual games, remote repair, and remote surgery. However, to guarantee users get a vivid-enough experience, HT needs strict QoS requirements.

Traditional communication methods often result in a waste of bandwidth resources and computing resources due to the blindness of the content. To meet the requirements of HT communication, the SC system, which transmits data based on semantics, emerges as a promising solution to make full use of bandwidth and computing resources.

1.5.4 Intelligently Transport System

Intelligently connected vehicle (ICV) networks are the integration of vehicles with advanced communication technologies and intelligent systems that enable them to interact with each other, infrastructure, and external environments. Employing sensors, radar, cameras, and other technologies, ICVs collect driving-related information, such as the vehicle's surroundings and road conditions [Ye et al., 2021]. This network facilitates the sharing of information among vehicles, such as data on vehicle locations, potholes, and water puddles, ultimately enhancing safety, optimizing traffic flow, and improving the overall driving experience.

The rich information within the system has inherent meaning, indicating they do not need to be transmitted bit by bit. Essential semantic information such as the traffic signs and road conditions can be extracted from the raw information flow [Yang et al., 2023a]. Therefore, the introduction of SC helps the ICV networks reduce the delay while improving the quality of communication. Moreover, SC can help to extract semantics from the information, reducing the time and effort consumed in the information processing part. In addition to considering transmission efficiency, SC also plays an important role in differentiating the priority of the information, directing the attention of the intelligent transport system toward safety considerations, as noted by Nanda et al. [2019].

1.5.5 Smart Cities

In the development of smart cities, the quality of communication is an important aspect. A smart city includes the ICV networks and IoT systems such as smart factories and intelligent medical services. For different applications within smart cities, carefully tailored SC systems can be designed to satisfy specific requirements. For example, intelligent medical services usually consider the semantics in medicine, requiring the design of a specialized SC system.

Smart cities call for seamless and immediate interaction between different devices. In the face of the limited available wireless spectrum resources, SC helps to enhance the overall quality of communication within these dynamic urban environments.

References

Eirina Bourtsoulatze, David Burth Kurka, and Deniz Gündüz. Deep joint source-channel coding for wireless image transmission. *IEEE Transactions on Cognitive Communications and Networking*, 5:567–579, 2018. URL https://api.semanticscholar.org/CorpusID:52167612.

Yanyu Cheng, Dusit Niyato, Hongyang Du, Jiawen Kang, Chunyan Miao, and Dong In Kim. Interest-based semantic information transmission with RSMA in smart cities. In *ICC 2023 – IEEE International Conference on Communications*, pages 82–87, 2023a. doi: 10.1109/ICC45041.2023.10279270.

Yanyu Cheng, Dusit Tao Niyato, Hongyang Du, Jiawen Kang, Zehui Xiong, Chunyan Miao, and Dong In Kim. Resource allocation and common message selection for task-oriented semantic information transmission with RSMA. *IEEE Transactions on Wireless Communications*, 2023b. URL https://api.semanticscholar.org/CorpusID:264888379.

Diego Collarana, Mikhail Galkin, Ignacio Traverso-Ribón, Christoph Lange, Maria-Esther Vidal, and Sören Auer. Semantic data integration for knowledge graph construction at query time. In *2017 IEEE 11th International Conference on Semantic Computing (ICSC)*, pages 109–116, 2017. doi: 10.1109/ICSC.2017.85.

Nariman Farsad, Milind Rao, and Andrea J. Goldsmith. Deep learning for joint source-channel coding of text. In *2018 IEEE International Conference on Acoustics, Speech and Signal Processing (ICASSP)*, pages 2326–2330, 2018. doi: 10.1109/ICASSP.2018.8461983.

Jia Guo and Chenyang Yang. Learning precoding for semantic communications. In *2022 IEEE International Conference on Communications Workshops (ICC Workshops)*, pages 163–168, 2022. doi: 10.1109/ICCWorkshops53468.2022.9814464.

Samira Hayat, Evşen Yanmaz, and Raheeb Muzaffar. Survey on unmanned aerial vehicle networks for civil applications: A communications viewpoint. *IEEE Communications Surveys & Tutorials*, 18:2624–2661, 2016. URL https://api.semanticscholar.org/CorpusID:424353.

Han Hu, Xingwu Zhu, Fuhui Zhou, Wei Wu, Rose Qingyang Hu, and Hongbo Zhu. One-to-many semantic communication systems: Design, implementation, performance evaluation. *IEEE Communications Letters*, 26(12):2959–2963, 2022. doi: 10.1109/LCOMM.2022.3203984.

Yan Huang, Qi Wu, Wei Wang, and Liang Wang. Image and sentence matching via semantic concepts and order learning. *IEEE Transactions on Pattern Analysis and Machine Intelligence*, 42(3):636–650, 2020. doi: 10.1109/TPAMI.2018.2883466.

Danlan Huang, Xiaoming Tao, Feifei Gao, and Jianhua Lu. Deep learning-based image semantic coding for semantic communications. In *2021 IEEE Global Communications Conference (GLOBECOM)*, pages 1–6, 2021. doi: 10.1109/GLOBECOM46510.2021.9685667.

Peiwen Jiang, Chao-Kai Wen, Shi Jin, and Geoffrey Y. Li. Deep source-channel coding for sentence semantic transmission with HARQ. *IEEE Transactions on Communications*, 70(8):5225–5240, 2022a. URL https://api.semanticscholar.org/CorpusID:235358364.

Shengteng Jiang, Yueling Liu, Yichi Zhang, Peng Luo, Kuo Cao, Jun Xiong, Haitao Zhao, and Jibo Wei. Reliable semantic communication system enabled by knowledge graph. *Entropy*, 24(6), 2022b. ISSN 1099-4300. doi: 10.3390/e24060846. URL https://www.mdpi.com/1099-4300/24/6/846.

Feibo Jiang, Yubo Peng, Li Dong, Kezhi Wang, Kun Yang, Cunhua Pan, and Xiaohu You. Large AI model empowered multimodal semantic communications. *arXiv e-prints*, art. arXiv:2309.01249, September 2023a. doi: 10.48550/arXiv.2309.01249.

Peiwen Jiang, Chao-Kai Wen, Shi Jin, and Geoffrey Ye Li. Wireless semantic communications for video conferencing. *IEEE Journal on Selected Areas in Communications*, 41(1):230–244, 2023b. doi: 10.1109/JSAC.2022.3221968.

Qiao Lan, Dingzhu Wen, Zezhong Zhang, Qunsong Zeng, Xu Chen, Petar Popovski, and Kaibin Huang. What is semantic communication? A view on conveying meaning in the era of machine intelligence. *Journal of Communications and Information Networks*, 6(4):336–371, 2021. doi: 10.23919/JCIN.2021.9663101.

Ang Li, Xin Wei, Dan Wu, and Liang Zhou. Cross-modal semantic communications. *IEEE Wireless Communications*, 29(6):144–151, 2022. doi: 10.1109/MWC.008.2200180.

Xuewen Luo, Hsiao-Hwa Chen, and Qing Guo. Semantic communications: Overview, open issues, and future research directions. *IEEE Wireless Communications*, 29(1):210–219, 2022a. doi: 10.1109/MWC.101.2100269.

Xuewen Luo, Ruobin Gao, Hsiao-Hwa Chen, Shu-Wen Chen, Qing Guo, and Ponnuthurai Nagaratnam Suganthan. Multi-modal and multi-user semantic communications for channel-level information fusion. *IEEE Wireless Communications*, 31(2):117–125, 2022b. URL https://api.semanticscholar.org/CorpusID:253309006.

Yijie Mao and Bruno Clerckx. Beyond dirty paper coding for multi-antenna broadcast channel with partial CSIT: A rate-splitting approach. *IEEE Transactions on Communications*, 68(11):6775–6791, 2020. doi: 10.1109/TCOMM.2020.3014153.

Yijie Mao, Bruno Clerckx, and Victor O. K. Li. Rate-splitting multiple access for downlink communication systems: Bridging, generalizing, and outperforming SDMA and NOMA. *EURASIP Journal on Wireless Communications and Networking*, 2018:133, 2017. URL https://api.semanticscholar.org/CorpusID:29163706.

Yijie Mao, Bruno Clerckx, and Victor O. K. Li. Rate-splitting for multi-antenna non-orthogonal unicast and multicast transmission: Spectral and energy efficiency analysis. *IEEE Transactions on Communications*, 67(12):8754–8770, 2019. doi: 10.1109/TCOMM.2019.2943168.

Yijie Mao, Onur Dizdar, Bruno Clerckx, Robert Schober, Petar Popovski, and H. Vincent Poor. Rate-splitting multiple access: Fundamentals, survey, and future research trends. *IEEE Communications Surveys & Tutorials*, 24(4):2073–2126, 2022. doi: 10.1109/COMST.2022.3191937.

Samaneh Movassaghi, Mehran Abolhasan, Justin Lipman, David Smith, and Abbas Jamalipour. Wireless body area networks: A survey. *IEEE Communications Surveys & Tutorials*, 16(3):1658–1686, 2014. doi: 10.1109/SURV.2013.121313.00064.

Xidong Mu and Yuanwei Liu. Semantic communications in multi-user wireless networks. *ArXiv*, abs/2211.08932, 2022. URL https://api.semanticscholar.org/CorpusID:253553318.

Ashish Nanda, Deepak Puthal, Joel José P. C. Rodrigues, and Sergei A. Kozlov. Internet of autonomous vehicles communications security: Overview, issues, and directions. *IEEE Wireless Communications*, 26:60–65, 2019. URL https://api.semanticscholar.org/CorpusID:201622757.

Chae-Hoon Park, Jinhyuk Choi, Jihong Park, and Seong-Lyun Kim. Federated codebook for multi-user deep source coding. In *2022 13th International Conference on Information and Communication Technology Convergence (ICTC)*, pages 994–996, 2022. doi: 10.1109/ICTC55196.2022.9952500.

Zhijin Qin, Huiqiang Xie, and Xiaoming Tao. Mem-DeepSC: A semantic communication system with memory. In *ICC 2023 – IEEE International Conference on Communications*, pages 3854–3859, 2023. doi: 10.1109/ICC45041.2023.10279664.

Yulin Shao and Deniz Gunduz. Semantic communications with discrete-time analog transmission: A PAPR perspective. *arXiv e-prints*, art. arXiv:2208.08342, August 2022. doi: 10.48550/arXiv.2208.08342.

Jonathan Shen, Ruoming Pang, Ron J. Weiss, Mike Schuster, Navdeep Jaitly, Zongheng Yang, Zhifeng Chen, Yu Zhang, Yuxuan Wang, Rj Skerrv-Ryan, Rif A. Saurous, Yannis Agiomvrgiannakis, and Yonghui Wu. Natural TTS synthesis by conditioning wavenet on MEL spectrogram predictions. In *2018 IEEE International Conference on Acoustics, Speech and Signal Processing (ICASSP)*, pages 4779–4783, 2018. doi: 10.1109/ICASSP.2018.8461368.

Guangming Shi, Yong Xiao, Yingyu Li, and Xuemei Xie. From semantic communication to semantic-aware networking: Model, architecture, and open problems. *IEEE Communications Magazine*, 59(8):44–50, 2021. doi: 10.1109/MCOM.001.2001239.

Emilio Calvanese Strinati and Sergio Barbarossa. 6G networks: Beyond Shannon towards semantic and goal-oriented communications. *Computer Networks*, 190:107930, 2020. URL https://api.semanticscholar.org/CorpusID:227227910.

Haonan Tong, Zhaohui Yang, Sihua Wang, Ye Hu, Walid Saad, and Changchuan Yin. Federated learning based audio semantic communication over wireless networks. In *2021 IEEE Global Communications Conference (GLOBECOM)*, pages 1–6, 2021. doi: 10.1109/GLOBECOM46510.2021.9685654.

Tze-Yang Tung and Deniz Gündüz. DeepWiVe: Deep-learning-aided wireless video transmission. *IEEE Journal on Selected Areas in Communications*, 40(9):2570–2583, 2022. URL https://api.semanticscholar.org/CorpusID:244709579.

Quan Wang, Zhendong Mao, Bin Wang, and Li Guo. Knowledge graph embedding: A survey of approaches and applications. *IEEE Transactions on Knowledge and Data Engineering*, 29(12):2724–2743, 2017. doi: 10.1109/TKDE.2017.2754499.

Sixian Wang, Jincheng Dai, Zijian Liang, Kai Niu, Zhongwei Si, Chao Dong, Xiaoqi Qin, and Ping Zhang. Wireless deep video semantic transmission. *IEEE Journal on Selected Areas in Communications*, 41(1):214–229, 2023. doi: 10.1109/JSAC.2022.3221977.

Zhenzi Weng and Zhijin Qin. Semantic communication systems for speech transmission. *IEEE Journal on Selected Areas in Communications*, 39:2434–2444, 2021. URL https://api.semanticscholar.org/CorpusID:232046408.

Zhenzi Weng, Zhijin Qin, and Geoffrey Ye Li. Semantic communications for speech signals. In *ICC 2021 – IEEE International Conference on Communications*, pages 1–6, 2021. doi: 10.1109/ICC42927.2021.9500590.

Zhenzi Weng, Zhijin Qin, Xiaoming Tao, Chengkang Pan, Guangyi Liu, and Geoffrey Ye Li. Deep learning enabled semantic communications with speech recognition and synthesis. *IEEE Transactions on Wireless Communications*, 22(9):6227–6240, 2023. doi: 10.1109/TWC.2023.3240969.

Huiqiang Xie and Zhijin Qin. A lite distributed semantic communication system for internet of things. *IEEE Journal on Selected Areas in Communications*, 39(1):142–153, 2021. doi: 10.1109/JSAC.2020.3036968.

Huiqiang Xie, Zhijin Qin, Geoffrey Ye Li, and Biing-Hwang Juang. Deep learning based semantic communications: An initial investigation. In *GLOBECOM 2020 – 2020 IEEE Global Communications Conference*, pages 1–6, 2020. doi: 10.1109/GLOBECOM42002.2020.9322296.

Huiqiang Xie, Zhijin Qin, and Geoffrey Y. Li. Task-oriented multi-user semantic communications for VQA. *IEEE Wireless Communications Letters*, 11:553–557, 2021a. URL https://api.semanticscholar.org/CorpusID:245130898.

Huiqiang Xie, Zhijin Qin, Geoffrey Ye Li, and Biing-Hwang Juang. Deep learning enabled semantic communication systems. *IEEE Transactions on Signal Processing*, 69:2663–2675, 2021b. doi: 10.1109/TSP.2021.3071210.

Huiqiang Xie, Zhijin Qin, Xiaoming Tao, and Khaled Ben Letaief. Task-oriented multi-user semantic communications. *IEEE Journal on Selected Areas in Communications*, 40:2584–2597, 2022. URL https://api.semanticscholar.org/CorpusID:245334909.

Mingyu Yang, Chenghong Bian, and Hun-Seok Kim. Deep joint source channel coding for wireless image transmission with OFDM. In *ICC 2021 – IEEE International Conference on Communications*, pages 1–6, 2021. doi: 10.1109/ICC42927.2021.9500996.

Wanting Yang, Hongyang Du, Zi Qin Liew, Wei Yang Bryan Lim, Zehui Xiong, Dusit Niyato, Xuefen Chi, Xuemin Shen, and Chunyan Miao. Semantic communications for future internet: Fundamentals, applications, and challenges. *IEEE Communications Surveys & Tutorials*, 25(1):213–250, 2023a. doi: 10.1109/COMST.2022.3223224.

Zhaohui Yang, Mingzhe Chen, Zhaoyang Zhang, and Chongwen Huang. Energy efficient semantic communication over wireless networks with rate splitting. *IEEE Journal on Selected Areas in Communications*, 41:1484–1495, 2023b. URL https://api.semanticscholar.org/CorpusID:255440611.

Qiang Ye, Weisen Shi, Kaige Qu, Hongli He, Weihua Zhuang, and Xuemin Shen. Joint ran slicing and computation offloading for autonomous vehicular networks: A learning-assisted hierarchical approach. *IEEE Open Journal of Vehicular Technology*, 2:272–288, 2021. doi: 10.1109/OJVT.2021.3089083.

Won Joon Yun, Byungju Lim, Soyi Jung, Young-Chai Ko, Jihong Park, Joongheon Kim, and Mehdi Bennis. Attention-based reinforcement learning for real-time UAV semantic communication. In *2021 17th International Symposium on Wireless Communication Systems (ISWCS)*, pages 1–6, 2021. doi: 10.1109/ISWCS49558.2021.9562230.

Cheng Zeng, Jun-Bo Wang, Ming Xiao, Changfeng Ding, Yijian Chen, Hongkang Yu, and Jiangzhou Wang. Task-oriented semantic communication over rate splitting enabled wireless control systems for URLLC services. *IEEE Transactions on Communications*, 2023. URL https://api.semanticscholar.org/CorpusID:264362823.

Guangyi Zhang, Qiyu Hu, Zhijin Qin, Yunlong Cai, and Guanding Yu. A unified multi-task semantic communication system with domain adaptation. In *GLOBECOM 2022 – 2022 IEEE Global Communications Conference*, pages 3971–3976, 2022. doi: 10.1109/GLOBECOM48099.2022.10000850.

Hongwei Zhang, Shuo Shao, Meixia Tao, Xiaoyan Bi, and Khaled B. Letaief. Deep learning-enabled semantic communication systems with task-unaware transmitter and dynamic data. *IEEE Journal on Selected Areas in Communications*, 41(1):170–185, 2023a. doi: 10.1109/JSAC.2022.3221991.

Ping Zhang, Xiaodong Xu, Chen Dong, Kai Niu, Haotai Liang, Zijian Liang, Xiaoqi Qin, Mengying Sun, Hao Chen, Nan Ma, Wenjun Xu, Guangyu Wang, and Xiaofeng Tao. Model division multiple access for semantic communications. *Frontiers of Information Technology & Electronic Engineering*, 24:801–812, 2023b. URL https://api.semanticscholar.org/CorpusID:259054404.

Wenyu Zhang, Kaiyuan Bai, Sherali Zeadally, Haijun Zhang, Hua Shao, Hui Ma, and Victor C. M. Leung. DeepMA: End-to-end deep multiple access for wireless image transmission in semantic communication. *IEEE Transactions on Cognitive Communications and Networking*, 10(2):387–402, 2023c. doi: 10.1109/TCCN.2023.3326302.

Zhouxiang Zhao, Zhaohui Yang, Quoc-Viet Pham, Qianqian Yang, and Zhaoyang Zhang. Semantic communication with probability graph: A joint communication and computation design. *arXiv e-prints*, art. arXiv:2310.00015, September 2023. doi: 10.48550/arXiv.2310.00015.

Conghao Zhou, Wen Wu, Hongli He, Peng Yang, Feng Lyu, Nan Cheng, and Xuemin Sherman Shen. Deep reinforcement learning for delay-oriented IoT task scheduling in SAGIN. *IEEE Transactions on Wireless Communications*, 20:911–925, 2021. URL https://api.semanticscholar.org/CorpusID:265039938.

Fuhui Zhou, Yihao Li, Xinyuan Zhang, Qihui Wu, Xianfu Lei, and Rose Qingyang Hu. Cognitive semantic communication systems driven by knowledge graph. In *ICC 2022 – IEEE International Conference on Communications*, pages 4860–4865, 2022. URL https://api.semanticscholar.org/CorpusID:247083930.

Fuhui Zhou, Yihao Li, Ming Xu, Lu Yuan, Qihui Wu, Rose Qingyang Hu, and Naofal Al-Dhahir. Cognitive semantic communication systems driven by knowledge graph: Principle, implementation, and performance evaluation. *IEEE Transactions on Communications*, 72(1):193–208, 2023. doi: 10.1109/TCOMM.2023.3318605.

2

Joint Cell Association and Spectrum Allocation in Semantic Communication Networks

Le Xia, Yao Sun, and Muhammad Ali Imran

James Watt School of Engineering, University of Glasgow, Glasgow, Lanarkshire, UK

2.1 Introduction

In contemporary wireless networks, there is a notable surge in traffic demands driven by the emergence of pervasive application intelligence and the need for extensive content delivery services encompassing text, images, speech, and video [Saad et al., 2019; Shen et al., 2021]. This surge, coupled with the imperative for low latency and high reliability, forewarns of a forthcoming scarcity of communication resources such as spectrum and energy. In conjunction with the inherent constraints posed by the Shannon limit, these impending bottlenecks underscore the necessity for radical reconfigurations in the design of future networks. This necessitates a paradigm shift from conventional bit-based communication to context-driven *semantic communication* (SemCom) [Carnap and Bar-Hillel, 1952; Weaver, 1953; Bao et al., 2011; Basu et al., 2014; Xie and Qin, 2020; Strinati and Barbarossa, 2021; Weng and Qin, 2021; Xie et al., 2021; Liu et al., 2022; Luo et al., 2022; Xia et al., 2023]. The concept of SemCom, originally posited by Weaver in his seminal work [Weaver, 1953], delineates communication challenges into three tiers: the technical quandary at the bit level, the semantic complexity at the meaning level, and the efficacy issue at the information exchange level. While substantial strides have been made in addressing the technical quandary through classical Shannon information theory [Shannon, 1948], the transition toward SemCom is still in its nascent stages, with a pivotal emphasis on the conveyance of meaning as opposed to conventional bit transmission.

In SemCom, the transmitter initiates the communication process by harnessing pertinent background knowledge associated with the source messages. This facilitates the filtration of extraneous content and the refinement of semantic features, thereby reducing the requisite bits for transmission – a procedure referred to as "semantic encoding." Subsequently, upon receiving the requisite knowledge, the destination receiver employs local semantic interpreters to accurately reconstruct the original meaning from the transmitted bits, even in the presence of intolerable bit errors at the syntactic level. This retrieval process is termed "semantic decoding." Consequently, SemCom enables efficient information exchanges with minimal semantic ambiguity, particularly when there is parity in background knowledge between the source and destination. Moreover, SemCom serves to significantly mitigate resource scarcity concerns [Bao et al., 2011; Strinati and Barbarossa, 2021; Luo et al., 2022].

Wireless Semantic Communications: Concepts, Principles, and Challenges, First Edition.
Edited by Yao Sun, Lan Zhang, Dusit Niyato, and Muhammad Ali Imran.

Indeed, several noteworthy studies have contributed significantly to the advancement of SemCom. For instance, leveraging sophisticated natural language processing (NLP) algorithms, researchers in Xie et al. [2021] and Xie and Qin [2020] have devised transformer-based text sentence similarity metrics, crucial for evaluating semantic performance within end-to-end SemCom systems. Simultaneously, in Weng and Qin [2021], two semantic metrics linked to speech distortion ratios have been employed to assess speech signals transmitted via SemCom. Moreover, the study by Xia et al. [2023] explores the application of SemCom in wireless virtual reality video delivery. Their work aims to achieve high-performance feature extraction and semantic recovery in this domain. In addition to these advancements in semantic-transceiver design, research focusing on the information-theoretic characterization of SemCom is also of paramount importance. For instance, in Bao et al. [2011] and Basu et al. [2014], authors have quantitatively measured semantic entropy by proposing a semantic channel coding theorem, rooted in the logical probability of messages, as initially suggested by Carnap and Bar-Hillel in Carnap and Bar-Hillel [1952]. Furthermore, in a recent study by Liu et al. [2022], the semantic rate-distortion function of information sources has been investigated, considering both intrinsic state and extrinsic observation in the memoryless source.

Building upon the foundational research in link-level SemCom, we advocate for a shift toward exploring wireless SemCom from a networking perspective, particularly focusing on the upper layer. Our primary objective is to investigate optimal wireless resource management strategies within the SemCom-enabled network (SC-Net) to enhance overall network performance in a semantics-aware manner. This entails addressing the unique challenge of ensuring background knowledge alignment between multiple mobile users (MUs) and multitier base stations (BSs). Efficient resource management remains paramount within the SC-Net, offering numerous benefits such as guaranteeing high-quality SemCom services and enhancing spectrum utilization. By devising effective resource allocation strategies, we aim to optimize network performance and facilitate seamless communication within the SC-Net ecosystem.

In light of the novel paradigm of SC-Net, we are confronted with three fundamental networking challenges:

- *Challenge 1: How to construct a reasonable semantic channel model well when considering the characteristics of SemCom?* Unlike conventional bit-based channel models, the primary focus of the semantic channel model is to establish a mathematical framework for characterizing semantic information transmission from a source to its destination [Liu et al., 2022]. In particular, discrepancies in background knowledge between the semantic encoder and decoder can introduce semantic ambiguity and distort the transmitted information [Bao et al., 2011]. Therefore, the initial challenge lies in devising a coherent semantic channel model that accounts for varying degrees of knowledge alignment, approached from the perspective of semantic information theory.
- *Challenge 2: How to devise an appropriate metric to measure the semantic-related network performance?* In SemCom, where the emphasis shifts from transmitted bits to the conveyed meaning of messages, conventional performance metrics rooted in Shannon's framework, such as system throughput measured in bits, are rendered obsolete for assessing the network performance of SC-Net. Given the distinct characteristics of the semantic channel model, determining a suitable metric relevant to SemCom poses another significant challenge.
- *Challenge 3: How to find the optimal resource management solution to maximize the semantic-related performance?* Within the cellular network architecture, cell association (CA) and spectrum allocation (SA) stand as pivotal mechanisms for resource management [Xu et al.,

2021]. Yet, in the context of SC-Net, alongside practical limitations such as constrained spectrum resources and the imposition of single-BS association, the existence of varying degrees of knowledge alignment between MUs and BSs introduces additional stringent criteria for CA and SA. Of particular note is the connection between SemCom-related network performance and the stochastic nature of source information generation. This dynamic underscores the necessity for efficiently devising a joint optimal CA and SA strategy, which constitutes the third challenge in SC-Net.

To our understanding, none of the existing literature has comprehensively tackled the array of challenges outlined previously. In this chapter, our primary focus is on delving into the resource management intricacies within the downlink of SC-Net. Acknowledging the distinctive knowledge matching dynamics inherent in SemCom, we identify two distinct SC-Net scenarios, each presenting unique joint optimization challenges concerning CA and SA. Consequently, we propose two distinct and effective solutions tailored to optimize the semantic-level performance of SC-Net. In summary, our main contributions can be encapsulated as follows:

- Our initial step involves the formal identification and definition of two overarching SC-Net scenarios, systematically categorized based on all feasible knowledge matching states between MUs and BSs. These scenarios are termed as "perfect knowledge matching" (PKM)-based SC-Net and imperfect knowledge matching (IKM)-based SC-Net. Subsequently, we embark on a mathematical elucidation of the unique semantic channel capacity model pertinent to the PKM-based SC-Net scenario, approached from a semantic information-theoretical standpoint. Building upon this foundation, we methodically construct the semantic channel model applicable to the IKM-based SC-Net scenario. This comprehensive approach effectively addresses the aforementioned Challenge 1, pertaining to the establishment of a reasonable semantic channel model.
- Utilizing the distinctive semantic channel models inherent in SC-Net, we employ a bit-rate-to-message-rate (B2M) transformation function to quantify the message rate for each SemCom-enabled link. This approach leads to the introduction of a novel metric known as "system throughput in messages" (STM), facilitating a precise assessment of the overall network performance at the semantic level. This development effectively addresses the previously mentioned Challenge 2. Furthermore, we formulate two joint optimization problems aimed at maximizing STM, encompassing both CA and SA, tailored to the specific characteristics of the two SC-Net scenarios.
- To address resource management challenges under both scenarios, we develop distinct solutions. For the deterministic optimization problem in PKM-based SC-Net, we directly apply a primal–dual decomposition method coupled with a Lagrange multiplier approach to derive the optimal CA and SA strategy. In contrast, for the stochastic optimization problem inherent in IKM-based SC-Net, we devise a two-stage solution. Initially, we employ a chance-constrained model to transform the primal stochastic problem into a deterministic counterpart, introducing a predefined semantic confidence level. Subsequently, in the second stage, we utilize an interior-point method in conjunction with a heuristic algorithm to finalize the joint optimal solution for CA and SA. Consequently, Challenge 3, pertaining to the efficient derivation of a joint optimal CA and SA strategy, is effectively addressed under both SC-Net scenarios.
- We conduct comprehensive simulations for both SC-Net scenarios to assess the efficacy of the proposed solutions. Through comparative analysis against two baseline approaches, our numerical results underscore the significant superiority of our solutions in terms of STM performance. Furthermore, our simulations unveil the critical importance of ensuring adequate

knowledge matching, highlighting its pivotal role in minimizing semantic ambiguity and maximizing message rates in SemCom.

The subsequent sections of this chapter are structured as follows: The following subsection introduces the semantic channel models for both PKM-based and IKM-based SC-Nets. Sections 2.3 and 2.4 delineate the joint CA and SA optimization problems for the two distinct SC-Nets, respectively. Corresponding solutions are proposed thereafter. Section 2.5 presents and discusses the numerical results obtained from extensive simulations. Finally, conclusions are drawn in Section 2.6.

2.2 Semantic Communication Model

2.2.1 Background Knowledge Matching Model

Consider the depicted scenario of an SC-Net, illustrated in Figure 2.1, where all communication entities, including BSs and MUs, possess the capability of engaging in SemCom with one another. It is noteworthy that the accuracy of SemCom critically hinges upon the alignment of background knowledge between the transceiver pairs, wherein a higher degree of matching is presumed to result in reduced semantic ambiguity and enhanced efficiency in information exchange [Bao et al., 2011; Strinati and Barbarossa, 2021; Xie et al., 2021]. To illustrate, in the context of a singular downlink depicted in Figure 2.1, consider the transmission of a message pertaining to an individual's preferred music genre from a BS to an associated MU. For accurate SemCom to occur, both parties must possess congruent background knowledge within the musical domain. In essence, the MU must ensure that its affiliated BS possesses background knowledge that closely aligns with its own before soliciting SemCom services.

Building upon this premise, a fundamental concept known as "auxiliary *knowledge base*" (KB) is introduced within SemCom. The KB serves as a compact information repository housing background knowledge specific to particular application domains (e.g., music or sports) relevant to distinct SemCom services [Luo et al., 2022; Shi et al., 2021; Strinati and Barbarossa, 2021]. Leveraging the computational prowess and storage capabilities of BSs, we posit that each BS maintains a cluster of KBs, encompassing random amounts and types of knowledge pertinent to various SemCom services. Consequently, MUs can access diverse SemCom services by associating with different BSs that possess the requisite KBs. However, it is imperative to acknowledge that messages received by MUs may encompass varying degrees of background knowledge simultaneously, thereby resulting in disparities in knowledge alignment between MUs and their associated BSs during the CA process. In light of these considerations, we present our initial definition as follows.

Definition 2.1 *For all knowledge-matching cases, we define two different SC-Net scenarios as shown in Figure 2.2.*

- *PKM-based SC-Net: Each MU within the network is ensured to have access to at least one BS that possesses all the requisite KBs necessary for achieving perfect alignment, thereby facilitating SemCom.*
- *IIKM-based SC-Net: No single BS within the network is equipped with all the requisite KBs for a given MU. However, various BSs associated with the MU may exhibit differing degrees of imperfect alignment, thereby enabling SemCom with varying levels of accuracy.*

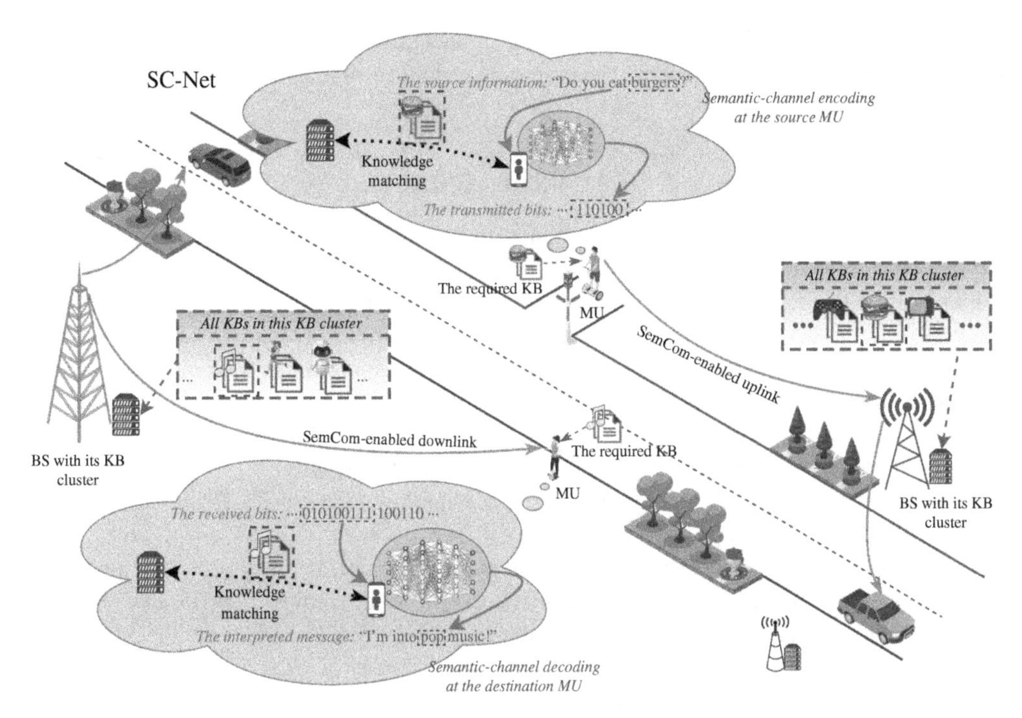

Figure 2.1 An overview of SC-Net.

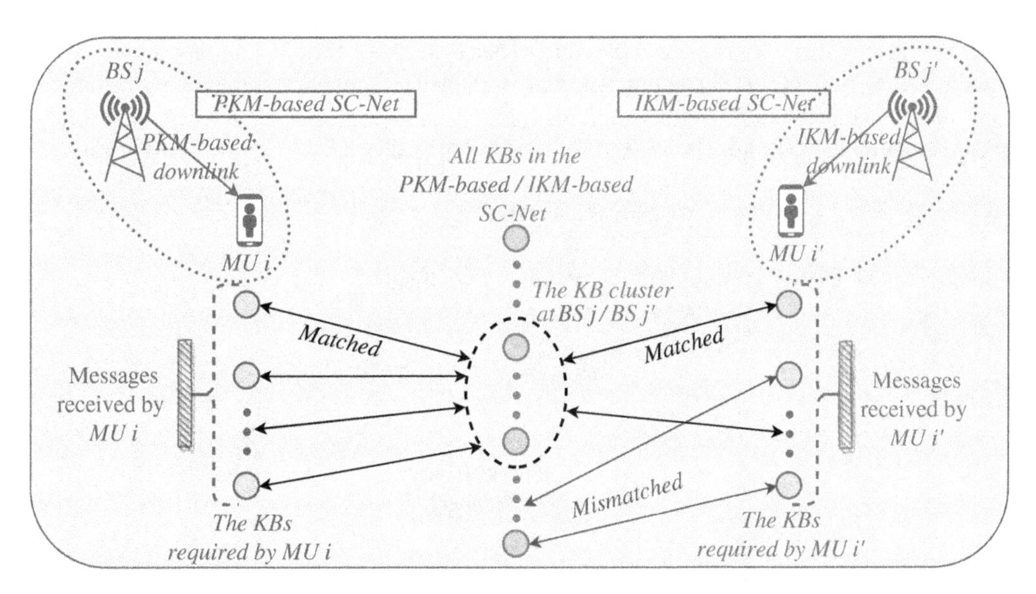

Figure 2.2 The PKM-based SC-Net and the IKM-based SC-Net for a single SemCom-enabled link.

From the depiction in Figure 2.2, it is evident that MU i on the left is categorized within the PKM scenario, wherein its associated BS j precisely holds all the KBs relevant to its received messages. Conversely, MU i' on the right falls under the IKM scenario, as its associated BS j' possesses only a subset of the required KBs. Consequently, only a portion of the received messages can be accurately interpreted, defining the IKM scenario. In Sections 1.2.2 and 1.2.3, we will provide detailed elaboration on these distinct SC-Net scenarios and their corresponding semantic channel models, respectively.

2.2.2 Semantic Channel Model in the PKM-Based SC-Net

Let us commence by examining a SemCom diagram illustrated in Figure 2.3. To facilitate our discussion, we model the source information, representing the intended meaning to be conveyed, as a random variable denoted by W. The observable message, which could be a sentence or a speech signal conveying the desired meaning, is represented by the random variable X. Both W and X are defined over a suitable product alphabet $\mathcal{W} \times \mathcal{X}$. Correspondingly, $\hat{X} \in \hat{\mathcal{X}}$ denotes the received message at the destination, such as the reconstructed sentence or speech signal. From $\hat{X}$, the destination interprets the information and extracts $\hat{W} \in \hat{\mathcal{W}}$. In this setup, a one-bit encoder and one-bit decoder are interconnected via a bit pipe, which could represent the wireless physical channel in conventional communications, enabling the transmission of the codeword $Y \in \mathcal{Y}$ at a certain code rate [Liu et al., 2022].

It is pertinent to note that in Figure 2.3, the message generator at the source is responsible for generating X from W based on a specific semantic encoding strategy. Here, we model this semantic encoding strategy as a conditional probabilistic distribution $P(\mathcal{X}|\mathcal{W})$, akin to approaches presented in previous works such as Bao et al. [2011] and Liu et al. [2022]. Furthermore, it is observed that different coding strategies represented by $P(\mathcal{X}|\mathcal{W})$ may result in varying degrees of semantic ambiguity at a statistical level [Strinati and Barbarossa, 2021]. This ambiguity arises because a given observable message may semantically correspond to multiple meanings, albeit only some of them being true with respect to the source [Bao et al., 2011; Strinati and Barbarossa, 2021; Luo et al., 2022]. Ensuring sufficient efficiency and accuracy in semantic coding thus necessitates a comprehensive understanding of both the background knowledge and the inference capability of each coding model, as highlighted in Bao et al. [2011] and Basu et al. [2014]. Moreover, the

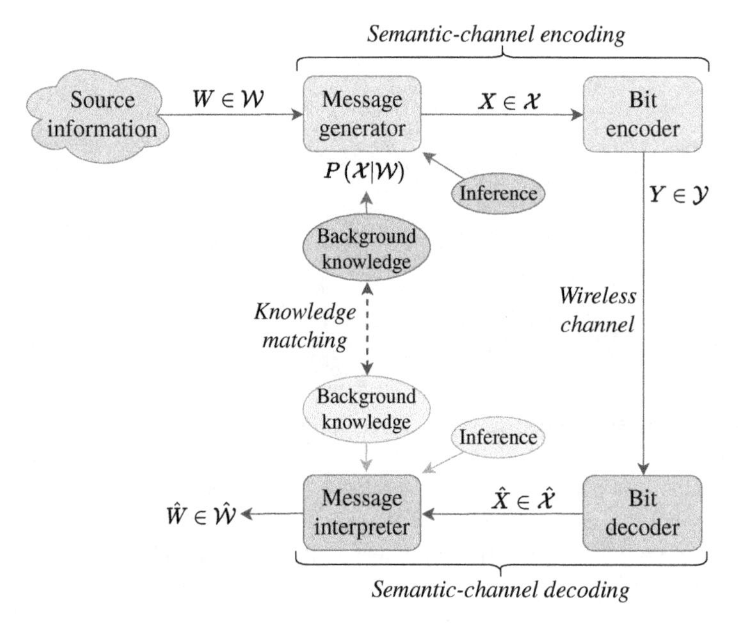

Figure 2.3 The SemCom diagram with the information source and destination.

inference capability of semantic coding models plays a crucial role, particularly in the context of deep learning-driven SemCom. This capability encompasses the feature compression and meaning interpretation abilities of coding models, which are intricately linked to the specific structures and compositions of neural networks utilized [Strinati and Barbarossa, 2021]. For instance, in the realm of SemCom driven by NLP, attention-based models often exhibit superior inference capabilities compared to traditional recurrent (e.g., LSTM) or convolutional models (e.g., TextCNN), particularly in tasks related to context prediction or sequence transduction [Vaswani et al., 2017].

Continuing along the lines of [Bao et al., 2011], if we assume that the message generator and the message interpreter possess identical inference capabilities and perfectly matched background knowledge, then the conditions outlined in the semantic-channel coding theorem given in Bao et al. [2011] are fully satisfied. This theorem asserts that the semantic channel capacity, denoted in units of messages per unit time (msg/s), for a discrete memoryless channel should be

$$C^s = \sup_{P(\mathcal{X}|\mathcal{W})} \left\{ \sup_{P(\mathcal{Y}|\mathcal{X})} \{I(\mathcal{X};\hat{\mathcal{X}})\} - H(\mathcal{W}|\mathcal{X}) + \overline{H_s(\hat{\mathcal{X}})} \right\}. \tag{2.1}$$

Here, $I(\mathcal{X};\hat{\mathcal{X}})$ represents the mutual information between $\mathcal{X}$ and $\hat{\mathcal{X}}$ under the traditional bit encoding strategy, which is modeled as $P(\mathcal{Y}|\mathcal{X})$. On the other hand, $H(\mathcal{W}|\mathcal{X})$ quantifies the semantic ambiguity of coding at the source with respect to $P(\mathcal{X}|\mathcal{W})$. Both of these quantities are expressed in the form of classical Shannon entropy. In contrast, $\overline{H_s(\hat{\mathcal{X}})}$ evaluates the semantic entropy of received messages, calculated using the logical probability denoted as $P_s(\hat{X})$ [Bao et al., 2011]. Specifically, $\overline{H_s(\hat{\mathcal{X}})}$ is computed as $-\sum_{\hat{X}\in\hat{\mathcal{X}}} P(\hat{X})\log_2 P_s(\hat{X})$. This metric captures the degree of semantic uncertainty in the received messages, offering insights beyond traditional Shannon entropy measures.[1]

Here, if we denote the optimal semantic encoding solution as $P^*(\mathcal{X}|\mathcal{W})$, we can substitute it into (2.1) and get

$$\begin{aligned} C^s &= \sup_{P(\mathcal{Y}|\mathcal{X})} \{I^*(\mathcal{X};\hat{\mathcal{X}})\} - H^*(\mathcal{W}|\mathcal{X}) + \overline{H_s^*(\hat{\mathcal{X}})} \\ &\triangleq \sup_{P(\mathcal{Y}|\mathcal{X})} \{I^*(\mathcal{X};\hat{\mathcal{X}})\} + H^s \\ &\triangleq C^b + H^s, \end{aligned} \tag{2.2}$$

where C^b represents the traditional Shannon channel capacity, measured in units of bits per unit time (bit/s). On the other hand, H^s is a semantic-relevant term, calculated as $H^s = \overline{H_s(\hat{\mathcal{X}})} - H(\mathcal{W}|\mathcal{X})$. This term can either be positive or negative, contingent upon the background knowledge (i.e., the aforementioned knowledge bases) and the inference capability of specific semantic coding models employed at both the source and the destination.

Given Eq. (2.2), let us now consider a single downlink channel between MU i and BS j within the PKM-based SC-Net. Referring to Definition 2.1, we establish that MU i and BS j must possess perfectly matched knowledge bases. Assuming further that BS j's semantic encoder (i.e., message generator) exhibits equivalent inference ability to MU i's semantic decoder (i.e., message interpreter), we can apply Theorem 3 as outlined in Bao et al. [2011] to this link. In accordance with Eq. (2.2), let C_{ij}^s denote the achievable message rate for this link, C_{ij}^b denote the achievable bit rate, and H_{ij}^s denote the given semantic-relevant term specific to this link. Consequently, we can formalize their relationship through the following definition.

1 The notion of logical probability was initially introduced by Carnap and Bar-Hillel [1952]. It characterizes the likelihood of an observable message being true, which differs significantly from the conventional statistical probability represented by $P(\hat{X})$. For further technical elucidation on $P_s(\hat{X})$, interested readers can refer to Bao et al. [2011], Basu et al. [2014], and Strinati and Barbarossa [2021], and the seminal work by Carnap and Bar-Hillel [1952].

Definition 2.2 *In the PKM-based SC-Net, we define $S_{ij}^P(\cdot)$ as the bit-rate-to-message-rate (B2M) transformation function of the wireless link between MU i and BS j, such that*

$$S_{ij}^P\left(C_{ij}^b\right) \triangleq C_{ij}^s = C_{ij}^b + H_{ij}^s. \tag{2.3}$$

In accordance with Shannon's theorem, we recognize that C_{ij}^b can be directly computed based on the spectrum and the signal-to-interference-plus-noise ratio (SINR) of the link. Consequently, leveraging the channel conditions and the semantic coding models, we possess the capability to adjust the spectrum (i.e., the input of $S_{ij}^P(\cdot)$) allocated to this link, thereby optimizing the corresponding achievable message rate (i.e., the output of $S_{ij}^P(\cdot)$). Furthermore, it is evident that $S_{ij}^P(\cdot)$ exhibits linearity with respect to the bit rate C_{ij}^b, thereby offering a straightforward pathway toward solving the subsequent PKM-based resource optimization problem.

2.2.3 Semantic Channel Model in the IKM-Based SC-Net

In the context of the IKM-based SC-Net, each communication link no longer adheres to the perfect knowledge matching condition, rendering the aforementioned semantic-channel coding theorem inapplicable to any IKM-based case. Unfortunately, as far as our current knowledge extends, no existing work has put forth a SemCom-related information theory with rigorous derivations to ascertain the semantic channel capacity under mismatched background knowledge between the transceivers, as observed in the IKM scenario. Nevertheless, within the scope of this study, we acknowledge that for a given IKM-based link, there still exists a discernible relationship between the achievable message rate and the degree of knowledge matching. Specifically, the better the knowledge matching between the source and the destination, the greater the number of messages the destination can accurately interpret and vice versa [Strinati and Barbarossa, 2021]. This rationale is speculative in nature. For instance, as outlined in Definition 2.2, the message rate can attain an upper bound C_{ij}^s if MU i and BS j are in the PKM state. Conversely, no source information can be correctly interpreted if they lack matched knowledge bases [Bao et al., 2011].

Given the earlier discussion, the following definition introduces our semantic channel modeling for the IKM case.

Definition 2.3 *In the IKM-based SC-Net, we define $S_{ij}^I(\cdot)$ as the B2M transformation function of the physical link between MU i and BS j. This function is correlated with its PKM-based counterpart $S_{ij}^P(\cdot)$ as follows:*

$$S_{ij}^I(\cdot) = \beta_{ij} \cdot S_{ij}^P(\cdot). \tag{2.4}$$

Here, β_{ij} represents the knowledge matching coefficient, modeled as a random variable with values ranging from 0 to 1. A value of $\beta_{ij} = 0$ signifies a completely mismatched state between MU i and BS j, whereas $\beta_{ij} = 1$ denotes a perfectly matched state.

Clearly, the upper bound of the message rate in the IKM case must coincide with the message rate obtained in its PKM case (i.e., $\beta_{ij} = 1$). Conversely, it is also possible for the message rate to reach zero when there is no common background knowledge between the transceivers (i.e., $\beta_{ij} = 0$), as previously mentioned. Moreover, since the information source is typically modeled as a stochastic process [Liu et al., 2022], the specific quantity of messages generated corresponding to matched or mismatched knowledge bases becomes uncertain, even given the knowledge matching state. Consequently, in comparison to the PKM case, only a random proportion β_{ij} of messages can be accurately interpreted in the IKM scenario, ultimately resulting in a random message rate with respect to $S_{ij}^{I}(\cdot)$. In light of the aforementioned observations, we propose the following proposition.

Proposition 2.1 Given the knowledge matching degree, denoted as τ_{ij}, between MU i and BS j in the IKM case, the random knowledge matching coefficient β_{ij} follows a Gaussian distribution with a mean of τ_{ij} and a variance of $\sigma^2 ij$, i.e., $\beta ij \sim \mathcal{N}\left(\tau_{ij}, \sigma^2 ij\right)$, where $\sigma^2 ij = \tau_{ij}\left(1 - \tau_{ij}\right)$.

Proof: Consider a downlink scenario between MU i and BS j. Let us first assume that there exists a set of source information, modeled as a stochastic process $\left\{W_m\right\}$ $(m = 1, 2, \ldots, M)$, at the BS j side, where each W_m is independent of each other [Liu et al., 2022]. Furthermore, let K_m denote the knowledge base required by source information W_m for SemCom, and let $\mathcal{K}_{ij}$ denote the set of knowledge bases matched between MU i and BS j. Hence, we can define a KB matching indicator for source information as follows:

$$Z_m = \begin{cases} 1, & \text{if } K_m \in \mathcal{K}_{ij} \\ 0, & \text{otherwise} \end{cases}. \tag{2.5}$$

Given the knowledge matching degree τ_{ij} of the link between MU i and BS j, the probability of successful matching of W_m, i.e., the probability of $Z_m = 1$, becomes τ_{ij}. Moreover, based on the same link, its different source information $\left\{W_m\right\}$ should have the same probability of successful matching, which is entirely independent of the knowledge matching situations of other links. Therefore, $\left\{Z_m\right\}$ follows an identical binomial distribution with respect to τ_{ij}, such that

$$\begin{cases} \Pr\left(Z_m = 1\right) = \tau_{ij} \\ \Pr\left(Z_m = 0\right) = 1 - \tau_{ij} \end{cases}, \tag{2.6}$$

where $\Pr(\cdot)$ is the probability measure.

With these considerations, the random knowledge matching coefficient β_{ij} can now be expressed as the mean of the sum of $\left\{Z_m\right\}$ from a statistical average point of view as M approaches infinity. Mathematically, this can be formulated as

$$\beta_{ij} = \lim_{M \to +\infty} \frac{1}{M} \sum_{m=1}^{M} Z_m. \tag{2.7}$$

Based on (2.6) and (2.7), the classical central limit theorem [Fischer, 2011] can be directly applied to determine the distribution of β_{ij}, leading to $\beta_{ij} \sim \mathcal{N}\left(\tau_{ij}, \tau_{ij}\left(1 - \tau_{ij}\right)\right)$.

From Proposition 2.1, it is observed that in the IKM-based SC-Net case, each BS j is only capable of obtaining the deterministic information of τ_{ij} (i.e., the distribution of β_{ij}) from its link associated with MU i. This circumstance consistently leads to a stochastic optimization problem for IKM-based resource management. Consequently, the IKM problem is inevitable, and its solution should be distinct from that of the PKM problem due to the stochastic nature of each β_{ij}.

2.2.4 Basic Network Topology of SC-Net

Let us consider the network topology of both PKM-based and IKM-based SC-Nets. As illustrated in Figure 2.1, suppose there are a total of U MUs randomly distributed within the coverage area of

B BSs. Each MU $i \in \mathcal{U} = \{1, 2, \ldots, U\}$ can only be associated with one BS $j \in \mathcal{B} = \{1, 2, \ldots, B\}$ at a given time. Specifically, by Definition 2.1, let $\mathcal{B}i^P$ ($\mathcal{B}i^P \subseteq \mathcal{B}, \forall i \in \mathcal{U}$) denote the set of BSs holding all the KBs required by MU i in the PKM-based SC-Net. For the IKM-based SC-Net, assuming a minimum threshold for the knowledge matching degree, denoted as τ_0, is required to ensure the minimum quality of SemCom. Thus, let $\mathcal{B}i^I$ represent the set of BSs that MU i is eligible for CA in the IKM-based SC-Net. Mathematically, $\mathcal{B}i^I = \left\{j \mid j \in \mathcal{B}, \tau_{ij} \geqslant \tau_0\right\}, \forall i \in \mathcal{U}$. Given the previous considerations, if we define the binary CA indicator for both scenarios as $x_{ij} \in \{0, 1\}$, where $x_{ij} = 1$ indicates that MU i is associated with BS j and $x_{ij} = 0$ otherwise, the CA constraints for MU i in the PKM-based and IKM-based SC-Nets can be formulated as follows:

$$\sum_{j \in \mathcal{B}_i^P} x_{ij} = 1 \quad \text{and} \quad \sum_{j \in \mathcal{B}_i^I} x_{ij} = 1, \ \forall i \in \mathcal{U}, \tag{2.8}$$

respectively.

In the meantime, the total budget for SA of BS j is denoted as N_j, and the amount of spectrum that the BS j assigns to MU i is denoted as n_{ij}. Let γ_{ij} be the SINR experienced by the link, so the achievable bit rate can be found by $C_{ij}^b = n_{ij} \log_2\left(1 + \gamma_{ij}\right)$. Further, according to Definitions 2.2 and 2.3, the corresponding achievable message rate is $S_{ij}^P\left(C_{ij}^b\right)$ in the PKM-based SC-Net and $S_{ij}^I\left(C_{ij}^b\right)$ in the IKM-based SC-Net. With these, considering the uniqueness and significance of message rate in SemCom (i.e., where the conveyed message itself becomes the sole focus for correct reception in SemCom, rather than traditional transmitted bits [Bao et al., 2011; Strinati and Barbarossa, 2021]), we define a new performance metric herein, namely*STM*, to specifically measure the overall message rates obtained by all MUs in the network. Consequently, the STM of PKM-based SC-Net is given as

$$STM^P = \sum_{i \in \mathcal{U}} \sum_{j \in \mathcal{B}} x_{ij} S_{ij}^P \left(C_{ij}^b\right) = \sum_{i \in \mathcal{U}} \sum_{j \in \mathcal{B}} x_{ij} S_{ij}^P \left(n_{ij} \log_2\left(1 + \gamma_{ij}\right)\right). \tag{2.9}$$

Likewise, the STM of IKM-based SC-Net is

$$STM^I = \sum_{i \in \mathcal{U}} \sum_{j \in \mathcal{B}} x_{ij} S_{ij}^I \left(C_{ij}^b\right) = \sum_{i \in \mathcal{U}} \sum_{j \in \mathcal{B}} x_{ij} \beta_{ij} S_{ij}^P \left(n_{ij} \log_2\left(1 + \gamma_{ij}\right)\right). \tag{2.10}$$

With the metrics STM^P and STM^I in hand, we can proceed to jointly optimize the CA and SA from the perspective of SemCom. This optimization aims to maximize the overall network performance for both the PKM-based and IKM-based SC-Net scenarios.

2.3 Optimal CA and SA Solution in the PKM-Based SC-Net

2.3.1 Problem Formulation

In order to ensure the delivery of high-quality SemCom services to all MUs in the PKM-based SC-Net, it is crucial to maximize the semantic throughput metric (STM^P) while adhering to various SemCom-related and practical system constraints. To achieve this objective, we formulate an STM-maximization problem that jointly optimizes the CA variable x_{ij} and the SA variable n_{ij}. For clarity, we introduce matrices $\boldsymbol{x}$ and $\boldsymbol{n}$, which respectively contain all CA and SA variables associated with MUs and BSs. It is noteworthy that both $\boldsymbol{x}$ and $\boldsymbol{n}$ are intricately linked to semantic aspects: $\boldsymbol{x}$ is subject to constraints imposed by PKM-based links, while $\boldsymbol{n}$ determines the upper bound not only for the bit-based channel capacity but also for the semantic channel capacity. Specifically, the joint optimization problem for the PKM-based SC-Net is formulated as follows:

$$\mathbf{P1}: \max_{\boldsymbol{x},\boldsymbol{n}} \sum_{i\in\mathcal{U}}\sum_{j\in\mathcal{B}} x_{ij} S_{ij}^{P}\left(n_{ij}\log_2\left(1+\gamma_{ij}\right)\right) \tag{2.11}$$

$$\text{s.t.} \sum_{j\in\mathcal{B}_i^P} x_{ij} = 1, \ \forall i \in \mathcal{U}, \tag{2.11a}$$

$$\sum_{i\in\mathcal{U}} x_{ij} n_{ij} \leqslant N_j, \ \forall j \in \mathcal{B}, \tag{2.11b}$$

$$x_{ij} \in \{0,1\}, \ \forall (i,j) \in \mathcal{U} \times \mathcal{B}. \tag{2.11c}$$

Constraint (2.11a) enforces the single-BS constraint for CA, ensuring that only BSs belonging to $\mathcal{B}_i^P$ can associate with MU i to achieve the perfect knowledge matching (PKM) state. Constraint (2.11b) restricts the total spectrum allocated to MUs, ensuring it does not exceed the SA budget of each BS. Constraint (2.11c) ensures the binary nature of $\boldsymbol{x}$, indicating the association or non-association of MUs with BSs.

2.3.2 Optimal Solution for UA

The main challenge in solving **P1** lies in the binary constraint (2.11c). To address this, we initially relax $\boldsymbol{x}$ to continuous variables between 0 and 1. However, directly solving the relaxed problem while ensuring the recovery of the binary property of $\boldsymbol{x}$ without compromising performance significantly remains challenging. To overcome this hurdle, we introduce a minimum SA, denoted as n_{ij}^T, that BS j must allocate to its associated MU i to ensure basic signal quality given the channel conditions. By setting each n_{ij} to n_{ij}^T, **P1** can be reformulated as

$$\mathbf{P1.1}: \max_{\boldsymbol{x}} \sum_{i\in\mathcal{U}}\sum_{j\in\mathcal{B}} x_{ij}\xi_{ij}^{T} \tag{2.12}$$

$$\text{s.t.} \sum_{j\in\mathcal{B}_i^P} x_{ij} = 1, \forall i \in \mathcal{U}, \tag{2.12a}$$

$$\sum_{i\in\mathcal{U}} x_{ij} n_{ij}^T \leqslant N_j, \forall j \in \mathcal{B}, \tag{2.12b}$$

$$0 \leqslant x_{ij} \leqslant 1, \forall (i,j) \in \mathcal{U} \times \mathcal{B}, \tag{2.12c}$$

where

$$\xi_{ij}^T \triangleq S_{ij}^P\left(n_{ij}^T\log_2\left(1+\gamma_{ij}\right)\right). \tag{2.13}$$

It is worth noting that ξ_{ij}^T is treated as a constant in the objective function (2.12), given that n_{ij}^T, γ_{ij}, and H_{ij}^s in the B2M function $S_{ij}^P(\cdot)$ are all constants for the specified link between MU i and BS j.

In the context of **P1.1**, we utilize the Lagrange dual method (Low, 1999) to derive its dual optimization problem. Introducing a Lagrange multiplier $\boldsymbol{\mu} = \left\{\mu_j \mid j \in \mathcal{B}\right\}$, we can merge the inequality constraint (2.12b) into (2.12). Consequently, the Lagrange function is formulated as follows:

$$L(x,\mu) = \sum_{i\in\mathcal{U}}\sum_{j\in\mathcal{B}} x_{ij}\xi_{ij}^T + \sum_{j\in\mathcal{B}} \mu_j\left(N_j - \sum_{i\in\mathcal{U}} x_{ij} n_{ij}^T\right). \tag{2.14}$$

Hence, the Lagrange dual problem of **P1.1** becomes

$$\mathbf{D1.1}: \quad \min_{\boldsymbol{\mu}} D(\boldsymbol{\mu}) = g_{\boldsymbol{x}}(\boldsymbol{\mu}) + \sum_{j\in\mathcal{B}} \mu_j N_j \tag{2.15}$$

$$\text{s.t.} \quad \mu_j \geqslant 0, \ \forall j \in \mathcal{B}, \tag{2.15a}$$

where we have

$$g_{\boldsymbol{x}}(\boldsymbol{\mu}) = \sup_{\boldsymbol{x}} \sum_{i\in\mathcal{U}}\sum_{j\in\mathcal{B}} x_{ij}\left(\xi_{ij}^T - \mu_j n_{ij}^T\right) \quad \text{s.t. (2.12a), (2.12c).} \tag{2.16}$$

It is noteworthy that strong duality is preserved in this primal–dual transformation. This is because the objective function (2.12) of **P1.1** exhibits convexity, and all constraints are linear and affine inequalities, thereby satisfying Slater's condition [Boyd et al., 2004].

Given the initial dual variable $\boldsymbol{\mu}$, the optimal $\boldsymbol{x}$ can be determined first, denoted as $\boldsymbol{x} = \left\{x_{ij} \mid i \in \mathcal{U}, j \in \mathcal{B}\right\}$. Subsequently, a gradient descent method can be employed to update $\boldsymbol{\mu}$ iteratively to solve **D1.1** [Boyd et al., 2004]. Upon careful examination of (2.16), it is evident that under the fixed n_{ij}^T, MU i can be served by its optimal BS j if and only if it satisfies the following condition:

$$x_{ij}^* = \begin{cases} 1, & \text{if } j = \arg\max\limits_{j\in\mathcal{B}_i^P}\left(\xi_{ij}^T - \mu_j n_{ij}^T\right) \\ 0, & \text{otherwise} \end{cases}, \quad \forall i \in \mathcal{U}. \tag{2.17}$$

Following the derivation of $\boldsymbol{x}^*$, the gradient with respect to $\boldsymbol{\mu}$ in the objective function $D(\boldsymbol{\mu})$ is computed and employed as the gradient for each iteration. Consequently, μ_j ($\forall j \in \mathcal{B}$) is updated according to the following expression:

$$\mu_j(t+1) = \left[\mu_j(t) - \delta(t)\cdot\left(N_j - \sum_{i\in\mathcal{U}} x_{ij}(t)\, n_{ij}^T\right)\right]^+. \tag{2.18}$$

The operator $[\cdot]^+$ is utilized to compute the maximum value between its argument and zero, ensuring that $\boldsymbol{\mu}$ remains nonnegative as prescribed in constraint (2.15a). The stepsize in iteration t, denoted by $\delta(t)$, is crucial for ensuring convergence in gradient descent methods, as noted in previous work [Wang et al., 2016]. Additionally, to uphold the spectral availability constraint (2.12b), it is imperative to assess the total spectrum utilization at each BS based on the solution $\boldsymbol{x}^*$. For BSs that breach constraint (2.12b), a strategy is adopted to redistribute their most spectrum-consuming MUs to other BSs, following the guideline outlined in (2.17) until all BSs adhere to the spectral budget requirements. To summarize, by iteratively updating $\boldsymbol{x}$ and $\boldsymbol{\mu}$ until convergence, the resource allocation problem within the PKM-based SC-Net can be effectively resolved.

2.3.3 Optimization Solution for SA

With the CA solution $\boldsymbol{x}^*$ at hand and a predetermined spectrum threshold n_{ij}^T, we can readily construct the SA problem for each BS j ($\forall j \in \mathcal{B}$) as follows:

$$\mathbf{P1.2}^{(j)}: \quad \max_{\boldsymbol{n}} \sum_{i\in\mathcal{U}_j^P} S_{ij}^P\left(n_{ij}\log_2(1+\gamma_{ij})\right) \tag{2.19}$$

$$\text{s.t.} \quad \sum_{i\in\mathcal{U}_j^P} n_{ij} = N_j, \tag{2.19a}$$

$$n_{ij} \geqslant n_{ij}^T, \ \forall i \in \mathcal{U}_j^P, \tag{2.19b}$$

where

$$\mathcal{U}_j^P \triangleq \left\{ i \mid x_{ij}^* = 1 \right\}. \tag{2.20}$$

In this context, $\mathcal{U}j^P$ denotes the set of MUs associated with BS j in the preceding CA phase. Given the linear nature of $Sij^P(\cdot)$, it is evident that within each $\mathbf{P1.2}^{(j)}$ subproblem, both the objective function and all constraints exhibit convexity. Consequently, leveraging efficient optimization toolboxes such as CVXPY [Diamond and Boyd, 2016] enables the direct derivation of the optimal SA solution for the PKM-based SC-Net.

2.4 Optimal CA and SA Solution in the IKM-Based SC-Net

2.4.1 Problem Formulation

Similar to the rationale underlying **P1**, ensuring the optimality of STM^I is imperative for enhancing the overall SemCom-related network performance within the IKM-based SC-Net. Drawing from the CA indicator $\boldsymbol{x}$ and the SA indicator $\boldsymbol{n}$,[2] the joint optimization problem of the IKM-based SC-Net can be formulated as

$$\mathbf{P2}: \max_{\boldsymbol{x},\boldsymbol{n}} \sum_{i\in\mathcal{U}} \sum_{j\in\mathcal{B}} x_{ij}\beta_{ij}S_{ij}^P\left(n_{ij}\log_2\left(1+\gamma_{ij}\right)\right) \tag{2.21}$$

$$\text{s.t.} \sum_{j\in\mathcal{B}_i^I} x_{ij} = 1, \quad \forall i \in \mathcal{U}, \tag{2.21a}$$

$$\sum_{i\in\mathcal{U}} x_{ij}n_{ij} \leqslant N_j, \quad \forall j \in \mathcal{B}, \tag{2.21b}$$

$$x_{ij} \in \{0,1\}, \quad \forall (i,j) \in \mathcal{U} \times \mathcal{B}. \tag{2.21c}$$

In contrast to **P1**, the CA constraint (2.21a) in **P2** serves to restrict the association between MU i and only those BSs enabled with IKM capabilities within $\mathcal{B}i^I$. This association is determined by a predefined minimum knowledge matching threshold $\tau 0$ within the network. Similarly, constraints (2.21b) and (2.21c) impose limitations on the SA for each BS and enforce the binary nature of $\boldsymbol{x}$, respectively.

It is crucial to highlight that the introduction of the random knowledge matching coefficient β_{ij} marks the primary distinction between **P1** and **P2**. In **P1**, we encounter a deterministic optimization problem, whereas **P2** presents a stochastic optimization problem. Specifically, in **P1**, the solution $(\boldsymbol{x}, \boldsymbol{n})$ directly determines the numerical value of STM^P. However, in **P2**, these variables impact the pdf of STM^I with respect to β_{ij}. Consequently, the primary challenge in solving **P2** lies in addressing the stochastic nature of β_{ij}. In our study, we have developed a dedicated two-stage methodology to determine the optimal $\boldsymbol{x}$ and $\boldsymbol{n}$. Firstly, we convert the nondeterministic problem **P2** into a deterministic one by employing a chance-constrained optimization model. Subsequently, we propose an effective heuristic algorithm in the second stage to finalize the solution for CA and SA within the IKM-based SC-Net.

2 To streamline notation and avoid redundancy, we employ the same symbols (e.g., $\boldsymbol{x}$ and $\boldsymbol{n}$) in the IKM-based SC-Net as in the PKM-based SC-Net, signifying identical physical interpretations.

2.4.2 Problem Transformation with Semantic Confidence Level

Upon thorough examination of **P2**, it becomes apparent that the set $\boldsymbol{\beta} = \{\beta_{ij} \mid i \in \mathcal{U}, j \in \mathcal{B}\}$ solely appears within its objective function (2.21). Considering the distributional nature of (2.21), in our initial solution phase, we integrate Kataoka's model [Kataoka, 1963] to introduce a novel objective function alongside an additional constraint. This adjustment ensures the primal problem's suitability for stochastic optimization while preserving the original intent. Designating the new objective function as $\overline{F}(\boldsymbol{x}, \boldsymbol{n})$ (with its explicit expression to be provided later), in accordance with Kataoka's model [Kataoka, 1963], **P2** can be equivalently reformulated as

$$\mathbf{P2.1}: \max_{\boldsymbol{x},\boldsymbol{n}} \quad \overline{F}(\boldsymbol{x}, \boldsymbol{n}) \tag{2.22}$$

$$\text{s.t.} \quad \Pr\left\{STM^I \geqslant \overline{F}(\boldsymbol{x}, \boldsymbol{n})\right\} \geqslant \alpha, \tag{2.22a}$$

$$(2.21a), (2.21b), (2.21c). \tag{2.22b}$$

Constraint (2.22a) represents the newly introduced probabilistic (chance) constraint, governed by a specified confidence level α ($0 < \alpha < 1$, typically set to a high value in practical applications [Charnes and Cooper, 1959]). To elaborate, given the stochastic nature of β_{ij}, the objective of **P2.1** is to achieve optimality in $(\boldsymbol{x}, \boldsymbol{n})$ to ascertain the optimal pdf of STM^I. This involves maximizing its lower bound $\overline{F}(\boldsymbol{x}, \boldsymbol{n})$ based on the predetermined confidence level α. In this context, α is referred to as "a semantic confidence level preset" for the IKM-based SC-Net.[3]

Furthermore, it is apparent that upon attaining the optimal solution to **P2.1**, STM^I exhibits a nondegenerate distribution, indicating that it does not converge to a constant value [Prékopa, 2013]. This observation implies

$$\Pr\left\{STM^I \geqslant \overline{F}(\boldsymbol{x}, \boldsymbol{n})\right\} = \alpha. \tag{2.23}$$

Based on our Proposition 2.1, it is evident that the sufficient condition outlined in Theorem 10.4.1 presented in Prékopa [2013] is entirely met. This condition entails determining the specific expression of $\overline{F}(\boldsymbol{x}, \boldsymbol{n})$ from (2.23). Hence, we get

$$\begin{aligned}\overline{F}(\boldsymbol{x}, \boldsymbol{n}) = &\sum_{i\in\mathcal{U}}\sum_{j\in\mathcal{B}} x_{ij}\tau_{ij}S_{ij}^{P}\left(n_{ij}\log_2\left(1+\gamma_{ij}\right)\right) \\ &- \Phi^{-1}(\alpha)\sqrt{\sum_{i\in\mathcal{U}}\left(\sum_{j\in\mathcal{B}} x_{ij}\sigma_{ij}S_{ij}^{P}\left(n_{ij}\log_2\left(1+\gamma_{ij}\right)\right)\right)^2},\end{aligned} \tag{2.24}$$

where $\Phi^{-1}(\cdot)$ denotes the inverse function of the standard normal probability distribution. Upon examining (2.23) and (2.24), it becomes apparent that even the maximum value of $\overline{F}(\boldsymbol{x}, \boldsymbol{n})$ (or equivalently, all possible combinations of $(\boldsymbol{x}, \boldsymbol{n})$) satisfies the confidence constraint in (2.22a). Consequently (2.22a), can now be eliminated from **P2.1**.

Building upon this foundation, we employ a similar strategy as in (2.13) to render **P2.1** tractable. Here, each n_{ij} in $\boldsymbol{n}$ is constrained by the predetermined spectrum threshold n_{ij}^T within the IKM-based SC-Net. Simultaneously, we relax the CA variable $\boldsymbol{x}$ into a continuous form to address the NP-hard challenge. As a result, we transform **P2.1** into a deterministic optimization problem,

---

3 While an expected value of the optimization goal could also serve as a measure of optimality criterion [Charnes and Cooper, 1959], opting for a specific probability mitigates the risk associated with obtaining significantly low profits under the given expectation, as elucidated in Kataoka [1963]. Hence, we prioritize setting a predetermined probability to enhance practicality.

expressed as

$$\mathbf{P2.2}: \max_{\boldsymbol{x}} \quad \overline{F}(\boldsymbol{x}) \triangleq \overline{F}(\boldsymbol{x}, \boldsymbol{n})|_{n_{ij}=n_{ij}^T} \tag{2.25}$$

$$\text{s.t.} \quad \sum_{j \in \mathcal{B}_i^I} x_{ij} = 1, \quad \forall i \in \mathcal{U}, \tag{2.25a}$$

$$\sum_{i \in \mathcal{U}} x_{ij} n_{ij}^T \leqslant N_j, \quad \forall j \in \mathcal{B}, \tag{2.25b}$$

$$0 \leqslant x_{ij} \leqslant 1, \quad \forall (i,j) \in \mathcal{U} \times \mathcal{B}. \tag{2.25c}$$

As emphasized in Kataoka [1963] and Charnes and Cooper [1959], the convexity of the objective function $\overline{F}(\boldsymbol{x})$ can be ensured under the assumption $\alpha > 1/2$, equivalently $\Phi^{-1}(\alpha) > 0$. This assumption holds practical significance, as an excessively small α imposes an overly stringent limit on the solution space $(\boldsymbol{x}, \boldsymbol{n})$ dictated by constraint (2.22a). This could potentially lead to the nonexistence of feasible solutions when combined with other constraints in **P2.1**. Moreover, it is noteworthy that in **P2.2**, n_{ij}^T, τ_{ij}, σ_{ij}, and α should all be regarded as known constants associated with the link between MU i and BS j during the problem-solving process.

2.4.3 Solution Finalization for CA and SA

In our second-stage solution, we initially employ the interior-point method [Potra and Wright, 2000] to transform the inequality-constrained problem **P2.2** into an equality-constrained problem. This approach facilitates efficient approximation toward optimality. To elaborate, let $\varphi(\boldsymbol{x})$ represent a logarithmic barrier function corresponding to the SA constraint (2.25b), defined as

$$\varphi(\boldsymbol{x}) = \sum_{j \in \mathcal{B}} \log \left(N_j - \sum_{i \in \mathcal{U}} x_{ij} n_{ij}^T \right). \tag{2.26}$$

That way, **P2.2** can be rephrased as

$$\mathbf{P2.3}: \max_{\boldsymbol{x}} \quad \overline{F}(\boldsymbol{x}) + r \cdot \varphi(\boldsymbol{x}) \tag{2.27}$$

$$\text{s.t. (2.25a), (2.25c).} \tag{2.27a}$$

Here, r represents a small positive scalar that determines the accuracy of the approximation. As r diminishes toward zero, the maximum value of the new objective function, as depicted in (2.27), converges to the optimal solution of the primal problem [Potra and Wright, 2000]. It is crucial to note that (2.27) retains convexity, as both $\overline{F}(\boldsymbol{x})$ and $\varphi(\boldsymbol{x})$ are convex functions. Consequently, we can readily identify a set $\boldsymbol{x}(r)$ containing all optimal x_{ij} values with respect to a given r for **P2.3**. Furthermore, employing the sequential unconstrained minimization technique [Fiacco and McCormick, 1990], the optimal solution $\boldsymbol{x}$ to **P2.2** (denoted as $\hat{\boldsymbol{x}}$) can be iteratively attained by updating the descent value of r until convergence.[4]

However, it is important to note that $\hat{\boldsymbol{x}}$ may not ensure binary values for each $\hat{x}_{ij}$. Thus, we introduce a heuristic algorithm to determine the optimal solution to **P2** (i.e., $\boldsymbol{x}^*$) based on the provided $\hat{\boldsymbol{x}}$. In this algorithm, each x_{ij}^* is determined according to the following rule:

$$x_{ij}^* = \begin{cases} 1, & \text{if } j = \arg\max_{j \in \mathcal{B}_i^I} \hat{x}_{ij} \\ 0, & \text{otherwise} \end{cases}, \quad \forall i \in \mathcal{U}. \tag{2.28}$$

4 The initialization and update rule of r are appropriately determined at the outset of the barrier method. Further technical details are available in Boyd et al. [2004].

An implicit interpretation of (2.28) suggests that each MU in the IKM state possesses multiple potentially associated BSs alongside the corresponding optimal weights, i.e., $\hat{x}_{ij}$s. These weights are highly correlated with the performance of STM. Consequently, each user has the liberty to select the BS with the maximum weight during CA to optimize the overall network performance within the SC-Net.

However, even after applying (2.28), the resulting spectrum consumption may still exceed the budget of some BSs. In such cases, we adopt the same strategy as in the PKM scenario by reallocating those MUs who consume the most spectrum from these BSs to other BSs, based on their weight list provided in $\hat{\boldsymbol{x}}$. This process continues until the SA constraint (2.25b) is satisfied for all BSs. Subsequently, akin to the approach for solving $\mathbf{P1.2^{(j)}}$, we can further formulate the SA optimization problem for each BS j ($j \in \mathcal{B}$) based on the obtained $\boldsymbol{x}^*$ and the fixed n_{ij}^T. This entails

$$\mathbf{P2.4^{(j)}}: \max_{\boldsymbol{n}} \quad \overline{F}(\boldsymbol{x}^*, \boldsymbol{n}) \tag{2.29}$$

$$\text{s.t.} \quad \sum_{i \in \mathcal{U}_j^I} n_{ij} = N_j, \tag{2.29a}$$

$$n_{ij} \geqslant n_{ij}^T, \quad \forall i \in \mathcal{U}_j^I, \tag{2.29b}$$

where

$$\mathcal{U}_j^I = \left\{ i \mid i \in \mathcal{U}, x_{ij}^* = 1 \right\}. \tag{2.30}$$

In each $\mathbf{P2.4^{(j)}}$, the objective function (2.29) is evidently convex, given the convexity of $\overline{F}(\boldsymbol{x}, \boldsymbol{n})$, while both constraints (2.29a) and (2.29b) are linear. This renders the problem amenable to optimization using the CVXPY toolbox [Diamond and Boyd, 2016]. Conclusively, both the CA and SA problems have been effectively optimized within the IKM-based SC-Net, despite the presence of the random knowledge matching coefficient $\boldsymbol{\beta}$.

2.5 Numerical Results and Discussions

In this section, we assess the performance of our proposed CA and SA solutions for PKM-based and IKM-based SC-Nets, respectively. In our baseline network configuration, we randomly deploy 5 picocell BSs (PBSs), 10 femtocell BSs (FBSs), and 200 MUs within a circular area with a radius of 500 m. Additionally, a macrocell BS (MBS) is positioned at the center of the circle. Furthermore, the transmit power of the MBS, PBSs, and FBSs is set to 43, 35, and 20 dBm, respectively, with each having a spectrum budget of 2 MHz. For the wireless propagation model, we adopt the path loss models $L(d) = 34 + 40\log(d)$ and $L(d) = 37 + 30\log(d)$ to represent the path loss for the MBS/PBSs and FBSs, respectively. Additionally, we assume a fixed noise power of -111.45 dBm [Boostanimehr and Bhargava, 2014].

In our SemCom-related model, we simulate a general text transmission-enabled SC-Net environment to evaluate the proposed solutions for accurate demonstration purposes. It is worth noting that while we focus on text transmission scenarios, the proposed solutions can be applied to other types of content (e.g., images or videos) for performance testing. We choose text-based scenarios due to the existence of well-established NLP-driven SemCom models. Specifically, we adopt a transformer model with the same structure as proposed in Xie et al. [2021] as a unified semantic coding model for all SemCom-enabled links. We employ the PyTorch-based Adam optimizer for network training with an initial learning rate of 1×10^{-3}. Furthermore, all source information utilized for transmission is derived from a public dataset extracted from the proceedings of the European Parliament [Koehn, 2005]. Initial sentences are pruned to a specified word-counting range from 4 to 30 to enhance subsequent computational efficiency and mitigate potential issues related to gradient vanishing or explosion. With these considerations, the corresponding PKM-based B2M function $S_{ij}^{P}(\cdot)$ can be approximated through model testing and will be presented in the results section. During the solution simulation of the PKM-based case, we set a dynamic step size of $\delta(t) = 0.8/t$ to update the Lagrange multipliers in (2.18). This ensures the convergence of each trial. In the IKM-based case, the knowledge matching degree τ_{ij} (with respect to β_{ij} in Proposition 2.1, $\forall(i,j) \in \mathcal{U} \times \mathcal{B}$) is uniformly set to 0.5 for all possible links in the SC-Net. Henceforth, we omit the subscript ij from τ_{ij} for brevity. Moreover, the semantic confidence level is fixed at $\alpha = 95\%$ in the two-stage solution of the IKM case.

For comparison purposes, we employ two baseline algorithms for CA and SA in both the PKM-based and IKM-based SC-Nets: (i) A *max-SINR plus water-filling* algorithm [He et al., 2013], which associates each MU with the BS providing the strongest SINR during the CA phase, and the water-filling SA method is applied; (ii) A *max-SINR plus evenly distributed* algorithm [Ye et al., 2013] that adopts the same max-SINR strategy for CA but utilizes an evenly distributed SA method. Additionally, we set a bit rate threshold (denoted by n_{ij}^{T}) of 0.01 Mbit/s in both the proposed and baseline solutions to ensure a basic quality of SemCom services for all MUs. It is important to note that all parameter values are set to default unless otherwise specified. Moreover, all subsequent simulation results are obtained by averaging over a significantly large number of trials.

2.5.1 Performance Evaluations in the PKM-Based SC-Net

We begin by evaluating the performance of the bilingual evaluation understudy (BLEU) metric in the PKM-based SC-Net. BLEU is a classical metric in the NLP field, ranging between 0 and 1 [Papineni et al., 2002]. It measures the similarity between the source and restored texts, with higher scores indicating better text recovery. By assessing BLEU, we gain insights into the accuracy of semantic interpretation, which is closely linked to the MUs' ability to correctly interpret messages in the network. To this end, we present the BLEU scores (1 gram) at different bit rates (C_{ij}^{b}) in Figure 2.4, considering four different SINRs in the link. In Figure 2.4, we observe that the BLEU score initially increases as the bit rate improves, stabilizing at a consistent score after approximately 0.03 Mbit/s. Moreover, higher SINRs correspond to higher BLEU scores, with scores plateauing when the SINR exceeds 6 dB. This behavior is expected, as varying channel conditions can lead to different degrees of signal attenuation, ultimately affecting the received bits. Notably, higher received bits result in lower semantic ambiguity. Overall, these observations underscore the importance of providing a minimum bit rate for MUs under favorable channel conditions in the SC-Net to achieve high-quality SemCom.

According to (2.2), if the optimal semantic encoding strategy is ensured, it is anticipated that each MU can achieve a higher message rate as the corresponding bit rate improves, a trend that correlates with its BLEU score. To examine this conjecture, we depict the B2M transformation relationship (with respect to $S_{ij}^{P}(\cdot)$) under the same four SINRs in Figure 2.5. It is important to note that the message rate (i.e., C_{ij}^{s}) obtained here is calculated based on the number of messages correctly interpreted within a given time unit. As anticipated, we observe that the transformed message rate steadily increases with the increment of the bit rate. Additionally, a higher SINR corresponds to a higher transformation rate of B2M, affirming the expected relationship between bit rate, semantic interpretation accuracy (as indicated by BLEU), and the B2M transformation rate.

The effectiveness of our CA and SA solution is demonstrated in the following two simulations. In these simulations, the PKM-enabled BS implies that each associated MU utilizes a well-trained

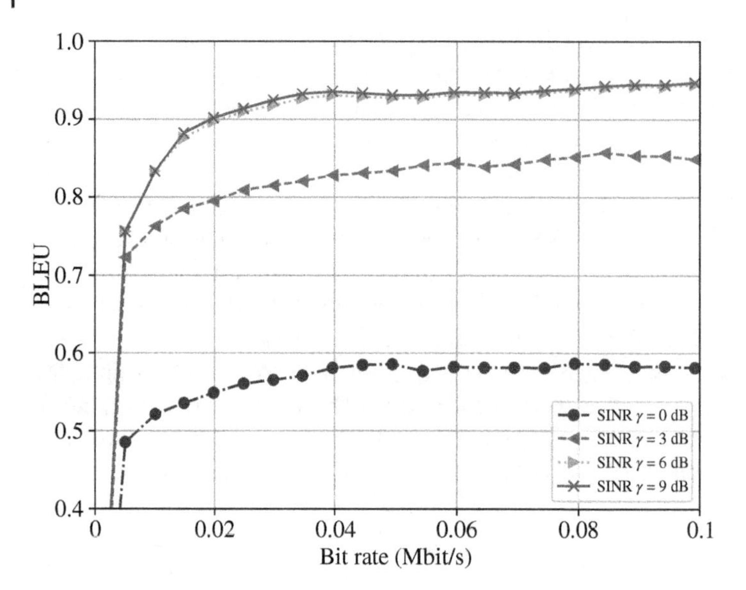

Figure 2.4 The BLEU score (1-gram) versus bit rates.

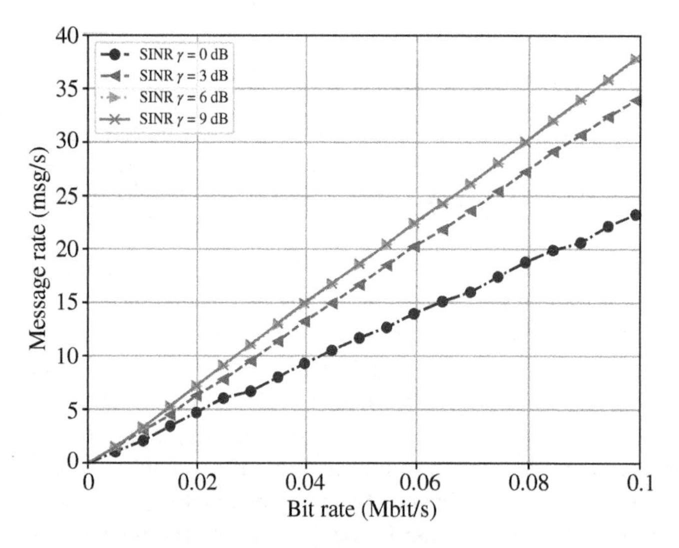

Figure 2.5 Demonstration of B2M transformation in the PKM-based SC-Net.

transformer decoding model with perfectly matched training data. Figure 2.6 compares the proposed solution with the two baseline algorithms by evaluating the STM performance under varying numbers of MUs ranging from 100 to 200. It is evident that the STM achieved by our solution consistently outperforms the two baselines. Specifically, the proposed solution maintains an average STM of 42.5 kmsg/s, which is approximately 6 kmsg/s higher than the max-SINR plus water-filling baseline and 13 kmsg/s higher than the max-SINR plus evenly distributed baseline. Additionally, the stable STM trend observed across all methods is due to the fact that the spectrum budget of all BSs is exhausted during the SA phase. Consequently, enhancing the STM performance solely by increasing the number of MUs becomes challenging.

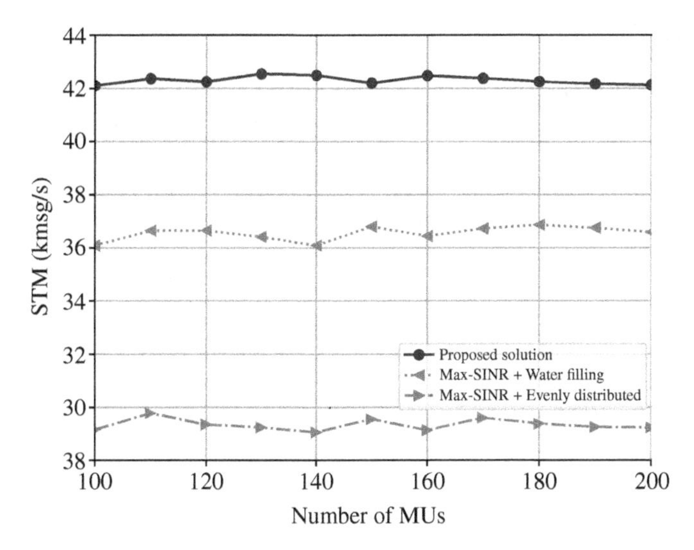

Figure 2.6 Comparison of the PKM-based STM performance under different numbers of MUs.

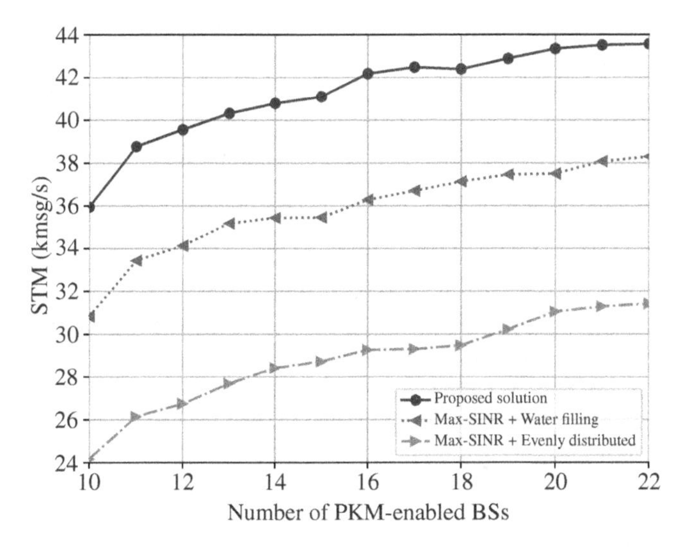

Figure 2.7 Comparison of the PKM-based STM performance under different numbers of BSs.

Furthermore, similar comparisons are conducted with different numbers of PKM-enabled BSs ranging from 10 to 20, as illustrated in Figure 2.7. Consistent with the results observed in Figure 2.6, our proposed solution achieves an additional average STM of around 6 kmsg/s compared to the max-SINR plus water-filling baseline and approximately 12 kmsg/s compared to the max-SINR plus evenly distributed baseline. Moreover, we observe that the STM performance of our solution initially improves and then gradually stabilizes after exceeding 20 BSs. With more BSs capable of providing PKM-based SemCom services to MUs, the MUs have access to increased spectrum resources, enabling higher message rates. However, when the number of BSs surpasses a certain threshold, the STM performance is expected to saturate or even degrade. This is attributed to the heightened channel interference resulting from the surplus of BSs.

2.5.2 Performance Testing in the IKM-Based SC-Net

To evaluate the effectiveness of the proposed two-stage solution in the IKM-based SC-Net, Figure 2.8 presents comparisons of the STM with varying numbers of MUs. Three different semantic confidence levels of $\alpha = 55\%, 75\%$, and 95% are considered. From Figure 2.8, it is evident that the obtained STM initially increases with the number of MUs, stabilizing after surpassing 130 MUs. This stabilization occurs because the spectrum budget of some BSs is exhausted after serving a high number of MUs, resulting in a plateau in STM performance, as discussed earlier. Moreover, an upward trend in STM performance is observed with a decrease in α. This observation can be elucidated from two perspectives. Mathematically, a higher value of α corresponds to a lower achievable bound on STM, leading to overall inferior network performance, as described in (2.23). Semantically, the randomness of the knowledge matching degree in the IKM case necessitates a moderate decrease in the preset semantic confidence to mitigate the risk of falling below the expected message rate for each MU. Consequently, a certain performance compromise in terms of STM is deemed acceptable, aligning with the high preset semantic confidence level. Furthermore, it is noteworthy that even at the highest required semantic confidence level of $\alpha = 95\%$, the proposed solution consistently outperforms the two benchmark algorithms.

Similarly, in Figure 2.9, comparable performance improvements are observed. Each solution is executed under two mean knowledge matching degrees, $\tau = 0.3$ and 0.7, with respect to $\boldsymbol{\beta}$ as defined in Definition 2.3. Under each τ, our two-stage solution consistently achieves an additional STM performance gain ranging from 2 to 6 kmsg/s with an increasing number of MUs compared to the two baseline algorithms. Additionally, the SA constraint of each method is consistently met, leading to STM performance stabilization beyond 130 MUs, a trend consistent with the findings in Figure 2.9. Furthermore, the impact of different knowledge matching degrees reveals a decreasing trend in STM as τ decreases. Higher τ values signify a greater likelihood of achieving accurate knowledge matching for SemCom. Consequently, each MU can attain higher message interpretation accuracy, resulting in improved STM performance in the IKM-based SC-Net.

Next, we evaluate the STM performance with varying semantic confidence levels α and knowledge matching degrees τ in Figures 2.10 and 2.11, respectively, under different numbers of IKM-enabled BSs. Figure 2.10 depicts the STM at different α values, showcasing higher STM with

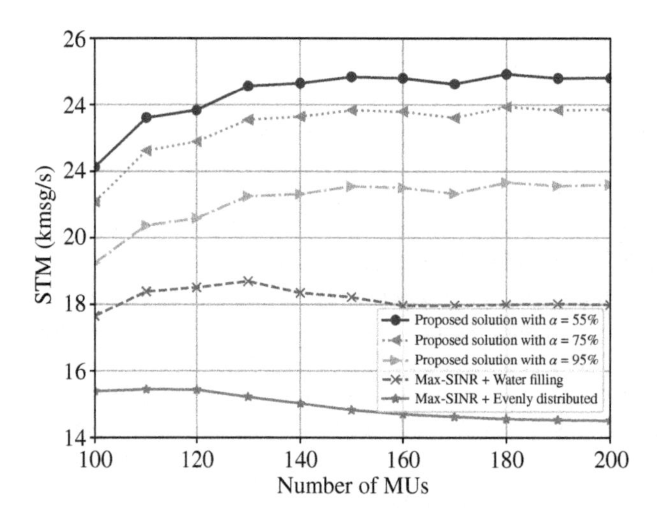

Figure 2.8 The IKM-based STM performance against varying number of MUs.

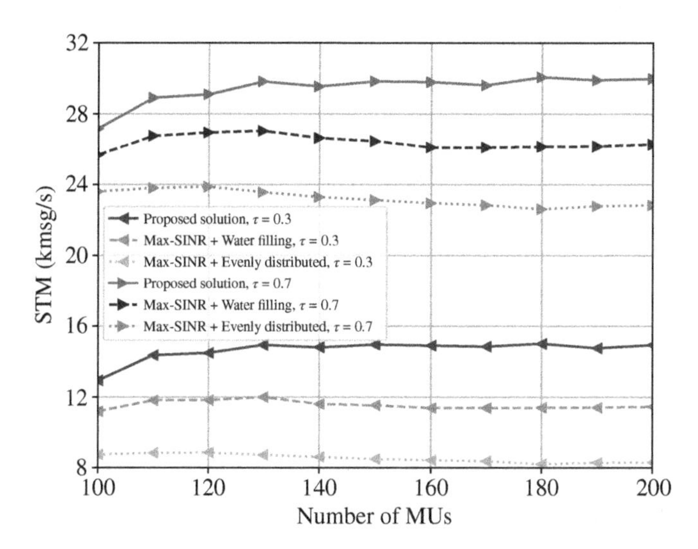

Figure 2.9 The IKM-based STM performance against varying number of MUs.

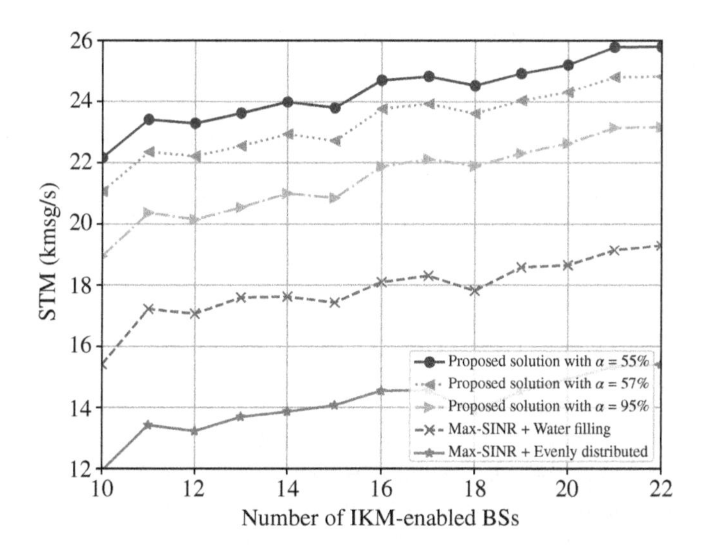

Figure 2.10 The IKM-based STM performance against different numbers of BSs.

lower semantic confidence levels, consistent with the observations in Figure 2.8. It is important to note that while lower semantic confidence levels may yield higher STM, they may not be practical in real-world scenarios. Consequently, in IKM-based SC-Nets, striking a balance between the preset risk level and desired STM becomes a crucial consideration. Regarding the effect of τ, Figure 2.11 demonstrates that higher knowledge matching degrees result in improved STM performance, as anticipated. Furthermore, our proposed solution consistently achieves higher STM performance compared to the two baseline algorithms. Additionally, a gradual increase in STM is observed across all solutions as the number of BSs increases, attributed to the increased spectrum availability at MUs.

Finally, upon a lateral comparison of the results between PKM-based and IKM-based SC-Nets, it is evident that transitioning from PKM to IKM leads to an inevitable penalty in STM performance.

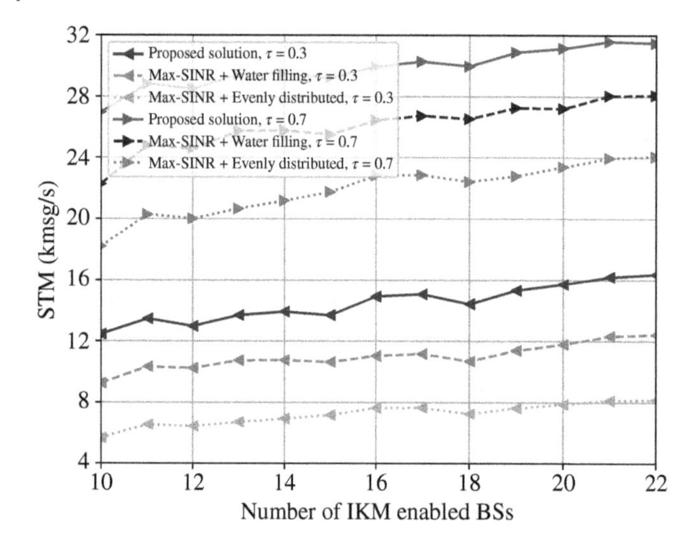

Figure 2.11 The IKM-based STM performance against different numbers of BSs.

For instance, considering the results from Figures 2.6 to 2.8, under the same simulation settings, our solution achieves an STM of approximately 42 kmsg/s in the PKM case, while only obtaining around 25 kmsg/s in the IKM case. This substantial reduction in STM performance can be attributed to the mismatching of partial KBs in the IKM-based semantic coding models. As a result, the message generation and interpretation abilities of IKM-based models cannot be fully harnessed, leading to a certain degree of semantic ambiguity compared to the PKM case. Therefore, ensuring adequate knowledge matching degrees in SemCom becomes paramount to achieving better network performance in IKM-based SC-Nets.

2.6 Conclusions

This chapter provides a comprehensive exploration of SemCom from a networking perspective. We begin by identifying two typical scenarios – PKM-based and IKM-based SC-Nets, and establish their semantic channel models in conjunction with existing works on semantic information theory. We introduce the concept of B2M transformation along with a novel network performance metric, STM, tailored to each SC-Net scenario. Subsequently, we formulate the joint optimization problems of CA and SA for both SC-Net scenarios, proposing solutions aimed at maximizing STM. Simulation results for each SC-Net scenario consistently demonstrate the superiority of our proposed solutions over two traditional benchmarks in terms of STM.

This chapter lays the groundwork for several future research avenues in wireless SemCom. One such avenue is the development of effective KB matching algorithms, which are crucial for ensuring high-quality SemCom service provisioning and STM performance in SC-Nets. Additionally, this work provides a foundation for extending resource management solutions to more complex SC-Net scenarios, such as PKM-IKM coexistence networks. Furthermore, insights gleaned from this study can inspire the design of new resource management strategies aligned with diverse SemCom-related objectives. These objectives may include enhancing the accuracy of message interpretation, minimizing end-to-end latency, and ensuring user fairness from a semantical standpoint.

References

Jie Bao, Prithwish Basu, Mike Dean, Craig Partridge, Ananthram Swami, Will Leland, and James A. Hendler. Towards a theory of semantic communication. In *2011 IEEE Network Science Workshop*, pages 110–117. IEEE, 2011.

Prithwish Basu, Jie Bao, Mike Dean, and James Hendler. Preserving quality of information by using semantic relationships. *Pervasive and Mobile Computing*, 11:188–202, 2014.

Hamidreza Boostanimehr and Vijay K. Bhargava. Unified and distributed QoS-driven cell association algorithms in heterogeneous networks. *IEEE Transactions on Wireless Communications*, 14(3):1650–1662, 2014.

Stephen Boyd, Stephen P. Boyd, and Lieven Vandenberghe. *Convex Optimization*. Cambridge University Press, 2004.

Rudolf Carnap and Yehoshua Bar-Hillel. An outline of a theory of semantic information, 1952.

Abraham Charnes and William W. Cooper. Chance-constrained programming. *Management Science*, 6(1):73–79, 1959.

Steven Diamond and Stephen Boyd. CVXPY: A Python-embedded modeling language for convex optimization. *The Journal of Machine Learning Research*, 17(1):2909–2913, 2016.

Anthony V. Fiacco and Garth P. McCormick. *Nonlinear Programming: Sequential Unconstrained Minimization Techniques*. SIAM, 1990.

Hans Fischer. *A History of the Central Limit Theorem: From Classical to Modern Probability Theory*. Springer, 2011.

Peter He, Lian Zhao, Sheng Zhou, and Zhisheng Niu. Water-filling: A geometric approach and its application to solve generalized radio resource allocation problems. *IEEE Transactions on Wireless Communications*, 12(7):3637–3647, 2013.

Shinji Kataoka. A stochastic programming model. *Econometrica: Journal of the Econometric Society*, 31(1/2):181–196, 1963.

Philipp Koehn. Europarl: A parallel corpus for statistical machine translation. In *MT Summit*, volume 5, pages 79–86. Citeseer, 2005.

Jiakun Liu, Shuo Shao, Wenyi Zhang, and H. Vincent Poor. An indirect rate-distortion characterization for semantic sources: General model and the case of Gaussian observation. *IEEE Transactions on Communications*, 70(9):5946–5959, 2022. doi: 10.1109/TCOMM.2022.3194978.

Xuewen Luo, Hsiao-Hwa Chen, and Qing Guo. Semantic communications: Overview, open issues, and future research directions. *IEEE Wireless Communications*, 29(1):210–219, 2022. doi: 10.1109/MWC.101.2100269.

Kishore Papineni, Salim Roukos, Todd Ward, and Wei-Jing Zhu. BLEU: A method for automatic evaluation of machine translation. In *Proceedings of the 40th Annual Meeting of the Association for Computational Linguistics*, pages 311–318, 2002.

Florian A. Potra and Stephen J. Wright. Interior-point methods. *Journal of Computational and Applied Mathematics*, 124(1-2):281–302, 2000.

András Prékopa. *Stochastic Programming*, volume 324. Springer Science & Business Media, 2013.

Walid Saad, Mehdi Bennis, and Mingzhe Chen. A vision of 6G wireless systems: Applications, trends, technologies, and open research problems. *IEEE Network*, 34(3):134–142, 2019.

Claude Elwood Shannon. A mathematical theory of communication. *The Bell System Technical Journal*, 27(3):379–423, 1948.

Xuemin Shen, Jie Gao, Wen Wu, Mushu Li, Conghao Zhou, and Weihua Zhuang. Holistic network virtualization and pervasive network intelligence for 6G. *IEEE Communications Surveys & Tutorials*, 24(1):1–30, 2021.

Guangming Shi, Yong Xiao, Yingyu Li, and Xuemei Xie. From semantic communication to semantic-aware networking: Model, architecture, and open problems. *IEEE Communications Magazine*, 59(8):44–50, 2021.

Emilio Calvanese Strinati and Sergio Barbarossa. 6G networks: Beyond Shannon towards semantic and goal-oriented communications. *Computer Networks*, 190:107930, 2021.

Ashish Vaswani, Noam Shazeer, Niki Parmar, Jakob Uszkoreit, Llion Jones, Aidan N. Gomez, Łukasz Kaiser, and Illia Polosukhin. Attention is all you need. *Advances in Neural Information Processing Systems 30* (NIPS 2017), 2017.

Ning Wang, Ekram Hossain, and Vijay K. Bhargava. Joint downlink cell association and bandwidth allocation for wireless backhauling in two-tier HetNets with large-scale antenna arrays. *IEEE Transactions on Wireless Communications*, 15(5):3251–3268, 2016.

Warren Weaver. Recent contributions to the mathematical theory of communication. *ETC: A Review of General Semantics*, 10(4):261–281, 1953.

Zhenzi Weng and Zhijin Qin. Semantic communication systems for speech transmission. *IEEE Journal on Selected Areas in Communications*, 39(8):2434–2444, 2021. doi: 10.1109/JSAC.2021.3087240.

Le Xia, Yao Sun, Chengsi Liang, Daquan Feng, Runze Cheng, Yang Yang, and Muhammad Ali Imran. WiserVR: Semantic communication enabled wireless virtual reality delivery. *IEEE Wireless Communications*, 30(2):32–39, 2023.

Huiqiang Xie and Zhijin Qin. A lite distributed semantic communication system for Internet of Things. *IEEE Journal on Selected Areas in Communications*, 39(1):142–153, 2020.

Huiqiang Xie, Zhijin Qin, Geoffrey Ye Li, and Biing-Hwang Juang. Deep learning enabled semantic communication systems. *IEEE Transactions on Signal Processing*, 69:2663–2675, 2021.

Yongjun Xu, Guan Gui, Haris Gacanin, and Fumiyuki Adachi. A survey on resource allocation for 5G heterogeneous networks: Current research, future trends and challenges. *IEEE Communications Surveys & Tutorials*, 23(2):668–695, 2021. doi: 10.1109/COMST.2021.3059896.

Qiaoyang Ye, Beiyu Rong, Yudong Chen, Mazin Al-Shalash, Constantine Caramanis, and Jeffrey G. Andrews. User association for load balancing in heterogeneous cellular networks. *IEEE Transactions on Wireless Communications*, 12(6):2706–2716, 2013.

3

An End-to-End Semantic Communication Framework for Image Transmission

Lei Feng, Yu Zhou, and Wenjing Li

The State Key Laboratory of Networking and Switching Technology, Beijing University of Posts and Telecommunications, Beijing, China

3.1 Introduction

In recent years, with the rapid increase in the demand for intelligent wireless communication, various emerging intelligent services based on wireless communication technologies (such as the industrial internet, connected vehicles, and augmented reality [AR]/virtual reality [VR]) have emerged in an endless stream. These emerging services are driving the transformation of future communication networks from a traditional paradigm that pursues high transmission rates to a new paradigm that is geared toward intelligent connectivity for all things. The traditional information theory is mainly concerned with the "bit" and "entropy" of information and how to transmit information effectively, especially in the communication channel with noise and interference [Dai et al., 2022]. The requirements of these emerging services for communication systems are not only to transmit data quickly and accurately but also to consider how to deal with and understand the content and meaning of data. Especially, a new communication paradigm called "semantic communication" has been stimulated as a whole new technology in 6G to break through the "Shannon bottleneck" [Yang et al., 2022]. Integrating users' information needs and semantics into the communication process is expected to greatly improve communication efficiency and enhance users' quality of experience (QoE).

The key to realizing semantic communication systems is the promotion of rapidly developing artificial intelligence (AI). Recently, various semantic communication technologies have emerged using deep learning methods to process different media such as text, images, and video [Xie et al., 2023]. Compared with text data explicitly represented by words, this chapter mainly focuses on research on semantic communication of image data because semantic information in images is implicit and knowledge based. We classify existing semantic communication work into two categories. The first category utilizes large-scale deep learning to extract global semantic information from the original data, embeds it into a low-dimensional space to compress the source information, and reconstructs the data at the semantic level based on the received semantic information. In Wang et al. [2022], they propose a distributed joint source-channel coding (JSCC) scheme that can encode and compress images for each device and jointly decode them at the receiving end to obtain a clear received image. [Huang et al., 2021] use an improved

Wireless Semantic Communications: Concepts, Principles, and Challenges, First Edition.
Edited by Yao Sun, Lan Zhang, Dusit Niyato, and Muhammad Ali Imran.

generative adversarial network (GAN) for semantic encoding and image transmission. Although these studies achieve better performance compared with traditional methods due to millions of parameters, the developed DNNs are very complex. Moreover, these works compress the semantic information of images into uninterpretable feature vectors that cannot be directly utilized and understood by human recipients, and they lack any physical meaning. The second category is task-oriented communication, which transmits only important and necessary information based on task metrics, deleting irrelevant information. It is undeniable that this semantic communication system achieves interpretable semantic compression and communication based on task-oriented metrics. However, due to its heavy reliance on the transmitted target, it is usually limited to certain specific wireless services or communication environments [Li et al., 2021; Xie et al., 2021]. Although the afore-mentioned semantic communication systems based on deep learning have demonstrated excellent performance compared to traditional methods, there are still some challenges in applying semantic communication to wireless networks for image data transmission, such as semantic information extraction, the impact of semantic noise, and the design of semantic metrics that can capture both semantic communication performance and wireless communication performance.

3.1.1 Related Works

Semantic noise is a type of noise that causes misinterpretation and decoding errors in semantic information. It leads to misguidance between the transmitted semantic symbols and the received semantic symbols, which can be introduced during semantic encoding, data transmission, and decoding [Hu et al., 2022]. During encoding, semantic noise may result from the incorrect recognition of entities and their relationships in the transmitted signal at the transmitter. Channel fading and noise may also cause the transmitted semantic messages to be lost and lead to semantic distortion. During decoding, the incorrect interpretation of semantic information and user misunderstanding at the receiver may also lead to the generation of semantic noise. For example, when there is a mismatch between the background knowledge bases (KBs) of the sender and receiver, the receiver may be unable to decode the correct semantic information. Therefore, in order to successfully interpret meaning at the semantic destination, we need to overcome not only physical channel noise but also semantic noise caused by encoding and decoding. In Hu et al. [2022], a masked vector quantization variational autoencoder (VQ-VAE) is designed, and samples with semantic noise are combined into the training dataset for adversarial training to achieve interference elimination. In Bourtsoulatze et al. [2019], JSCC for wireless image transmission is implemented using an autoencoder based on a deep convolutional neural network (CNN). Compared with traditional communication schemes, it significantly improves reliability performance against channel noise. In Güler et al. [2018], the authors propose an end-to-end semantic communication architecture that combines semantic reasoning with physical layer communication to achieve error correction at the semantic level. However, these works [Güler et al., 2018; Bourtsoulatze et al., 2019; Hu et al., 2022] all use model-free machine learning to implement encoding and decoding modules. The end-to-end operation is considered a black-box process, lacking interpretability. Semantic communication system designers must consider how to establish a connection between the complex and changing wireless environment and the precise semantic communication mechanism. Therefore, how to outline a mathematical model of semantic transmission that can capture the impact of semantic noise at the semantic level and quantitatively model the relationship between semantic transmission performance and semantic noise is currently an unresolved issue.

The key idea of semantic communication is to pursue maximum semantic fidelity using the minimum amount of communication resources rather than minimizing bit or symbol error rates. In semantic communication, the recipient should extract semantic information in a way that minimizes ambiguity about the transmitted message and use performance metrics to ensure the correct transmission of semantic information [Luo et al., 2022]. Therefore, appropriate metrics to measure semantic distance are crucial for optimizing image semantic communication systems. For images, image message performance can be measured using peak signal-to-noise ratio (PSNR) [Zhang et al., 2023b] or mean square error (MSE) [Luo et al., 2022], However, these traditional similarity metrics cannot well measure the semantic distance before and after transmission. They still focus on pixel-level reproduction rather than semantic-level reproduction. Generally speaking, image distortion necessarily means the failure of the transmission task. For example, in industrial workshops, high-definition camera images of products on the production line need to be transmitted to the cloud to complete various downstream tasks such as intelligent sorting and object contour monitoring. However, after applying semantic communication technology, information can be selectively compressed while preserving important information. Even if pixel-level distortion occurs during reconstruction, it will not cause significant performance losses when used for downstream tasks. Therefore, there is an urgent need to establish a reasonable measure of intention achievement efficiency on a true semantic level to optimize image semantic transmission performance.

3.1.2 Contribution

In this chapter, we establish a mathematical framework for end-to-end transmission models suitable for image semantic communication. We represent semantic information using knowledge graphs instead of feature vectors, which not only capture the meanings of various entities and their complex relationships but also consider personal preferences and background knowledge at both ends, ensuring accurate and interpretable semantic expression. Ignoring the impact of mismatched background knowledge and the wireless communication environment on both parties would greatly limit the efficiency of knowledge recognition and processing in semantic communication. Inspired by image communication systems, we innovatively propose the theory of semantic-spatial domain transformation. We project the extracted knowledge graph onto a three-dimensional (3D) tensor in the spatial domain, mapping the semantic ambiguity of entities to the intensity values of discrete points. At the receiving end, we reconstruct the knowledge graph based on background knowledge and the 3D tensor, addressing the limitations of traditional source–channel coding neural network structures that cannot visually reflect the interference of physical noise on the semantic level. To measure the differences in image semantic information before and after transmission, we propose a novel graph optimal transport (GOT) distance designed based on GOT theory to evaluate and calculate semantic similarity between two knowledge graphs graph-to-graph semantic similarity (GGSS). Additionally, we design an image-to-image semantic similarity (IISS) at the semantic level. We evaluate the quantitative accuracy of reconstructed images through visual question answering (VQA) and demonstrate consistency with human perception. Finally, we present a case study involving the intelligent task of VQA as the transmission objective for semantic communication systems.

3.2 The End-to-End Image Semantic Communication Framework Driven by Knowledge Graph

3.2.1 Overview

The proposed image semantic communication framework is shown in Figure 3.1. The original image at the source is $X \in \mathbb{R}^{H\times W\times 1}$, where H and W are the height and width of the original image, respectively. The original image is first compressed and encoded into semantic information G based on a knowledge graph, where the encoding process uses a priori knowledge stored in the source background knowledge base, including the parameter set $\boldsymbol{\theta}$ of the label distribution learning (LDL) and the relation tensor $\mathcal{R}$ that aids in entity and relationship detection. Then, as inspired by the theory of digital image processing, we project the knowledge graph G of semantic information into the spatial domain to obtain a 3D tensor $\mathcal{F}$ based on semantic ambiguity.

The projection process mainly regards the semantic ambiguity $\boldsymbol{w}$ of an entity in the semantic domain as the intensity vector $\boldsymbol{f}$ located at (x, y) by converting the entity bounding box position d of the entity extracted into spatial coordinates (x, y). From the point of view of digital image transmission, the random signal interference to which an image is subjected during ingestion or transmission manifests itself as random variations in image information or pixel brightness. From the perspective of semantic level, semantic information is affected by noise in the compilation, transmission, and decoding processes, and these interferences cause misrecognition of semantic information and distortion of interpretation, which is specifically manifested in the misinterpretation of entities and relationships. During the target recognition process, the size of the intensity values will affect the clarity of the image. If the difference in intensity values is too small, it will increase the difficulty of target recognition, and the probability of entity blur will also increase. Therefore, it is reasonable to associate the semantic ambiguity of entities in the semantic domain with pixel intensity vectors in the spatial domain, which is also consistent with human perception.

Consider that, with the exception of spatially periodic noise, channel noise is independent of spatial coordinates and uncorrelated with the image itself (i.e., there is no correlation between pixel values and noise component values). Therefore, the 3D tensor $\mathcal{F}$ will become $\hat{\mathcal{F}}$ after being transmitted over a wireless fading channel and noise interference. At the receiving end, we reduce the 3D tensor $\hat{\mathcal{F}}$ to get the received semantic information $\hat{G}$ through the mapping process. The mapping process mainly finds all the nonzero intensity values $\hat{\boldsymbol{f}}$ in the 3D tensor, whose spatial locations (x, y) are mapped to entity locations $\hat{d}$ in the semantic domain, with entity ambiguities $\hat{\boldsymbol{w}} = \hat{\boldsymbol{f}}$. The semantic decoder attempts to parse and generate the reconstructed image $\hat{X}$ from the received knowledge graph $\hat{G}$ by using its own background knowledge base.

Next, we introduce the details of semantic encoding, semantic information transfer, and semantic decoding. We then propose a new metric for the proposed image semantic communication framework, which can evaluate the semantic similarity between the extracted semantic information and the received semantic information. Furthermore, we propose an image-to-image semantic-level matching metric.

3.2.2 Semantic Encoding

We assume that the semantic information of an image consists of knowledge graph with entities and their relationships. Thus, the semantic information of each image is modeled by a knowledge graph defined by a set of nodes and edges, where a node denotes an entity and an edge denotes a

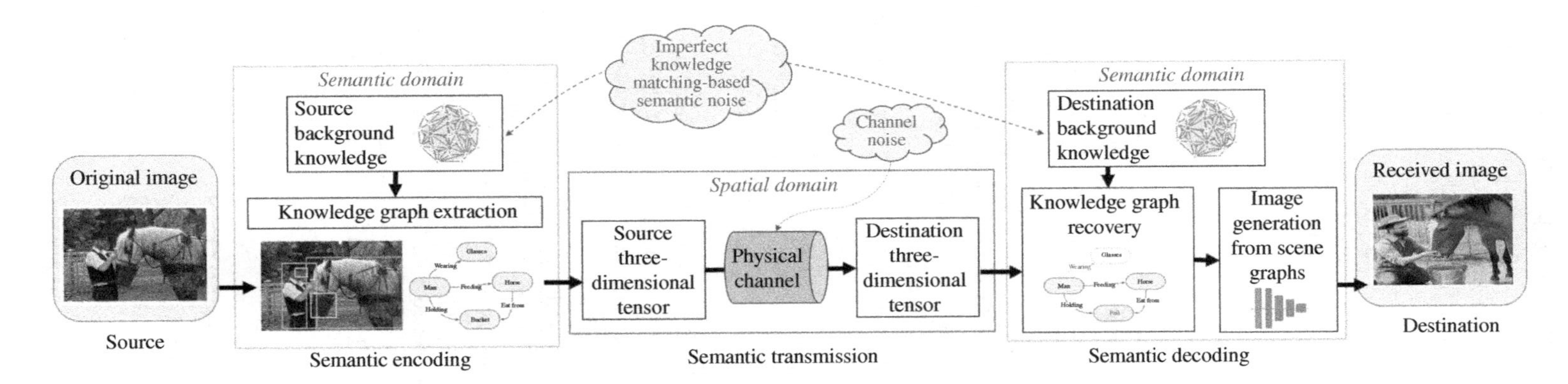

Figure 3.1 The end-to-end semantic communication framework with the knowledge graph for image transmission.

relationship between two entities. The semantic triple is the basic component of a knowledge graph. Each semantic triple corresponds to a combination of a head entity, a connecting relation, and a tail entity to represent subject–predicate–object information. For example, one of the semantic triples in Figure 3.1 is ["man"], ["feeding"], ["horse"]), where ["man"] and ["horse"] are entities and ["feeding"] is their relationship. The semantic information extraction process is divided into two steps: entity detection and relationship capture. Specifically, the user first detects and localizes all the entities in the image [Zhang et al., 2023a]. Then, all entities are sorted according to geometric and logical correlation to derive the relationship [Zellers et al., 2018] and output them in the form of a tuple.

3.2.2.1 Entity Detection

For entity detection, a CNN is usually used to obtain each entity or class instance present in the image and assign a bounding box to it based on its location. Due to variations in the performance of the detection model, semantic ambiguity can raise during the semantic extraction process, that is, the uncertainty between the real information in the image data and the extracted entities. For example, the identification of a "horse" in Figure 3.1 is classified as "donkey." Based on the LDL theory [Geng et al., 2013; Geng and Xia, 2014], we assume that for the input image X, instead of simply outputting a value, it quantizes the range of possible u values into several labels, and after each input image is divided into K blocks with bounding boxes, the block set is $\mathcal{K} = \{1, 2, \dots, K\}$. Each block is assigned a discrete label distribution $\boldsymbol{u}$. The ground-truth label distribution $\boldsymbol{u}$ is not available in most existing datasets, which must be generated under proper assumptions. The desirable label distribution $\boldsymbol{u} = (u_1, u_2, \dots, u_M)$ must satisfy some basic principles. (i) $\boldsymbol{u}$ should be a probability distribution. Thus, we have $u_i \in [0, 1]$ and $\sum_{i=1}^{M} u_i = 1$. (ii) The probability values u_i should have differences among all possible labels associated with an image. In other words, a less ambiguous category must be assigned a high probability, and those more ambiguous labels must have low probabilities.

This chapter mainly adopts the CNN to pretrain the LDL on a large number of images captured in this task [Gao et al., 2017], and stores the parameter $\boldsymbol{\theta}$ of the learned LDL framework in the source background knowledge base. Specifically, the label distribution vector $\boldsymbol{u} \in \mathbb{R}^{|\mathcal{O}|}$, where $\mathcal{O} = \left\{o_1, o_2, \dots, o_M\right\}$, is the label set defined for a specific task, where M is the number of all entities recorded in the source background knowledge base. We assume $\mathcal{O}$ is complete; i.e., any possible u value has a corresponding member in $\mathcal{O}$. The goal of LDL in pretraining is to learn a conditional probabilistic quality function directly from the sample dataset. Once $\boldsymbol{\theta}$ is learned, the label distribution $\boldsymbol{u}$ of any new instance can be generated by a forward run of the network. The original image is semantically encoded with the assistance of the parameters $\boldsymbol{\theta}$ of the known LDL framework in the source background knowledge base. The kth split block of the original image (i.e., $X_k, k \in \mathcal{K}$) is output by the LDL module as a series of category confidences.

$$\begin{aligned}\boldsymbol{w}_k &= [p(o_1|X_k;\boldsymbol{\theta}), p(o_2|X_k;\boldsymbol{\theta}), \dots, p(o_i|X_k;\boldsymbol{\theta}), \dots, p(o_M|X_k;\boldsymbol{\theta})] \\ &= [u_k^1, u_k^2, \dots, u_k^i, \dots, u_k^M],\end{aligned} \tag{3.1}$$

where the value of the probability u_k^i is defined as the probability that the bounding box detects and belongs to entity o_i, and the block X_k will be detected as the entity with the maximum u_k^i value. The semantic ambiguity of the detected entity from the kth block can be expressed as a one-dimensional vector $\boldsymbol{w}_k$.

In order to filter out redundant messages to improve transmission efficiency, if the semantic ambiguity of the block is lower than the threshold ξ_s, the patch will not be semantically encoded, where $\xi_s \in [0, 1]$ is the predefined threshold. After filtering, the input image X will be represented as a series of entities and a set of entity position coordinates.

$$E = \{(o_1, d_1), (o_2, d_2), \ldots, (o_k, d_k), \ldots\}, k \in \mathcal{K}_s = \{1, 2, \ldots, K_s\}, \tag{3.2}$$

where $K_s \leq K$ is the total number of entities detected and filtered from the original image X, which also means that K_s blocks have successfully detected entities.

3.2.2.2 Relationship Detection

The statistical analysis of the research [Zellers et al., 2018] shows that the type of the relationship between the two is easy to guess after knowing the ternary head entity and the tail entity. This shows that as long as the target detection is done well, then the relationship can be guessed based on the prior distribution of the relationship and the frequency of the statistics. This is the basic idea of visual relationship extraction in this chapter. In the background knowledge base, the prior distribution of the relationship is stored, which is the information obtained in advance by the method of factorization, i.e., the predicate distribution between two entities. This distribution is obtained by the tensor decomposition method. We can construct a relationship tensor $\mathcal{R} \in \mathbb{R}^{M \times M \times L}$ where $\mathcal{R}(i, j, l)$ contains the number of occurrences of the ith entity and jth entity having the lth predicate in the dataset (see Figure 3.2). If the relationship of man (entity index i) and horse (entity index j) described by feeding (predicate index l) has shown up c times, then we assign $\mathcal{R}(i, j, l) = c$.

The relationship tensor representation can reflect the intrinsic connection between entities and the distribution of relationships. The decomposition of the relationship tensor stored in the background knowledge base can obtain the a priori distribution of relationships. We can utilize the

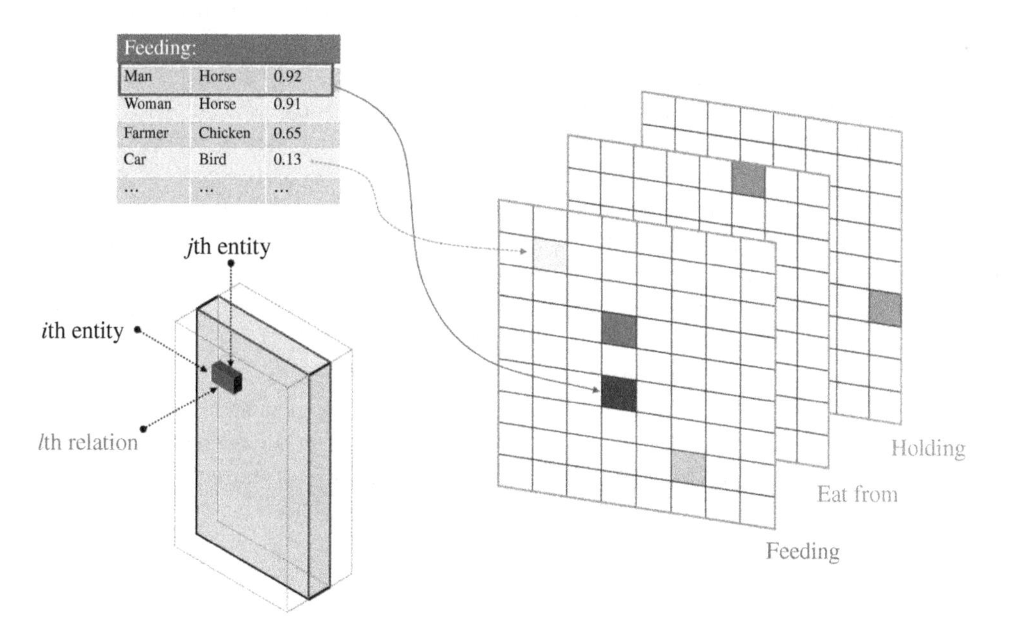

Figure 3.2 Relational tensor $\mathcal{R} \in \mathbb{R}^{M \times M \times L}$ given M entity categories and L possible predicates.

so-called a priori knowledge (e.g., statistical information, human common sense, and semantic information) to adjust the relationship prediction and improve the ability of zero-shot. Based on the statistical information of the relationship tensor, we can determine that the relation between the recognized entities i and j is

$$r_{ij} = \underset{l}{\operatorname{argmax}} \quad \mathcal{R}(i,j,l). \tag{3.3}$$

However, if $\max\{\mathcal{R}(i,j,l)\} < \zeta_s$, then there is no relationship between the entities i and j because ζ_s is the lowest threshold for the existence of a relationship between two entities. After the relation extraction, the relation among the whole image X can be described as a set of relations R consisting of different relations r_{ij}.

Therefore, after the process of entity detection and relation acquisition, the semantic information of the image X can be represented as knowledge graph $G = \{E, R\}$.

3.2.3 Semantic Transmission

Next, we introduce the model of semantic information transmission. Digital images are very sensitive to bit errors caused by channel noise. Some semantic methods are mostly trained on Gaussian white noise channels that are differentiable, while the actual channels of orthogonal frequency-division multiplexing (OFDM) systems are multiplicative and non-differentiable. The existing work uses neural networks to model the wireless channels but cannot fully utilize mathematical analysis of channel characteristics to help system design. The large-scale fading, small-scale fading, and noise interference in wireless transmission environments affect communication quality and increase the distortion of semantic information during transmission. Therefore, it is necessary to study the impact of wireless fading channels on wireless semantic communication.

First of all, let us first understand that the traditional image communication system model mainly includes several main parts such as image acquisition, preprocessing, coding, transmission, decoding, and display. In the image preprocessing process, the input image data are usually converted into a mathematical model or data structure through sampling and quantization to facilitate algorithm processing and operation. For example, for a color image, this may be represented as a 3D matrix, where each pixel corresponds to an element, and the value of the element represents the brightness of the red, green, and blue color channels of that pixel. For a grayscale image, which can usually be represented as a 2D matrix, it can also be extended as a 3D array, but the third dimension has a size of 1, i.e., it is considered to be an image with only one color channel, as shown in Figure 3.3, where each pixel corresponds to an element, and the value of the element denotes the luminance of that pixel (usually this value ranges from 0 to 255).

Inspired by the theory of digital image processing, this chapter uses a 3D tensor $\mathcal{F} \in \mathbb{R}^{\frac{H}{\kappa}\times\frac{W}{\kappa}\times M}$ to denote the knowledge graph G that should be transmitted. The input image is divided into a grid map whose size is equal to the size of the final feature map $\frac{H}{\kappa} \times \frac{W}{\kappa}$, and κ denotes the network step size. In other words, the knowledge graph characterizing semantic information in the semantic domain is projected onto a 3D tensor of representation images in the spatial domain (see Figure 3.4).

$$\mathcal{F}(x,y,i) = \boldsymbol{f}_i(x,y), \quad i \in \{1,2,\dots,M\}. \tag{3.4}$$

From the viewpoint of the purpose of wireless semantic transmission, the random change of pixel intensity values caused by the random signal interference during the ingestion or transmission of the image will inevitably lead to the misrecognition and semantic distortion of the decoding of the real semantic information content at the receiving end. Therefore, the semantic ambiguity

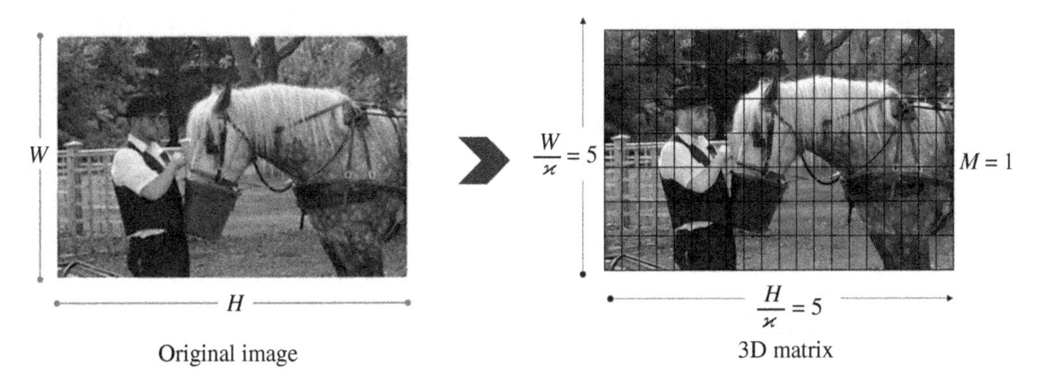

Figure 3.3 Pixel of an grayscale image can be formed to the corresponding pixel of single-channel 3D matrices using traditional image digital processing, where the size of the third dimension is 1.

of the entity $\boldsymbol{w}$ in the semantic domain is related to the intensity vector $\boldsymbol{f}$ in the spatial domain. The bounding box position d_k of the block, which can successfully detect an entity, corresponds to a spatial cell (x_k, y_k) in the spatial domain, and the intensity vector of the point is the semantic ambiguity of the entity detected by the block. For example, as shown in Figure 3.4, the entity set E in the knowledge graph has a total of $K_s = 4$ detected entities, and the 3D tensor $\mathcal{F} \in \mathbb{R}^{5\times5\times M}$. For the block $X_{k=1}$ where the entity "glasses" is identified, its projected coordinates (cell) in the spatial domain are $(2, 3)$, and its semantic ambiguity is associated with the intensity vector of the cell, $\boldsymbol{f}(2, 3) \triangleq \boldsymbol{w}_1$. In addition, the intensity vector at the corresponding cells of those filtered blocks is a zero-vector, and these cells are also called "zero cells."

$$\boldsymbol{f}(x_k, y_k) \triangleq \begin{cases} \boldsymbol{w}_k, & k \in \mathcal{K}_s \\ 0, & k \in \mathcal{K}, k \notin \mathcal{K}_s. \end{cases} \tag{3.5}$$

The semantic ambiguity-based 3D tensor array $\mathcal{F}$ obtained by projection needs to be channel coded and modulated to be fit for transmission in a wireless channel. Therefore, the received signal of a 3D tensor array $\mathcal{F}$, after transmission over a wireless fading channel and noise interference, can be expressed as follows [Jähne, 2005]:

$$\hat{\mathcal{F}} = \mathcal{F} * \mathcal{H} + \mathcal{N}, \tag{3.6}$$

where $\mathcal{H}$ is the channel coefficient tensor, $*$ is the Hadamard product, and $\mathcal{N}$ is the noise interference. In the discussion that follows, the noise tensor has the same size as a 3D tensor and can be viewed as a stack of M layers of 2D noise matrices. We assume that the elements in each layer of the 2D noise matrix are random numbers with a specified probability density function, independently sampled from a specified random distribution, such as Gaussian distribution [Jähne, 2005]. Based on the properties of independent random variables and random sampling, the one-dimensional vector composed of the elements in the same cell extracted from each noise matrix (that is, elements in the same row and column in each layer) still follows the specified random distribution. Note that here, $\mathcal{H}$ and $\mathcal{N}$ only affect nonzero cells, and the semantic ambiguity (intensity vector) for cells with filtered blocks remains zero before and after the wireless transmission.

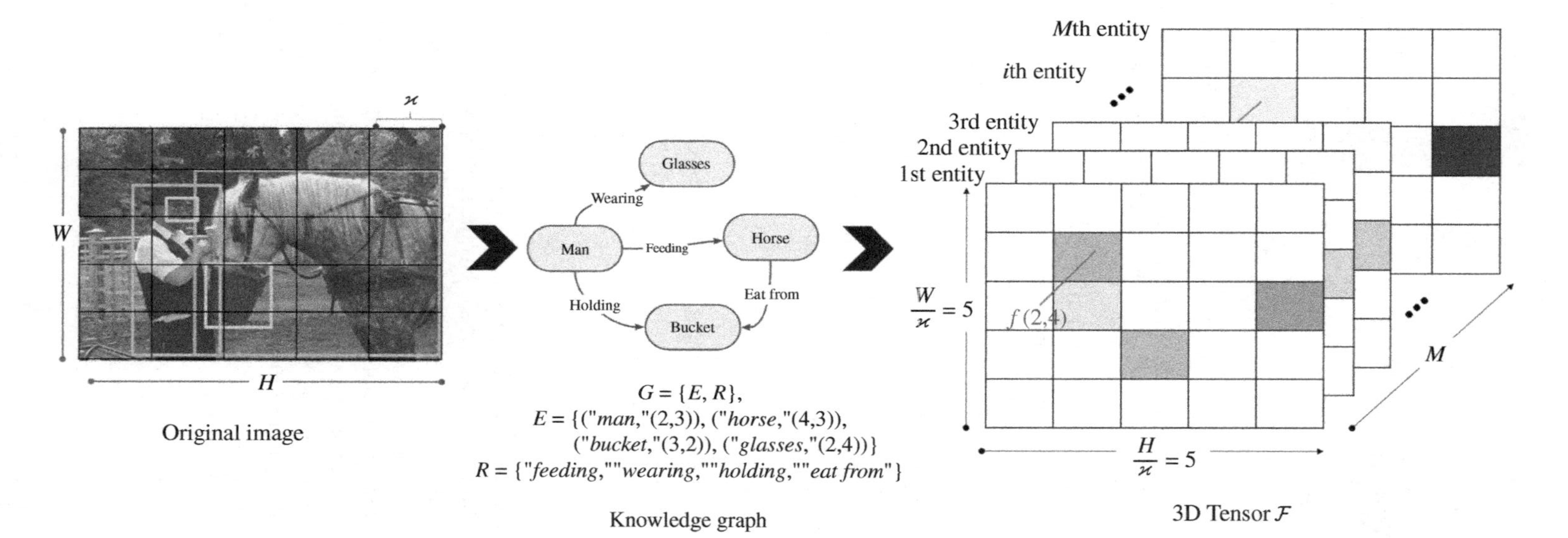

Figure 3.4 The semantic information based on the knowledge graph can be converted into a 3D tensor based on semantic ambiguity using the projection from the semantic domain to the spatial domain.

3.2.4 Semantic Decoding

3.2.4.1 Entity Mapping

At the destination, after demodulation and channel decoding, a new 3D tensor $\hat{\boldsymbol{f}}$ will be obtained. Through the mapping process, we restore the received semantic information according to the new 3D tensor. The mapping process mainly finds all the nonzero intensity vector $\hat{\boldsymbol{f}}$ in the 3D tensor. The spatial location (x_k, y_k) is mapped to the entity location d_k in the semantic domain, and the semantic ambiguity is equal to the intensity vector.

$$\begin{aligned}\hat{\boldsymbol{w}}_k &\triangleq \hat{\boldsymbol{f}} \\ &= [\hat{u}_k^1, \hat{u}_k^2, \dots, \hat{u}_k^i, \dots, \hat{u}_k^M].\end{aligned} \tag{3.7}$$

Due to channel fading and noise interference, the original category confidence of the kth block u_k^i may increase or decrease to $\hat{u}_k^i$, resulting in a change in the maximum category confidence of the block, so the block X_k is incorrectly identified as another entity by the destination. For example, the block in cell (2, 4) in Figure 3.5 before transmission is recognized as "bucket," and after going through the wireless channel and mapping, it is recognized as "pail."

Moreover, we need to re-rank the series of category confidence and filter out the probability that the confidence is lower than the threshold ξ_t.[1] If $\hat{u}_k^i \leq \xi_t, \forall i \in \{1, 2, \dots, M\}$, then the block X_k will be filtered, and this part of the semantic information will be distorted; i.e., the category confidence of the cell (2, 4) that originally contained the semantic information of "glasses" was too low, causing the block to be filtered and the semantic information of the cell to be lost.

When the mapping restoration process from the spatial domain to the semantic domain is completed, we will get the set of semantic entities at the receiver side.

$$\hat{E} = \{(o_1, d_1), (o_2, d_2), \dots, (o_k, d_k), \dots\}, k \in \mathcal{K}_t = \{1, 2, \dots, K_t\}, \tag{3.8}$$

where $\hat{K}_t$ is the number of entities obtained by mapping at the destination.

3.2.4.2 Relationship Acquisition

With the entity mapping in Section 3.2.4.1, we can obtain the semantic entities from the new 3D tensor $\hat{\mathcal{F}}$ and further guess the inter-entity relationships based on the prior distributions of the relationships and the frequencies of the statistics with the aid of the relationship tensor $\hat{\mathcal{R}}(i,j,l)$ from the destination background knowledge base. If $\max\{\hat{\mathcal{R}}(i,j,l)\} \leq \zeta_t$, then there is no relationship between entities o_i and o_j. If $\max\{\hat{\mathcal{R}}(i,j,l)\} > \zeta_t$, then a relationship exists between the entities o_i and o_j and the relationship is

$$\hat{r}_{ij} = \underset{l}{\operatorname{argmax}} \quad \hat{\mathcal{R}}(i,j,l). \tag{3.9}$$

Both $\hat{\mathcal{R}}(i,j,l)$ and ζ_t are determined by the background knowledge base of the receiver. After reduction by semantic relations, the set of relations $\hat{R}$ consisting of different relations $\hat{r}_{ij}$ can be used to represent the relations in the received knowledge graph $\hat{G}$.

The knowledge graph $\hat{G}$ can be constructed from the set of received semantic entities and their relationships, $\hat{G} = \{\hat{E}, \hat{R}\}$. The knowledge graph can be turned into a real image by the graph-to-image model [Johnson et al., 2018].

1 The mentioned threshold ξ_s and ξ_t are determined by their respective background knowledge, usually based on the perception ability of different observers to the image.

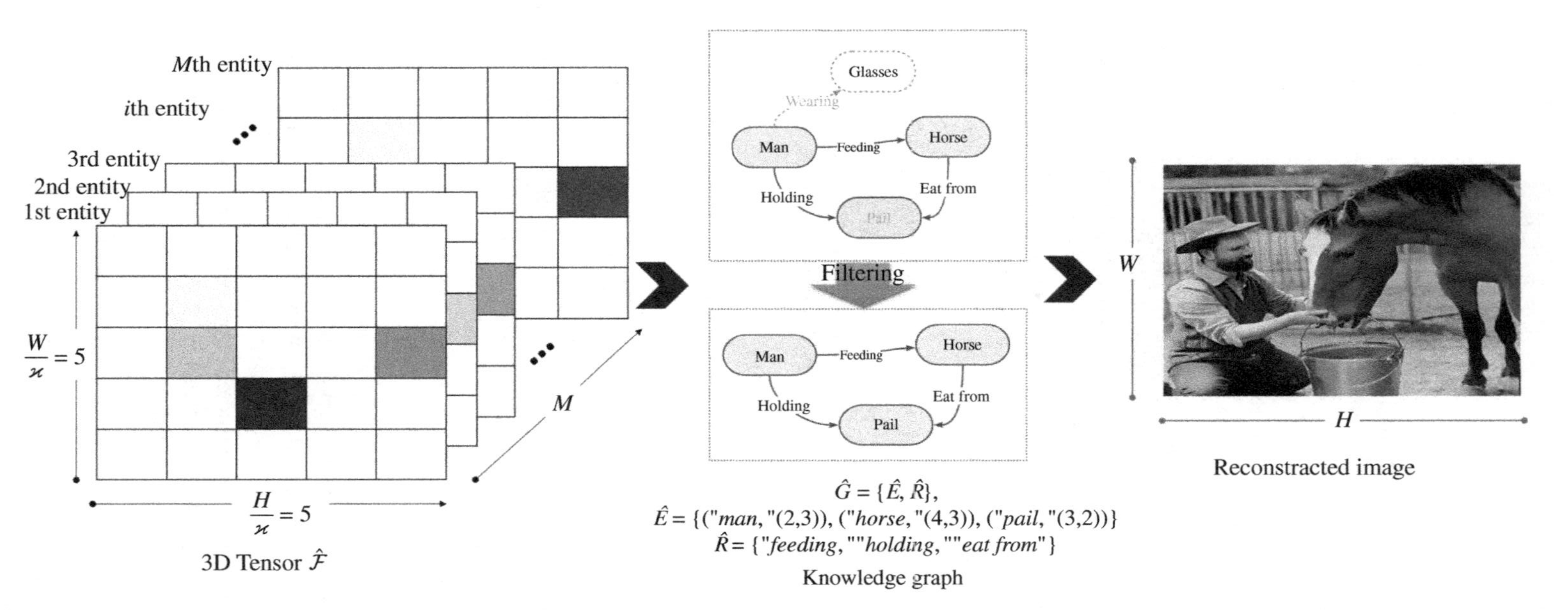

Figure 3.5 The new 3D tensor transmitted through the physical channel can be restored by mapping from the spatial domain to the semantic domain to obtain a new knowledge graph, and then the reconstructed image can be obtained through the graph-to-image model.

3.3 Semantic Similarity Measurement

3.3.1 Graph-to-Graph Semantic Similarity (GGSS)

To evaluate the effectiveness of image semantic communication, we introduce a new metric for semantic similarity between knowledge graphs based on GOT theory. The proposed GGSS captures the correlation of semantic information before and after wireless transmission. In traditional research on knowledge graph embedding, the cosine distance between embedded triple vectors is commonly used to calculate similarity. However, this distance only considers the similarity of entities and does not reflect the structure of the knowledge graph. For image semantic transmission, similarity requires understanding not only the meaning of regions in the image but also the relationships and spatial locations between regions.

Image and text data inherently contain rich sequential/spatial structures, and by representing them as knowledge graphs and calculating the similarity of knowledge graphs, it is possible to model not only inter-graph relationships but also leverage intra-graph relationships (e.g., semantic/spatial relationships between entities detected in an image [Li et al., 2019]). Therefore, we incorporate two types of pptimal transport (OT) distances to jointly measure the semantic similarity of two knowledge graphs based on the GOT framework, including the Wasserstein distance (WD) for calculating the similarity of entities (nodes) [Peyré and Cuturi, 2019] and the Gromov–Wasserstein distance (GWD) for capturing the similarity of relationships (edges) [Peyré et al., 2016].

As illustrated in Figure 3.6, WD measures only the distances between different knowledge graph entity embeddings without considering the topological information encoded in the graph. On the other hand, GWD compares graph structures by measuring the distance between a pair of nodes in each graph. The GSS that incorporates WD and GWD effectively considers the overall information of entities and relationships and can capture a more comprehensive semantic-level knowledge graph similarity.

3.3.1.1 Wasserstein Distance

Let $\boldsymbol{\mu}_s \in \mathbf{P}(\mathbb{X})$, $\boldsymbol{\mu}_t \in \mathbf{P}(\mathbb{Y})$ denote two discrete distributions, formulated as $\boldsymbol{\mu}_s = \sum_{i=1}^{n_s} p_i^s \delta(\boldsymbol{x}_i)$ and $\boldsymbol{\mu}_t = \sum_{j=1}^{n_t} p_j^t \delta(\boldsymbol{y}_j)$, where $\delta(\cdot)$ denotes the Dirac function, p_i^s and p_j^t are the probability masses to the ith sample and the jth sample, belonging to the n_s- and n_t-dimensional probability simplex, respectively, i.e., $\sum_{i=1}^{n_s} p_i^s = \sum_{j=1}^{n_t} p_j^t = 1$. The Wasserstein distance between the two discrete distributions $\boldsymbol{\mu}_s$, $\boldsymbol{\mu}_t$ is defined as

$$\begin{aligned} \mathcal{D}_w(\boldsymbol{\mu}_s, \boldsymbol{\mu}_t) &= \inf_{\boldsymbol{\pi}\in\Pi(\boldsymbol{\mu}_s,\boldsymbol{\mu}_t)} \mathbb{E}_{(\boldsymbol{x},\boldsymbol{y})\sim\boldsymbol{\pi}}[c(\boldsymbol{x},\boldsymbol{y})] \\ &= \min_{\mathbf{T}\in\Pi(\boldsymbol{\mu}_s,\boldsymbol{\mu}_t)} \sum_{i=1}^{n_s}\sum_{j=1}^{n_t} \mathbf{T}_{ij} \cdot c(\boldsymbol{x}_i, \boldsymbol{y}_j), \end{aligned} \tag{3.10}$$

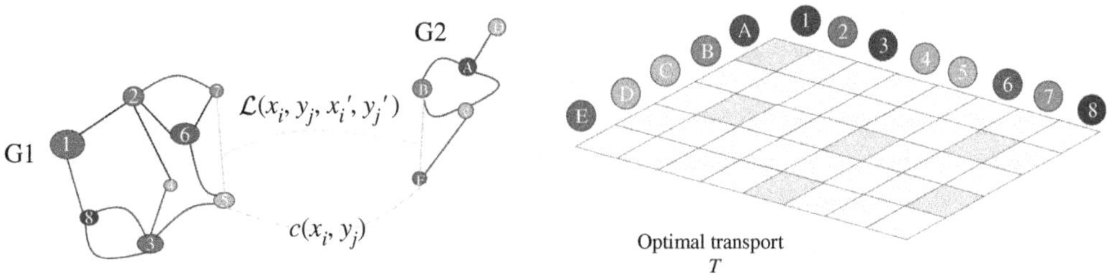

G1 G2 $\mathcal{L}(x_i, y_j, x_i', y_j')$ $c(x_i, y_j)$ Optimal transport T

Figure 3.6 The Wasserstein distance and Gromov–Wasserstein distance in graph optimal transport theory.

where $\Pi(\mu_s, \mu_t) = \left\{ \mathbf{T} \in \mathbb{R}_+^{n_s \times n_t} \mid \mathbf{T}\mathbf{1}_{n_t} = \mu_s, \mathbf{T}^\top \mathbf{1}_{n_s} = \mu_t \right\}$ is the set of all joint probability measures on $\mathbf{P}(\mathbb{X} \times \mathbb{Y})$ with marginals $\mu_s(x)$ and $\mu_t(y)$, and $\mathbf{1}_{n_s}$ denotes an n_t-dimensional all-one vector. We define a cost function $c(\mathbf{x}_i, \mathbf{y}_j)$ evaluating the distance between $\mathbf{x}_i$ and $\mathbf{y}_j$. $\mathbf{T}$ is denoted as the transport plan or transport matrix where $\mathbf{T}_{ij}$ represents the amount of mass to be shifted from $\mathbf{x}_i$ and $\mathbf{y}_j$ [Petric Maretic et al., 2019].

$\mathcal{D}_w(\mu_s, \mu_t)$ defines an optimal transport distance that measures the discrepancy of corresponding points in all possible couplings or matchings between the two distributions. In our study, this will represent the similarity distance of nodes (entities).

3.3.1.2 Gromov–Wasserstein Distance

Following the same notation as in WD, the Gromov–Wasserstein distance between μ_s and μ_t is defined as

$$\begin{aligned} \mathcal{D}_{gw}(\mu_s, \mu_t) &= \inf_{\pi \in \Pi(\mu_s, \mu_t)} \mathbb{E}_{(x,y)\sim\pi,(x',y')\sim\pi} \left[\mathcal{L}(\mathbf{x}, \mathbf{y}, \mathbf{x}', \mathbf{y}') \right] \\ &= \min_{\hat{\mathbf{T}} \in \Pi(\mu_s, \mu_t)} \sum_{i,i',j,j'} \hat{\mathbf{T}}_{ij} \hat{\mathbf{T}}_{i'j'} \mathcal{L}\left(\mathbf{x}_i, \mathbf{y}_j, \mathbf{x}'_i, \mathbf{y}'_j\right), \end{aligned} \tag{3.11}$$

where $\mathcal{L}(\cdot)$ is the cost function evaluating the intra-graph structural similarity between two pairs of nodes $(\mathbf{x}_i, \mathbf{x}'_i)$ and $(\mathbf{y}_j, \mathbf{y}'_j)$. The learned matrix $\hat{\mathbf{T}}$ now becomes a transport plan that relies on relational rather than positional similarities to analyze correspondences in different graphs.

Relying on WD alone to judge the similarity between knowledge graphs will result in a shift in knowledge graph alignment when there are duplicate entities represented by different nodes. On the contrary, using GWD alone to judge the similarity between knowledge graphs will result in an error in matching pairs of entities together; e.g., ("chair," "table") has a similar cosine similarity to the word pairs ("horse," "donkey"), but these two pairs of words have completely different semantic meanings. Therefore, by integrating WD and GWD and unifying these two distances in a mutually beneficial way, we propose the transportation plan $\mathbf{T}$ shared by WD and GWD.

Formally, the proposed GGSS metric based on GOT distance is defined as

$$\mathcal{D}_{GGSS}(\mu_s, \mu_t) = \min_{\mathbf{T} \in \Pi(\mu_s, \mu_t)} \sum_{i,i',j,j'} \mathbf{T}_{ij} \left(\lambda c(\mathbf{x}_i, \mathbf{y}_j) + (1 - \lambda) \mathbf{T}_{i'j'} \mathcal{L}\left(\mathbf{x}_i, \mathbf{y}_j, \mathbf{x}'_i, \mathbf{y}'_j\right) \right), \tag{3.12}$$

where λ is the hyper-parameter for controlling the importance of different cost functions. This leads to a structurally meaningful measure for comparing graphs, which is able to take into account the global structure of graphs, while most other measures merely observe local changes independently.

We consider two knowledge graphs, G and $\hat{G}$, with entities E and $\hat{E}$, respectively. We use a pretrained deep neural network as a vectorization function to transform the entities and relationships in the extracted knowledge graph into feature vectors. Each entity (node) $o_i \in E$ is represented by a feature vector $\mathbf{x}_i \in \mathbb{X}$; similarly, $\hat{o}_j \in \hat{E}$ corresponds to $\mathbf{y}_j \in \mathbb{Y}$. In particular, the GGSS metric proposed in this chapter can be extended to the text semantic transmission system, where entities can correspond to entities in images or words in sentences.

In our proposed model, we use GOT for the similarity measure of semantic information between knowledge graphs, where a transport plan $\mathbf{T} \in \mathbb{R}_+^{K_s \times K_t}$ is learned to find a minimum cost mapping between two point sets (formalized as discrete distributions μ_s and μ_t). In our case, we usually consider uniform weights, e.g., $p_i^s = \frac{1}{K_s}$ and $p_j^t = \frac{1}{K_t}$ [Alvarez-Melis and Jaakkola, 2018]. In Eq. (3.12), we define a cost function as

$$c(\mathbf{x}_i, \mathbf{y}_j) = 1 - \frac{\mathbf{x}_i^\top \mathbf{y}_j}{\|\mathbf{x}_i\|_2 \|\mathbf{y}_j\|_2}, \tag{3.13}$$

when evaluating the cosine distance between vectorized semantic entities $o_{k,i}$ and $\hat{o}_{k,j}$. We define the cost function as

$$\mathcal{L}\left(\boldsymbol{x}_i, \boldsymbol{y}_j, \boldsymbol{x}'_i, \boldsymbol{y}'_j\right) = 1 - \frac{\boldsymbol{r}_{i,i'}^{\mathrm{T}} \hat{\boldsymbol{r}}_{j,j'}}{\left\|\boldsymbol{r}_{i,i'}\right\|_2 \left\|\hat{\boldsymbol{r}}_{j,j'}\right\|_2}, \tag{3.14}$$

when evaluating edge similarity between the different knowledge graphs, in other words, to measure the cosine distance between the vectorized relation vectors $\boldsymbol{r}$ and $\hat{\boldsymbol{r}}$. Instead of using projected gradient descent or conjugated gradient descent as in Xu et al. [2019] and Titouan et al. [2019], we can calculate the GOT distance with the help of the improved Sinkhorn algorithm, which can be efficiently implemented in popular deep learning libraries such as PyTorch and TensorFlow [Chen et al., 2020].

3.3.2 Image-to-Image Semantic Similarity (IISS)

Furthermore, in addition to the measure between knowledge graphs, unlike the traditional pixel-level image communication performance metrics such as structural similarity index measure (SSIM) and PSNR, we propose a measure of semantic-level similarity between the original and recovered images, which is defined as the difference in the average accuracy of the transmitter and receiver answering a set of questions for each image.

To evaluate the image quality recovered from the received semantic information, the destination will use a series of problem sets $\mathcal{Q} = \left\{q_1, \dots, q_2, \dots, q_Q\right\}$ of Q questions for image $\hat{X}$. For example, the source wants to send the information related to the "chair" in image X. The questions used to evaluate the quality of the image in semantic level can be $\mathcal{Q}$ ="How many chairs are there?" and "What is on the chair?." Both the sender and receiver will answer each question based on the original image X and the restored image $\hat{X}$. If the answers of the two groups are similar, it is considered that the restored image has successfully obtained the semantic information of the original image, and the semantics of image X and Y are similar. The accuracy of answering a set of questions $\mathcal{Q}$ for image $\hat{X}$ is

$$\mathcal{D}_{AAQ}(\hat{X}) = \frac{1}{Q} \sum_{i=1}^{Q} A_i\left(\hat{X}\right), \tag{3.15}$$

where $A_i\left(\hat{X}\right) \in \{0, 1\}$ indicates whether the answer to question $q_i \in \mathcal{Q}$ obtained based on the original image $\hat{X}$ is correct. Here, $A_i\left(\hat{X}\right) = 1$ indicates that the obtained answer to question q_i is correct; otherwise, we have $A_i\left(Y_k\right) = 0$. Similarly,

$$\mathcal{D}_{AAQ}(X) = \frac{1}{Q} \sum_{i=1}^{Q} A_i(X). \tag{3.16}$$

We regard the answer to the question at the source as the standard answer, so $A_i(X)$ is regarded as 1 by default. Hence, the IISS can be defined as

$$\mathcal{D}_{IISS}\left(X_k, Y_k\right) = \mathcal{D}_{AAQ}\left(Y_k\right) - \mathcal{D}_{AAQ}\left(X_k\right). \tag{3.17}$$

The higher the similarity between the two images at the semantic level, the better the reliability of the image semantic communication system.

3.4 Simulation

The assessment of our proposed transmission framework's efficacy was conducted using the VG150 dataset [Xu et al., 2019] under the conditions of Rayleigh fading channels. We set the value of all elements of the average channel gain $\mathcal{H}$ to 1 and vary the noise variance to change the average signal-to-noise ratio (SNR). For enhanced clarity in our exposition, we only select the best-performing digital transmission scheme for the JPEG encoding method from all possible combinations of rate-1/2, rate-2/3, and rate-1/3 low-density parity-check (LDPC) codes with binary phase shift keying (BPSK), 4-QAM, 16-QAM, and 64-QAM modulation schemes. Furthermore, the DeepJSCC scheme was implemented by parameterizing both the encoder and decoder functions through two CNNs that were jointly trained. This approach aligns with the methodologies outlined in the existing literature [Bourtsoulatze et al., 2019].

Figure 3.7 illustrates the SSIM performance of both our proposed scheme and the benchmark scheme across various SNR levels. As depicted in Figure 3.7, our proposed framework demonstrates a growth trend in SSIM performance that is akin to that of the benchmark scheme. This similarity in trends underscores the validity of our theoretical model, which is based on the transformation between semantic and spatial domains. This model effectively encapsulates the impact of wireless channel fading on the perceived visual quality of reconstructed images, offering a convincing mathematical perspective. Notably, our proposed scheme exhibits slightly inferior performance at lower SNRs compared to the DeepJSCC approach. However, the training process for channel encoding and decoding in DeepJSCC usually is exceedingly time-intensive. Furthermore, in scenarios involving noisy channels, the neural network should be trained to map critical features to closely related representations instead of meaningless vector feature values so that similar images can be reconstructed even in the presence of noise.

In our continued evaluation, we delve into assessing the proposed GGSS metric, contrasting it with the benchmark schemes. This simulation involves extracting knowledge graphs from images

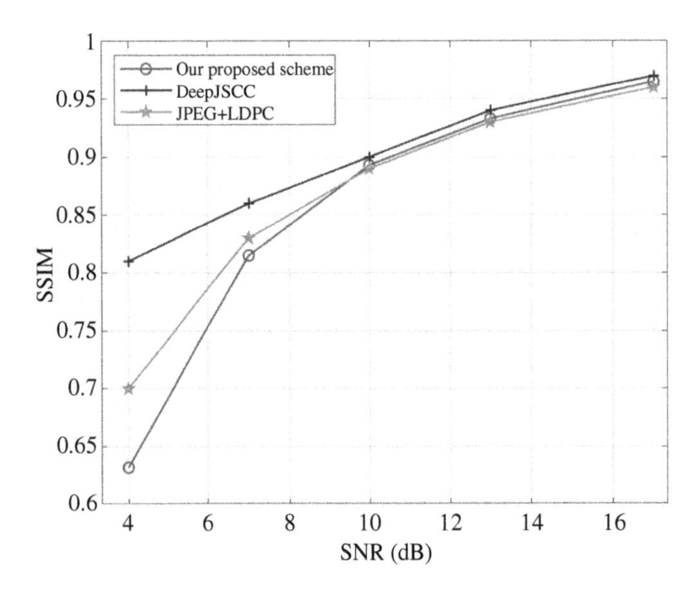

Figure 3.7 The impact of wireless channel fading on perceived visual performance of different wireless image transmission schemes.

Table 3.1 Performance evaluation results of the semantic similarity.

	PSNR	SSIM	GGSS
Our scheme	8.05	0.1425	3.2×10^3
JPEG	19.63	0.403	1.7×10^4
DeepJSCC	24.16	0.7537	4.1×10^3

reconstructed using the benchmark scheme and comparing them against knowledge graphs derived from the original images and calculating the semantic similarity between these pairs of knowledge graphs. Noted that we employ an entropy regularization parameter β in the Sinkhorn iterations to obtain the GOT distance, as outlined in Chen et al. [2020]. A larger β value results in a denser optimal coupling matrix $\mathbf{T}^*$, which facilitates faster runtime for each iteration. However, this comes at the expense of reduced performance accuracy. Conversely, a smaller β value yields a sparser optimal solution, potentially enhancing performance at the cost of increased computation time. Through experimental tuning, we have set β to 0.1, striking a balance between performance and computational efficiency.

These experiments in Table 3.1 are conducted in a low SNR environment, almost 1 dB. While our proposed scheme may lag behind the DeepJSCC approach in terms of traditional metrics like PSNR and SSIM, it demonstrates a significant advantage in fulfilling the objective of "express meaning" transmission. Specifically, our approach achieves approximately 21.9% improvement over DeepJSCC in terms of the GGSS metric. This improvement is also evident in the visual quality of the reconstructed images. Comparing the reconstructed image from our scheme (as seen in Figure 3.8a) with that of the DeepJSCC scheme (Figure 3.8b), against the original image (Figure 3.8c), clearly demonstrates this point. Additionally, as observed in Figure 3.8b and 3.8d, when channel quality is too low, channel codes cannot handle the increasing error rate, leading to a significant decline in the quality of reconstructed images. Unlike the JPEG approach, where image quality sharply declines with worsening channel conditions, the loss of semantic information in our proposed scheme is much more gradual. This resilience can be attributed to our focus on modeling how noise influences semantic ambiguity rather than just image clarity. In addition, knowledge graph reconstruction also has the assistance of background knowledge at the destination, enabling it to effectively resist semantic distortion caused by channel quality fluctuations. In terms of semantic similarity between images (IISS), relevant VQA research has made significant progress, which is reflected in many literatures [Wu et al., 2017]. In view of the extensiveness and depth of these studies, we will not repeat them in this chapter.

3.5 Conclusion

In this chapter, we focus on a knowledge graph-driven end-to-end image semantic communication system. We first propose a semantic representation method based on knowledge graphs. With the assistance of background knowledge of personal preferences and prior knowledge, knowledge graphs are extracted from original images through entity and relationship detection to ensure the accuracy and interpretability of semantic expressions. In addition, we also propose a projection

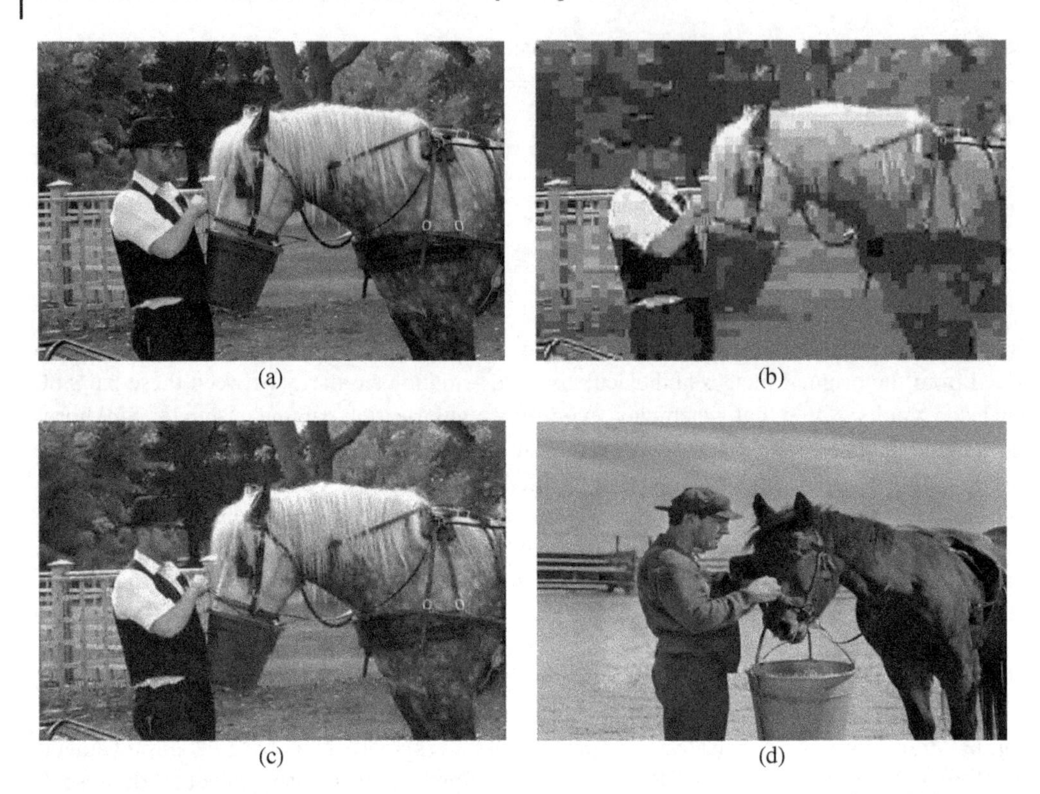

Figure 3.8 Examples of reconstructed images produced by our proposed scheme, the DeepJSCC, and the baseline digital schemes that use JPEG for image compression. (a) Original image, (b) JPEG, (c) DeepJSCC, and (d) our proposed scheme.

and mapping mechanism between the semantic domain and the spatial domain to mathematically model the impact of wireless fading channels on semantic communication. Finally, this chapter proposes GGSS and IISS metrics to measure the semantic-level similarity of knowledge graphs and images before and after wireless semantic communication and conducts some simulation experiments for verification.

References

David Alvarez-Melis and Tommi S. Jaakkola. Gromov-Wasserstein alignment of word embedding spaces. *arXiv preprint arXiv:1809.00013*, 2018.

Eirina Bourtsoulatze, David Burth Kurka, and Deniz Gündüz. Deep joint source-channel coding for wireless image transmission. *IEEE Transactions on Cognitive Communications and Networking*, 5(3):567–579, 2019.

Liqun Chen, Zhe Gan, Yu Cheng, Linjie Li, Lawrence Carin, and Jingjing Liu. Graph optimal transport for cross-domain alignment. In *International Conference on Machine Learning*, pages 1542–1553. PMLR, 2020.

Jincheng Dai, Ping Zhang, Kai Niu, Sixian Wang, Zhongwei Si, and Xiaoqi Qin. Communication beyond transmitting bits: Semantics-guided source and channel coding. *IEEE Wireless Communications*, 30(4):170–177, 2022.

Bin-Bin Gao, Chao Xing, Chen-Wei Xie, Jianxin Wu, and Xin Geng. Deep label distribution learning with label ambiguity. *IEEE Transactions on Image Processing*, 26(6):2825–2838, 2017.

Xin Geng and Yu Xia. Head pose estimation based on multivariate label distribution. In *Proceedings of the IEEE Conference on Computer Vision and Pattern Recognition*, pages 1837–1842, 2014.

Xin Geng, Chao Yin, and Zhi-Hua Zhou. Facial age estimation by learning from label distributions. *IEEE Transactions on Pattern Analysis and Machine Intelligence*, 35(10):2401–2412, 2013.

Başak Güler, Aylin Yener, and Ananthram Swami. The semantic communication game. *IEEE Transactions on Cognitive Communications and Networking*, 4(4):787–802, 2018.

Qiyu Hu, Guangyi Zhang, Zhijin Qin, Yunlong Cai, Guanding Yu, and Geoffrey Ye Li. Robust semantic communications against semantic noise. In *2022 IEEE 96th Vehicular Technology Conference (VTC2022-Fall)*, pages 1–6. IEEE, 2022.

Danlan Huang, Xiaoming Tao, Feifei Gao, and Jianhua Lu. Deep learning-based image semantic coding for semantic communications. In *2021 IEEE Global Communications Conference (GLOBECOM)*, pages 1–6. IEEE, 2021.

Bernd Jähne. *Digital Image Processing*. Springer Science & Business Media, 2005.

Justin Johnson, Agrim Gupta, and Li Fei-Fei. Image generation from scene graphs. In *Proceedings of the IEEE Conference on Computer Vision and Pattern Recognition*, pages 1219–1228, 2018.

Linjie Li, Zhe Gan, Yu Cheng, and Jingjing Liu. Relation-aware graph attention network for visual question answering. In *Proceedings of the IEEE/CVF International Conference on Computer Vision*, pages 10313–10322, 2019.

Xin Li, Jun Shi, and Zhibo Chen. Task-driven semantic coding via reinforcement learning. *IEEE Transactions on Image Processing*, 30:6307–6320, 2021.

Xuewen Luo, Hsiao-Hwa Chen, and Qing Guo. Semantic communications: Overview, open issues, and future research directions. *IEEE Wireless Communications*, 29(1):210–219, 2022.

Hermina Petric Maretic, Mireille El Gheche, Giovanni Chierchia, and Pascal Frossard. GOT: An optimal transport framework for graph comparison. *Advances in Neural Information Processing Systems 32* (NeurIPS 2019), 2019.

Gabriel Peyré and Marco Cuturi. Computational optimal transport: With applications to data science. *Foundations and Trends in Machine Learning*, 11(5-6):355–607, 2019.

Gabriel Peyré, Marco Cuturi, and Justin Solomon. Gromov-Wasserstein averaging of kernel and distance matrices. In *International Conference on Machine Learning*, pages 2664–2672. PMLR, 2016.

Vayer Titouan, Nicolas Courty, Romain Tavenard, and Rémi Flamary. Optimal transport for structured data with application on graphs. In *International Conference on Machine Learning*, pages 6275–6284. PMLR, 2019.

Sixian Wang, Ke Yang, Jincheng Dai, and Kai Niu. Distributed image transmission using deep joint source-channel coding. In *ICASSP 2022-2022 IEEE International Conference on Acoustics, Speech and Signal Processing (ICASSP)*, pages 5208–5212. IEEE, 2022.

Qi Wu, Damien Teney, Peng Wang, Chunhua Shen, Anthony Dick, and Anton Van Den Hengel. Visual question answering: A survey of methods and datasets. *Computer Vision and Image Understanding*, 163:21–40, 2017.

Huiqiang Xie, Zhijin Qin, and Geoffrey Ye Li. Task-oriented multi-user semantic communications for VQA. *IEEE Wireless Communications Letters*, 11(3):553–557, 2021.

Bingyan Xie, Yongpeng Wu, Yuxuan Shi, Derrick Wing Kwan Ng, and Wenjun Zhang. Communication-efficient framework for distributed image semantic wireless transmission. *IEEE Internet of Things Journal*, 10(24):22555–22568, 2023.

Hongteng Xu, Dixin Luo, and Lawrence Carin. Scalable Gromov-Wasserstein learning for graph partitioning and matching. *Advances in Neural Information Processing Systems 32* (NeurIPS 2019), 2019.

Wanting Yang, Hongyang Du, Zi Qin Liew, Wei Yang Bryan Lim, Zehui Xiong, Dusit Niyato, Xuefen Chi, Xuemin Sherman Shen, and Chunyan Miao. Semantic communications for future internet: Fundamentals, applications, and challenges. *IEEE Communications Surveys & Tutorials*, 25(1):213–250, 2022.

Rowan Zellers, Mark Yatskar, Sam Thomson, and Yejin Choi. Neural motifs: Scene graph parsing with global context. In *Proceedings of the IEEE Conference on Computer Vision and Pattern Recognition*, pages 5831–5840, 2018.

Wenjing Zhang, Yining Wang, Mingzhe Chen, Tao Luo, and Dusit Niyato. Optimization of image transmission in a cooperative semantic communication networks. *IEEE Transactions on Wireless Communications*, 23(2):861–873, 2023a.

Wenyu Zhang, Haijun Zhang, Hui Ma, Hua Shao, Ning Wang, and Victor C. M. Leung. Predictive and adaptive deep coding for wireless image transmission in semantic communication. *IEEE Transactions on Wireless Communications*, 22(8):5486–5501, 2023b.

4

Robust Semantic Communications and Privacy Protection

Xuefei Zhang

School of Information and Communication Engineering, Beijing University of Posts and Telecommunications, Beijing, China

4.1 Motivation and Introduction

Semantic communication, commonly depending on deep neural networks (DNNs) and knowledge bases, considers user's needs and the meanings of information into the communication process. It is expected to become a new basic paradigm for the future communication system. Although some progresses have proved that semantic communication can provide a better performance under low signal-to-noise ratio (SNR) conditions compared to the traditional bit-based communication, the security and privacy issues in semantic communication remain unclear and have not been discussed comprehensively.

(1) *Security issue*: DNNs, commonly adopted by semantic extraction/encoding and semantic decoding/recovery, are vulnerable to adversarial attacks. Most of the studies on semantic communication security regard the adversarial attacks as the semantic noise. Semantic noise are the perturbation signals that are tailored to focus on attacking the target semantics objects, while securing the transmission of other semantics. Such semantic-specific adversarial attacks are much more destructive than existing attacks crafted for conventional bit-based communication systems. Meanwhile, semantic noise may appear in the wireless channel, leading to the wrong interpretation of the semantic information at the receiver, which eventually leads to decoding errors.
(2) *Privacy issue*: Most of the existing semantic communication systems assume that the communicating nodes share the same public knowledge base, while ignoring the agnostic nature and diversity of private knowledge bases, which may lead to knowledge discrepancy. As a result, the receiver can infer sensitive information that the sender does not want to disclose.

Motivated by the challenges mentioned earlier, this chapter aims to explore robust semantic communication and privacy protection for semantic communication.

(1) In order to improve the security of semantic communication, a robust semantic communication model to defend semantic noise is proposed by adopting the idea of simulating the noise first and then defending. Specifically, semantic noise are produced by proposing an adversarial sample generation model. On this basis, a robust semantic representation approach is proposed to combat the generated semantic noise using a generative adversarial network (GAN). Through the aforementioned defense approach, the impact of semantic noise is eliminated, and the robustness of the semantic communication system is improved.

Wireless Semantic Communications: Concepts, Principles, and Challenges, First Edition.
Edited by Yao Sun, Lan Zhang, Dusit Niyato, and Muhammad Ali Imran.

(2) In order to achieve privacy protection for semantic communication, a knowledge difference-oriented privacy protection (KDPP) approach is proposed to reduce the privacy risk. One of the focuses in KDPP is knowledge inference to tackle the knowledge discrepancy resulting from the agnosticism and diversity of private knowledge bases. On this basis, a path truncation approach is proposed to reduce the risk of privacy leakage while retaining high data utility.

In the rest of this chapter, the proposed approaches and conclusions are described. Section 4.2 describes robust semantic communication techniques for image and speech transmission, respectively. Section 4.3 describes the KDPP approach. Section 4.4 gives the conclusions.

4.2 Robust Semantic Communication

Although semantic communication has achieved good performance in low SNR conditions, the impact of semantic noise and the robustness of the system have not been well studied. The weakness of DNNs that are vulnerable to adversarial attacks brings some new threats. In an adversarial attack, an attacker adds a perturbation to the input of a DNN, which can result in an incorrect output. Unlike traditional jamming attacks, such perturbations are not natural noise but deliberately optimized vectors in the input domain feature space that can deceive the DNN. Therefore, robust semantic communication to fight with the semantic noise can enhance the security of semantic communication.

There are two types of noise in semantic communication. The first type is physical channel noise, which also exists in traditional bit-based wireless communication system. The second is semantic noise, which may cause misrecognition and wrong interpretation of semantic information, and it exists in different positions in the semantic communication process [Luo et al., 2022]. First, semantic noise may attack the encoding process at the source, which is usually difficult to detect, and leads to feature extraction errors. Second, semantic noise can attack the signals over the air due to the openness of the wireless channel, which possibly leads to incorrect semantic recovery at the receiver.

Considering the diversity of the data modalities, e.g., image, text, and speech, the semantic noise in different modalities are generated as adversarial samples following their own characteristics. Some existing methods for generating semantic noise generation methods in different modalities are given in the following.

An attacker for image transmission generates semantic noise that can spoof DNNs but are not perceptible through human vision. Some methods for generating semantic noise in the image domain have been proposed, such as the fast gradient sign method (FGSM) [Goodfellow et al., 2015], projected gradient descent (PGD) [Yang et al., 2023], and universal adversarial attack (UAP) [Moosavi-Dezfooli et al., 2017]. Correspondingly, the approaches against the semantic noise are involved with adversarial training, network structure modification, and additional network structures [Akhtar et al., 2021].

In the text domain, a slight change is obviously detected by human beings. Therefore, semantic noise in the text domain occur through insertion, deletion, exchange, character replacement, and word replacement. It is noted that the modified part should be similar to the original text in terms of syntax and semantics, and the proportion of modification should not be too high. The defense approaches against semantic noise in the text domain are twofold. The first aprroach is adversarial training, and the second is data preprocessing, where the texts are processed before inputing into the model [Pruthi et al., 2019].

In the speech domain, the widely adopted semantic noise are produced by speech-to-label attacks and speech-to-text attacks. A speech-to-label attack is to generate adversarial samples that are similar to the original speech but enable the model to produce incorrect labels. A speech-to-text attack makes the transcription of the polluted speech different from the original text. The following methods are commonly used for semantic noise in the speech domain: end-to-end FGSM [Gong and Poellabauer, 2018], acoustic feature-based FGSM [Kreuk et al., 2018], and UAPs for speech recognition [Neekhara et al., 2019]. The defense approaches against semantic noise in the speech domain are divided into two main categories. One is active defense, achieved through adding or changing the structures of DNNs to enhance the robustness, e.g., input transformation and adversarial training. The other is adversarial detection, which does not require changes to the model but requires different methods to detect adversarial samples before they are fed into the model.

4.2.1 A Robust Semantic Communication Model for Image Transmission

As shown in Figure 4.1, semantic noise possibly appears in different positions of the semantic communication process. First, semantic noise may attack the encoding process at the source, which is usually difficult to detect and leads to feature extraction errors. Second, semantic noise can attack the signals over the air due to the openness of the wireless channel, which possibly leads to the wrong semantic recovery at the receiver.

To eliminate the impacts of the semantic noise on communication performance, a robust semantic communication model is proposed to support the idea of simulating an attack first and then defending. Specifically, GANs are added before semantic extraction or encoding to remove the effect of semantic noise at the transmitter, and a denoising autoencoder (DAE) is added before channel decoding to remove the effect of semantic noise attacking over the wireless channel. The design of the GAN and denoising is described in the following.

4.2.1.1 Generation of Semantic Noise

In order to characterize the effect of semantic noise on robust training and evaluation, we regard semantic noise as perturbation signals that attack a semantic communication system to meet the following criteria:

(1) *Wrong semantics*: Semantic noise added to the original signals are tailored to focus on attacking the target semantics, while securing the transmission of other semantics.
(2) *Imperceptible*: The perturbations to the signals cannot be detected by the transmitter and receiver.

The goal of a semantic communication system is to minimize the loss function for a particular task, e.g., cross-entropy for a classification task. In contrast, semantic noise aims to maximize the loss function. Therefore, semantic noise generation can be described as an optimization problem

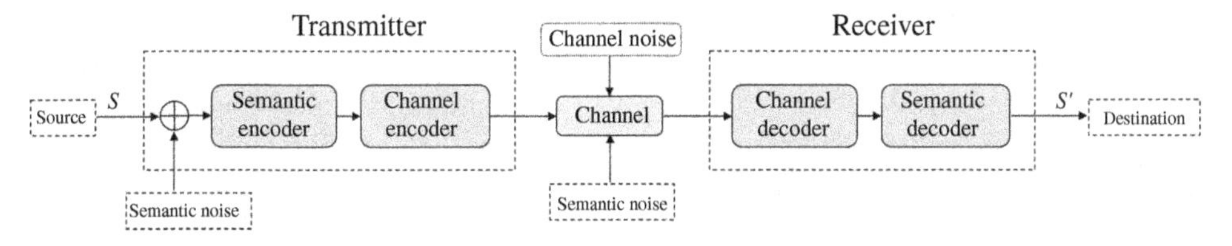

Figure 4.1 A semantic communication systems for image transmission with semantic noise.

to maximize a task-specific loss function given by

$$\max_{v_1} L(f_\theta(s+v_1), z), \tag{4.1}$$

$$\text{s.t. } \|v_1\|_p \le \alpha, \tag{4.2}$$

where s is the original signal, v_1 is the generated semantic noise, $f_\theta(\cdot)$ represents the outputs of the semantic encoding model with parameter θ, z is the real label of the image, α is used to control the amplitude of the perturbation, and $L(\cdot)$ represents the loss function of the model, and $\|\cdot\|_p$ is p-normal. The constraint (4.2) limits the scope of semantic noise to avoid being observed by humans.

First, we use the FGSM method to generate the semantic noise, where the loss function is as follows:

$$L(f_\theta(s+v_1), z) \approx L(f_\theta(s), z) + {v_1}^T \nabla_s L(f_\theta(s), z). \tag{4.3}$$

Then, we get the generated semantic noise as follows:

$$v_1 = \alpha \text{sign}(\nabla_v L(f_\theta(s), z)), \tag{4.4}$$

where sign($\cdot$) is a sign function. Similarly, we use PGD to generate the semantic noise. PGD is considered an iterative FGSM (K-FGSM), where K denotes the number of iterations. Compared to FGSM, which performs only one iteration with a larger step size, PGD builds the semantic noise through multiple iterations with a smaller step size.

Now, we turn to the semantic noise appearing over the air, denoted by v_2, as shown in Figure 4.1.

$$g(x+n+v_2) \neq g(x+n), \tag{4.5}$$

where x and n denote the output of the encoder and the channel noise, respectively, while $g(\cdot)$ represents the final output of the model. According to Eq. (4.6), we can find its approximate solution by utilizing the FGSM, which uses the Taylor expansion of the loss function as follows:

$$L(g(y_1+v_2), s) \approx L(g(y_1), s) + v_2^T \nabla_{y_1} L(g(y_1), s), \tag{4.6}$$

where $L(g(y_1+v_2), s)$ refers to the loss between the receiver's reconstructed and original samples after an attack at the channel. Our goal is to increase the loss function and provide an approximation of v_2 when the input is known. Since the attacker at the channel does not know the source's input, we utilize a UAP to imitate the semantic noise over the air, the goal of which is to generate perturbations to fool the DNN in the decoder. We first select some random samples, the number of which determines the number of iterations to create a universal adversarial perturbation. The iterative process is to compute a new perturbation based on the previous perturbation v_2, constantly updating the value of v_2 with the aim of misclassifying the input. The universal adversarial perturbation is then found by performing successive updates.

4.2.1.2 Robust Semantic Communication Model

A robust semantic communication model is proposed to defend the semantic noise mentioned earlier. As shown in Figure 4.2, we incorporate a GAN-based denoising module [Creswell et al., 2018] that is located before the semantic encoder to cope with the semantic noise possibly appearing at the source. This module consists of a generator and a discriminator that together minimize the noise in the training data through a feedback mechanism. The ultimate goal is to train a generative

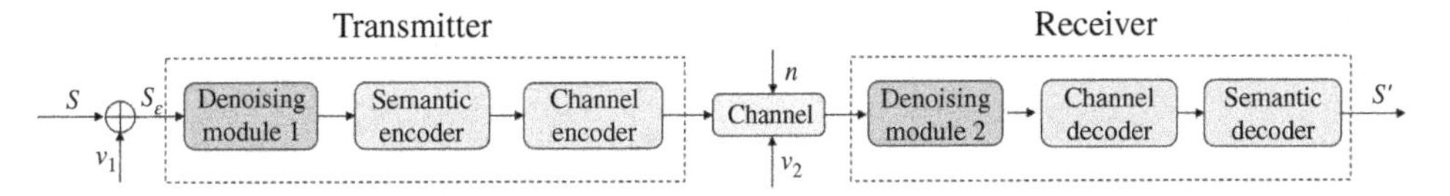

Figure 4.2 A robust semantic communication system.

function that is capable of eliminating imperceptible but intentional semantic noise from corrupted input images. To achieve this goal, we trained a generator network parameterized by θ_G, where θ_G represents the weights and biases of the generator network. The inputs to this generator are the adversarial samples s_ϵ generated by the adversarial attack and the original sample s. The objective function of the GAN model is as follows:

$$\min_{\theta_G} \max_{\theta_D} [E_{s \sim p_{data}(s)} \log D_{\theta_D}(s) + E_{s_\epsilon \sim p_G(s_\epsilon)} \log(1 - D_{\theta_D}(G_{\theta_G}(s_\epsilon)))], \tag{4.7}$$

where G_{θ_G} is the generator, D_{θ_D} is the discriminator, $E_{s \sim p_{data}(s)}[\log D_{\theta_D}(s)]$ is the expectation of $\log D_{\theta_D}(s)$ when all s are true samples, and the second expectation $E_{s_\epsilon \sim p_G(s_\epsilon)}[\log(1 - D_{\theta_D}(G_{\theta_G}(s_\epsilon)))]$ is the expectation of $\log(1 - D_{\theta_D}(G_{\theta_G}(s_\epsilon)))$ when all s_ϵ are adversarial samples. The main idea is that by fixing the generator G_{θ_G} to maximize the loss from the point of view of the discriminator D_{θ_D}'s point of view, followed immediately by fixing D_{θ_D} to minimize the loss from the point of view of the generator G_{θ_G}'s point of view, the discriminator and generator can be made to achieve confrontation with shared loss. During the training process, we use the Adam optimizer to continuously update the parameters and optimize the network to minimize the loss function. The purpose of this denoising module is to reduce the impacts of semantic noise at the source.

Now, we turn to eliminate the effect of semantic noise that possibly appears at the channel. Similar to tackling the semantic noise at the source, we add a denoising module before the channel decoder, which is structured as a DAE. A DAE is an autoencoder that injects semantic noise into the input and then uses the noisy "corrupted" samples to reconstruct the noise-free "clean" input. The goal of DAE is to minimize the loss between the input and the reconstructed signal. Due to the implicit layer representation of the DAE being obtained from a "corrupted" version of the original input, equivalent to introducing a proportion of "blank" elements into the original input, the DAE fills in the lost information by continuously adjusting its parameters to adapt to the dynamic changes in the input data and then learns the data structure. In this way, the extracted features can better reflect the characteristics of the original input [Ye et al., 2015]. As shown in Figure 4.2, the output x of the encoder is corrupted by semantic noise v_2 and channel noise n. The channel output is $y_1 = x + n + v_2$. y_1 is the input to the DAE, and the loss function of the DAE is given by

$$loss_{DE} = L_{DE}(f_D(y_1), x), \tag{4.8}$$

where $L_{DE}(\cdot)$ is the loss function and $f_D(\cdot)$ is the output function. The network parameters are obtained by iteratively learning the errors between the output of the DAE and x. The training objective is to minimize the reconstruction error.

4.2.1.3 Performance Analysis

In this section, we compare the proposed robust semantic communication model with the joint adversary-trained semantic communication model under AWGN, Rayleigh, and Rician channels. We use CIFAR-10 as a dataset, which consists of a total of 10 categories, each consisting 6000 images. Considering the image classification task, the performance of the following methods is provided:

- *Original*: The original semantic communication model [Bourtsoulatze et al., 2019] without the semantic noise.
- *Original + SN (FGSM)*: The semantic communication model with the semantic noise, where the semantic noise at the source is produced by an FGSM attack [Hu et al., 2022], and the semantic noise at the channel is produced by a UAP [Hu et al., 2023].
- *JOINT-AT + SN (FGSM)*: The semantic communication model with joint adversarial training defending with an FGSM attack at the source and a UAP at the channel.
- *Proposed-MD + SN (FGSM)*: The proposed robust semantic communication model with two denoising modules, defending with an FGSM attack at the source and a UAP at the channel.
- *Original + SN (PGD)*: The semantic communication model with the semantic noise, where the semantic noise at the source is produced by a PGD attack, and the semantic noise at the channel is produced by a UAP.
- *JOINT-AT + SN (PGD)*: The semantic communication model with joint adversarial training defending with a PGD attack at the source and a UAP at the channel.
- *Proposed-MD + SN (PGD)*: The proposed robust semantic communication model with two denoising modules defending with a PGD attack at the source and a UAP at the channel.

Figure 4.3 shows that the semantic communication system is highly sensitive to the semantic noise under the AWGN channel. Under the influence of semantic noise, the classification accuracy of the original semantic communication model is much lower than that of the original semantic communication model without the semantic noise. The classification accuracy of our proposed robust semantic communication model under the semantic noise at low SNR is close to the classification accuracy of the original semantic communication model without noise. Figure 4.4 shows that under the Rician channel, the overall trend is that the accuracy of our proposed robust semantic communication model slightly improves with the increasing SNR and gradually stabilizes. Our proposed robust semantic communication model has higher classification accuracy than the semantic communication model trained by joint adversarial training.

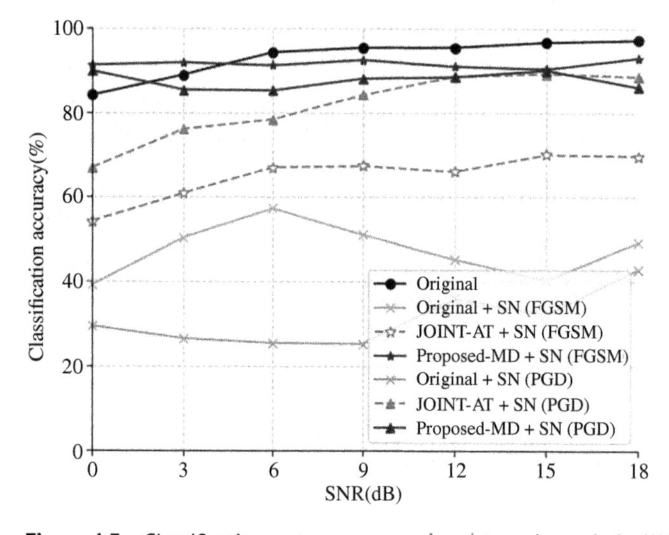

Figure 4.3 Classification accuracy versus signal-to-noise ratio in AWGN channel.

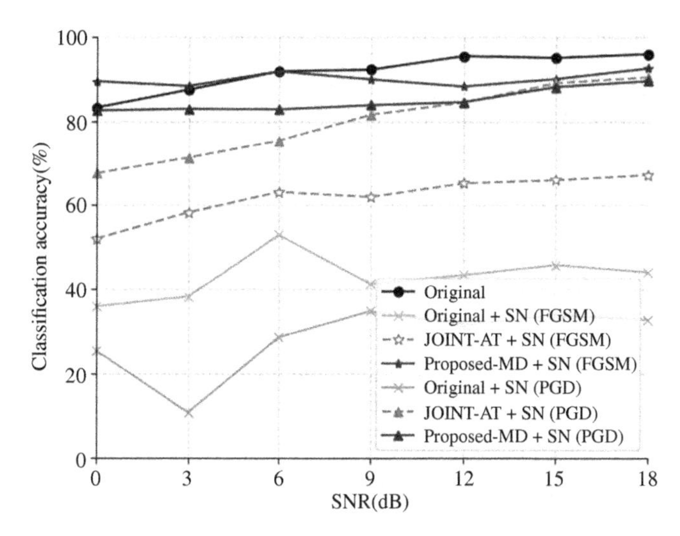

Figure 4.4 Classification accuracy versus signal-to-noise ratio in Rician channel.

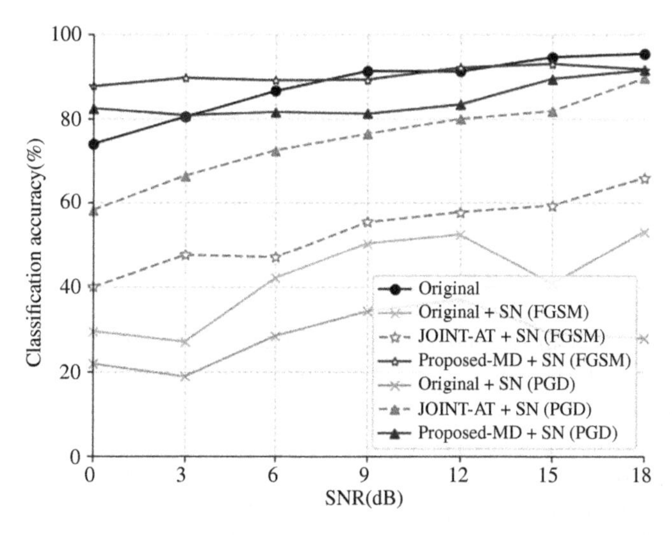

Figure 4.5 Classification accuracy versus signal-to-noise ratio in Rayleigh channel.

Figure 4.5 shows the classification accuracy versus SNR under the Rayleigh channel. We observe that the classification accuracy of the original semantic communication model gradually increases as the SNR increases. The effect of semantic noise on the classification accuracy of the original semantic communication system is more significant at lower SNR. Although the joint adversarial training improves the classification accuracy, it performs poorly if SNR is low. From the classification accuracy, it is easy to see that our proposed robust semantic communication model is significantly better than the semantic communication model with joint adversarial training. Meanwhile, the classification accuracy of our proposed robust semantic communication tends to be close to the classification accuracy of the original model at low SNR. We can also see that the classification accuracy of the Rayleigh channel is lower than that of the AWGN channel and the Rician channel, so the Rayleigh channel is more sensitive to semantic noise, resulting in lower classification accuracy.

4.2.2 A Robust Semantic Communication Model for Speech Transmission

If the speech signals are affected by semantic noise in semantic communication, the receiver may possibly not correctly understand or perceive the meaning of the original speech signals. Therefore, it is important to design a robust semantic communication system for speech transmission.

Semantic noise for images are usually performed through small and imperceptible pixel adjustments to deceive the DNNs into producing incorrect classification results. Meanwhile the presence of semantic noise for speech signals involves small modifications of the sound waveform to mislead the speech recognition system. Perceptually, semantic noise in images may be nearly imperceptible to the human visual system, while those in speech have relatively little impact on the human auditory system.

As shown in Figure 4.6, semantic noise may appear at different positions in the semantic communication process. We also adopt the idea of simulating the adversarial attack and then defending against it to improve the robustness of the model. We utilize the adversarial attacks for the audio domain to generate the semantic noise at the source. The commonly used method for generating adversarial samples is the Fast Gradient Sign Method (FGSM), which has been used for generating adversarial samples for both images and speeches. The adversarial attack for speech is given by

$$\delta = \varepsilon \text{sign}(\nabla_s L_{speech}(\theta, s, t')), \tag{4.9}$$

where δ denotes the adversarial perturbation, ε is used to control the amplitude of the perturbation, θ is the model parameter, s is the model input, t' is the classification label expected by the attacker, and $L_{speech}(\cdot)$ is the loss function of the neural network. By adjusting ε, the perturbation of the generated adversarial samples can be controlled. If ε is small, the generated adversarial samples may be more difficult to detect but have less impact on the performance of the model. A larger ε results in a more noticeable perturbation but also has a stronger impact on the model's performance. By directly perturbing the original waveform, this method avoids the extra loss caused by introducing perturbations in the feature domain. Moreover, we can imitate the semantic noise at the channel in the same way as we do for the images.

To defend against semantic noise for speech transmission, two common defense methods are adversarial sample detection and adversarial training. In particular, adversarial training, as a proactive defense approach, aims to combat the impacts of potential noise by improving the robustness of the neural network. Conversely, adversarial detection is to add additional detectors to identify adversarial samples before they are fed into the model. Considering the training cost, adversarial training is considered the main defense method to effectively eliminate the semantic noise in the semantic communication process, while significantly improving the overall robustness of the model.

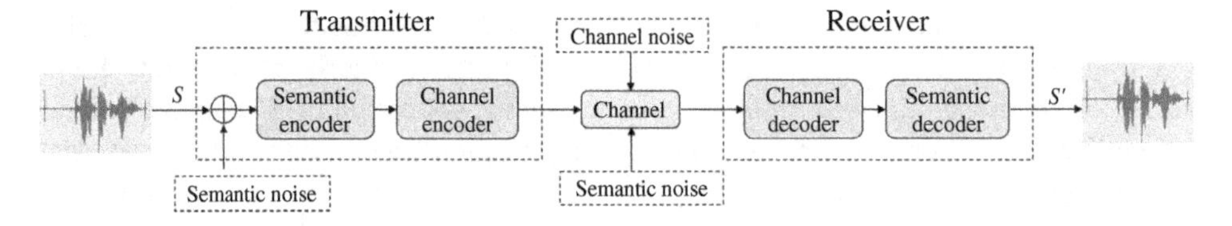

Figure 4.6 A semantic communication system for speech transmission with semantic noise.

4.3 Knowledge Discrepancy-Oriented Privacy Protection for Semantic Communication

4.3.1 Knowledge Discrepancy and Privacy Issue

Semantic communication, relying on DNNs and knowledge bases, can improve the efficiency of communication by transmitting only the key semantics rather than the whole data. Different from traditional communication where encoding and decoding can be done without understanding, semantic communication is characterized by understanding before transmission. Specifically, knowledge bases play a critical role in semantic understanding. On the sender side, knowledge bases assist in distilling core semantics by extracting the most critical content to reduce redundancy. On the receiver side, knowledge bases provide the extra inference ability for semantic reconstruction, enabling the receiver to recover the true meaning of the source message from received semantics even under harsh channel conditions.

Commonly, knowledge bases are divided into public and private ones [Shi et al., 2021]. Public knowledge bases are globally recognized knowledge resources (e.g., common knowledge and generic domain knowledge) that allow communication nodes to share, thereby building a foundation for semantic communication. Private knowledge bases, which are personalized resources closely tied to individuals' backgrounds (e.g., personal experience, professional background, and preferences), would not be shared widely. Currently, most semantic communication systems are on the premise that the sender and receiver share exactly the same knowledge base, i.e., the public knowledge base. However, it is difficult for the sender and receiver to hold a consistent background knowledge in reality [Goodfellow et al., 2015].

In this context, different knowledge bases are prone to introduce a *knowledge discrepancy*, which presents the different contents or levels of understanding between communication nodes. The knowledge discrepancy mainly comes from the agnostic and the diversity of private knowledge bases.

- *Agnostic*: Agnostic means that the sender has little knowledge about the private knowledge base of the receiver. Existing work mostly focuses on inferring private attributes based on public information from social platforms (e.g., personal interests [Chaabane et al., 2012] or posted content [Goga et al., 2013]). These involve the sender digging into and utilizing the personal information of the receiver, rather than inferring the private knowledge base directly. For knowledge inference, most work relies on the structural features of specific knowledge graph for inference [Yu et al., 2018]. However, the graph available to the public is usually anonymized through edge removal and node generalization [Qian et al., 2017]. Agnostic poses a challenge for the inference of anonymized knowledge graph with the auxiliary of prior knowledge graph that contains specific knowledge.
- *Diversity*: Diversity refers to the possibility that different knowledge bases may have different interpretations of the same object or fact. It is likely that the sender and receiver with different backgrounds have different understandings of the same thing [Xiao et al., 2022a]. For example, "Tweety" is always considered "an application" for adults, but kids recognize it as "a cartoon bird." Unfortunately, most existing knowledge inference methods ignore the possibility of different semantic interpretations between the inference results and prior knowledge [Ho et al., 2018; Krötzsch et al., 2018]. Although Xiao et al. [2022b] and Liang et al. [2022] proposed implicit

semantic reasoning mechanisms to resolve diverse understandings through collaboration, it is infeasible for the receiver to actively provide auxiliary information for the sender's inference. Thus, diverse understanding poses another challenge for narrowing down the knowledge discrepancy.

The knowledge discrepancy arising from these factors can lead to privacy risks. It enables the receiver to infer sensitive information that the sender does not want to disclose. For example, some sensitive information (e.g., the income of A) can be inferred based on the message "A and B are colleagues" and the knowledge about the company and occupation of B. To address the privacy risk arising from the knowledge discrepancy, it is imperative to delve into the agnostic and diversity of private knowledge bases.

4.3.2 Privacy Protection Model Against Knowledge Discrepancy

4.3.2.1 Overview

Semantic communication systems mainly consist of a semantic encoder and decoder, channel encoder and decoder, and knowledge bases of the sender and receiver. As shown in Figure 4.7, to address the privacy risk caused by the knowledge discrepancy, we propose a KDPP approach to decrease the privacy risk. In the semantic communication system, a privacy protector is added before the semantic encoder.

The protector consists of two modules: a knowledge inference module and a path cutting-off module. Commonly, by observing the receiver's public sources such as social platforms, the sender can obtain limited observed information O_B [Kosinski et al., 2013] (e.g., social relationships, personal preferences, and anonymous IDs). We assume that the observed information is an available graph data, similar to a published anonymous complex network graph. Based on the message X, the knowledge inference module $f(X, O_B, K_A)$ utilizes prior knowledge K_A to make inference on the observed information O_B. According to the output posterior knowledge $\hat{K}_B$, the path cutting-off module $g\left(X, \hat{K}_B\right)$ processes the message to avoid the receiver inferring sensitive information from the output message X'.

In the knowledge inference module, the initial stage focuses on knowledge mapping, which aims to match the prior knowledge with the observed information based on similarity. The second stage is disambiguation, relying on the semantic context to infer the receiver's interpretation of ambiguous concepts.

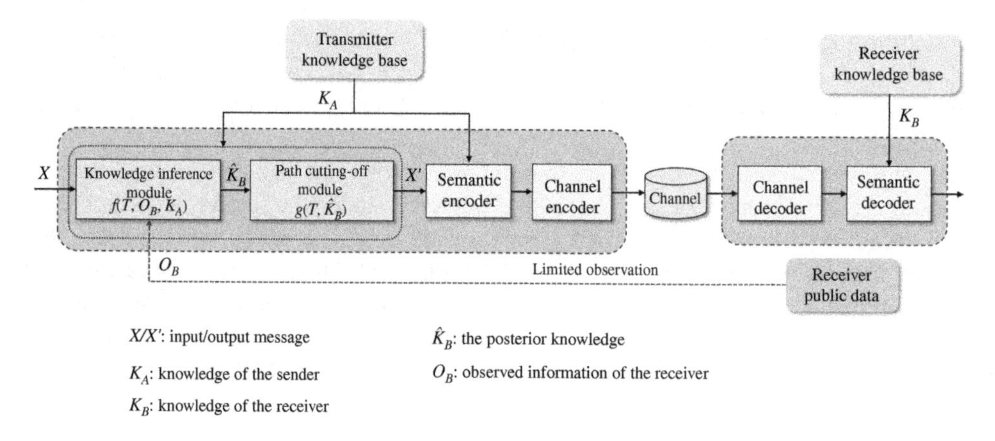

Figure 4.7 Semantic communication system with KDPP.

In the path cutting-off module, the main idea is to trace and cut off the receiver's privacy inference paths. Based on the knowledge, we simulate the receiver's inference of messages and ascertain the logic paths that contribute to privacy risk. Finally, we cut off these logical paths to prevent the receiver from inferring sensitive information.

4.3.2.2 Knowledge Inference Module

The knowledge of both communication nodes can be modeled as knowledge graphs. In this chapter, all we describe are undirected graphs. Each piece of knowledge is viewed as a triple (h, r, t), representing a fact composed of the head entity, relation, and tail entity. All the triples are linked to form a knowledge graph $G(V, E)$, where V is a set of entity nodes and E is a set of edges indicating the relationship between entities. We refer to the sender's knowledge as prior knowledge graph $G_S(V_S, E_S)$, the observed information from the receiver as limited knowledge graph $G_A(V_A, E_A)$, and the receiver's private background knowledge as $G_R(V_R, E_R)$. Here, the knowledge graph refers to a complex network that includes social relationships and personal attributes.

Definition 4.1 ***(Knowledge discrepancy).*** The difference set of triples between the sender's knowledge graph G_S and the receiver's knowledge graph G_R is called "the *knowledge discrepancy*," represented as the set $D = G_R \cap \overline{G_S}$, where $\overline{G_S}$ denotes the complementary set of G_S.

Definition 4.2 ***(Privacy risk).*** The privacy risk is defined as $P\,(S|X, G)_R$, where S denotes the sensitive information that the receiver infers from message X.

The privacy risk of semantic communication is caused by the knowledge discrepancy D, so the goal of privacy protection is to narrow D and reduce the risk $P\left(S|X, G_R\right)$.

It is possible that different knowledge backgrounds may hold inconsistent understanding of the same entity. Under such circumstances, the semantic of neighbor nodes according to the knowledge graphs can assist in more accurately deducing the interpretations. For example, as shown in Figure 4.8, the node "Tweety" can be interpreted as a cartoon character or a news application. To infer the receivers' understanding of "Tweety," we consider its neighbor nodes "TV show" and "cartoon character" as context, which can generate the semantic label of "cartoon." Meanwhile, in the sender's prior knowledge, we can generate the semantic label of "application" by its neighbor nodes "software" and "phone."

Therefore, our idea is to utilize the semantics of neighbor nodes to generate semantic labels for entities, and then we can quantify the difference in understanding through these labels. That is, the set of neighbor nodes $N(v) = \left\{n_1, n_2 \dots n_m\right\}$ reflects the context of v, where n_i is the ith neighbor node of v. To generate the semantic label for v based on $N(v)$, we adopt the FastText [Joulin et al., 2016] classifier, which is a typical deep learning representation of word vectors and has achieved outstanding results in classifying. Formally, the FastText classifier takes the neighbor set $N(v)$ as input and outputs the semantic label V_L.

A primary step of inference is to perform a node mapping $p : G_S \rightarrow G_A$ on two graphs, with the goal of specifying the anonymized graph. We consider two mapping criteria: the structural similarity and semantic similarity.

Structural similarity An essential criterion of node mapping is structural similarity. It is easy to understand that if two nodes are essentially the same, their structural features (e.g., social relationships and attribute connections) should be similar in G_A and G_S. To strengthen the

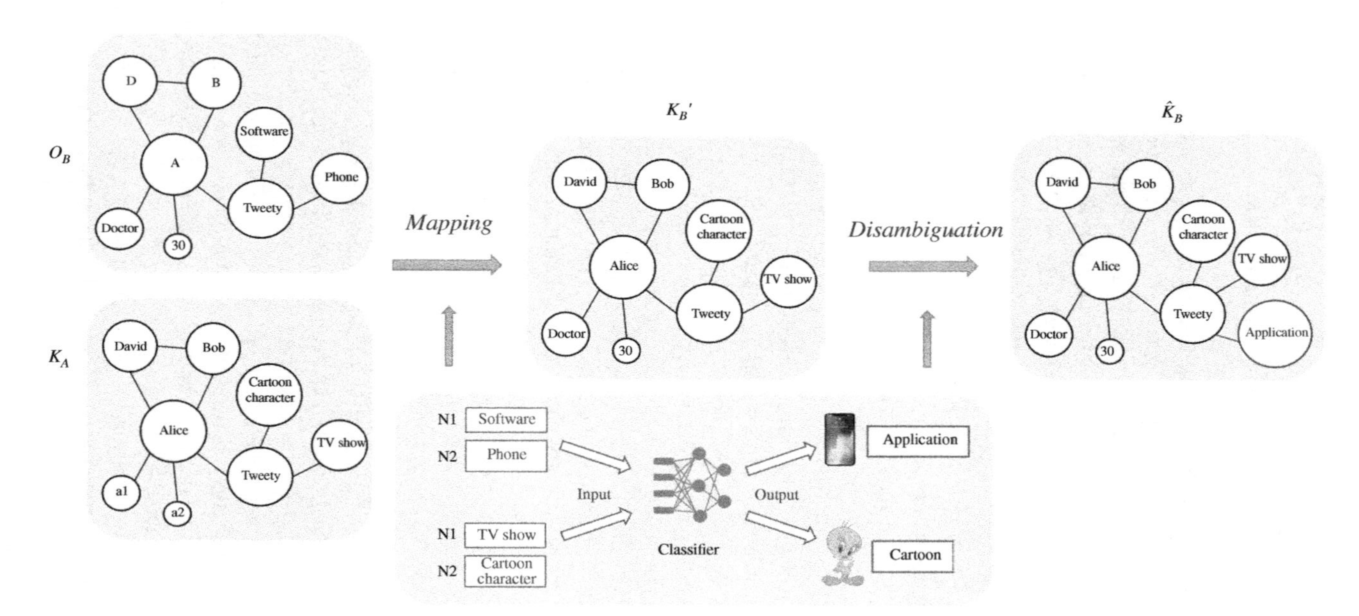

Figure 4.8 An example of proposed knowledge inference method with knowledge mapping and disambiguation.

criterion for accurate mapping, we define the following structural features in the knowledge graph: (i) The degree of the node $D_0(v)$, that is, the number of neighbors of the node, which is the most basic structural feature. (ii) The one-hop neighbor feature of the node $D_1(v)$, containing degree features of neighbors at a distance of 1 from the node, reflects the nearest neighbor similarity between nodes. (iii) The two-hop neighbor feature of the node $D_2(v)$, which contains degree features of neighbors at a distance of 2 from the node, reflects the second nearest neighbor similarity between nodes. Based on several definitions mentioned earlier, we can calculate the feature vector $f(v) = \left[D_0(v), D_1(v), D_2(v)\right]$ by cosine distance [Fang et al., 2019]. The formula for calculating the structural similarity $S_R(i,j)$, where $i \in G_S$ and $j \in G_A$, is shown here:

$$S_R\,(i,j) = \cos\left(f(i), f(j)\right). \tag{4.10}$$

Semantic Similarity Structural similarity is not sufficient for the mapping condition, as nodes with similar structures are not necessarily the same. Therefore, we incorporate semantic similarity as one of the mapping conditions. The trained Word2Vec model [Mikolov et al., 2013] is utilized to convert the generated semantic label V_L of node v into a vector $L(v) = (w_1, w_2 \ldots w_n)$, where w_i denotes the value of ith dimension for the word embedding. Thus, the semantic similarity S_A can be calculated by the following cosine formula:

$$S_A\,(i,j) = \cos\left(L(i), L(j)\right). \tag{4.11}$$

Thus, we assign weights to the structural similarity S_R and semantic similarity S_A to obtain the confidence between i and j:

$$\text{conf}\,(i,j) = (1-a)\cdot S_R\;(i,j) + a \cdot S_A\;(i,j), \tag{4.12}$$

where $a \in [0,1]$, $i \in G_S$, and $j \in G_A$. Equation (4.12) indicates that even if two entities have high structural similarity, confidence will be reduced due to a low S_A.

For each node pair between G_S and G_A, the goal is to find such a mapping p that maximizes the confidence.

$$\arg\max_{p}\;\text{conf}\,(i,j). \tag{4.13}$$

The process of mapping is to traverse through unmapped nodes between G_S and G_A, and then compute the mapping confidence. However, a global traversal will increase the time complexity significantly. We consider the nodes that exist in both G_S and G_A as seed nodes, which are featured by the same concept and semantic label. The intuition is that if two nodes match, their neighbors are also very likely to match. Thus, we divide the mapping process into two phases. In the first phase, we identify the seed nodes in the two graphs. In the second phase, the mapping spreads from the seed nodes and prioritizes the traversal of their neighbors, which is similar to breadth-first search. If the confidence conf(i,j) is higher than a predefined threshold value ε, the mapping is considered to be valid, and then the node i is added to the candidate set of j until the candidate count achieves K. It is obvious that the threshold ε and K will have a large impact on the results. Later, we will evaluate the impact of these parameters.

After the knowledge mapping, we consider the case of nodes with the same concept but different semantic labels as diverse understanding between two graphs. To address this diversity issue, we make further inference about the receiver through disambiguation, which is essentially an adding operation. Since the neighboring semantics conduce to the content interpretations of entities, we take the semantic labels V_L^i and V_L^j for nodes i and j with the same concept as understandings of the sender and receiver. As shown in Figure 1.7, we retain the sender's prior knowledge V_L^i and add the receiver's understanding V_L^j to the graph, i.e., connecting the node i to V_L^j by the edge E_L^j.

4.3.2.3 Path Cutting-Off Module

Privacy Extraction After the knowledge inference module based on G_S and G_A, the posterior knowledge graph G_P is obtained. We transform the sent message X into the knowledge graph G_T as well. In this way, the logic inside the message content can be corresponded to the connection of nodes for knowledge inference.

Before privacy protection, we first extract the message content that contributes to privacy risk. Privacy extraction is mainly divided into two steps: simulating the receiver's inference process and extracting the privacy relation path from simulation results. Based on G_P, we use the path ranking algorithm (PRA) [Lao et al., 2011] for simulation inference on G_T. The extracted relation path substantially corresponds to a kind of Horn clause, where path features can be interpreted as logical rules for inference.

Specifically, the inference process generates an n-dimensional feature vector $P = (P_1, P_2 \dots P_l \dots P_n)^T$ for each entity pair (h, t), where P_l is the feature value $P_l(h \to t)$ of the lth relational path π_l. Note that $\pi_l = \left(r_1, \dots r_i \dots r_n\right)$ is a multi-hop path with a certain length n, where r_i denotes the one-hop path. Let $\pi_l' = \left(r_1, \dots r_{i-1}\right)$, then the set of tail entities associated with π_l' is denoted as Range(π_l'). For any relational path, $h \xrightarrow{r_1} \dots \xrightarrow{r_{i-1}} e_{i-1} \xrightarrow{r_i} e_i \dots \xrightarrow{r_n} t$ where e_{i-1} and e_i are entities in the relational path, the feature value $P_l\left(h \to e_i\right)$ is obtained from iterations of the random walk probabilities:

$$P_l\left(h \to e_i\right) = \sum_{e_{i-1} \in \text{Range}(\pi_l')} P_l\left(h \to e_{i-1}\right) \cdot \Pr\left(e_i | e_{i-1}; r_i\right), \tag{4.14}$$

where $\Pr\left(e_i | e_{i-1}; r_i\right) = |r_i\left(e_{i-1}, e_i\right)| / |r_i\left(e_{i-1}, \cdot\right)|$ is the probability of reaching entity e_i through a single-hop path r_i from entity e_{i-1}, and $r_i(e_{i-1}, e_i)$ is a Boolean value representing the existence of the relation r_i between e_{i-1} and e_i. Using these feature vectors as training samples, the score function for the relationship to be predicted is calculated as follows:

$$s(h, t) = \sum P_l(h \to t) \cdot \theta_l, \tag{4.15}$$

where θ_l denotes the l_{th} path weight. A higher score $s(h, t)$ indicates a higher probability that the relationship exists between h and t.

To simulate the inference process is to train a classifier using path features between entities and use it to predict whether a certain relationship exists or not. Following the extraction of path features, we can train a logistic regression classifier for each relation r:

$$\Pr(y = 1 | P; \theta) = \frac{1}{1 + e^{-\theta^T \cdot P}}, \tag{4.16}$$

where $\theta^T \cdot P = s(h, t)$ is the score calculated earlier, and the approximate $\theta = (\theta_1, \theta_2 \dots \dots \theta_n)^T$ will be determined by classifier training.

In this chapter, we define privacy as a set of triples $\left(h_s, r_s, t_s\right)$ that contain private relationships or sensitive attributes, which are added to the privacy set S.

Path Cutting-Off As shown in Figure 4.9, the idea of privacy protection based on path cutting-off is to find out and cut off inference paths of privacy. The logical rules of inference are transformed into numerical vectors when simulating inference, which is the basis of the classifier's prediction in Eq. (4.16). Therefore, we analyze the numerical values of each feature vector and decide which logical rules should be cut off.

Equations (4.15) and (4.16) show that a larger feature value $P_l\left(h \to t; \pi_l\right) \cdot \theta_l$ indicates a greater possibility of the predicted relation. That is, π_l contributes more to predict r_s between h_s

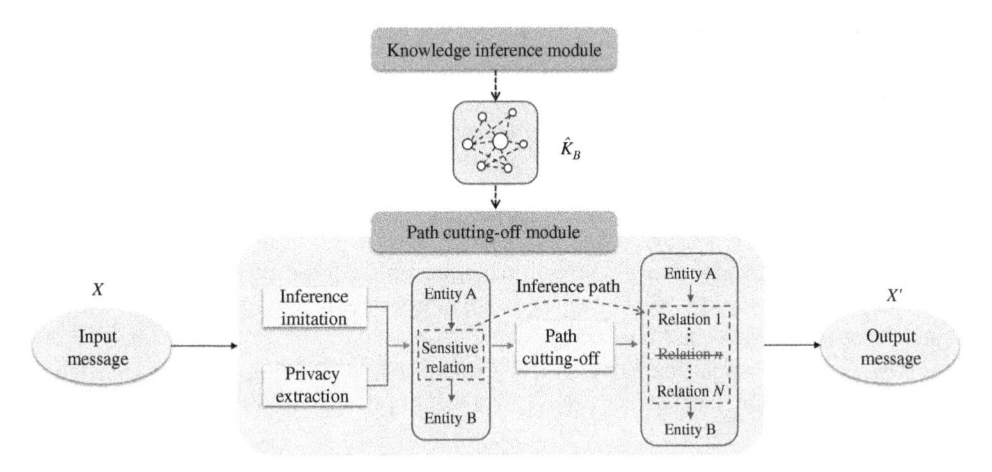

Figure 4.9 Schematic diagram of the path cutting-off module. It receives the posterior knowledge graph from the knowledge inference module along with messages as inputs, and output messages to meet the privacy protection criteria.

and t_s. To cut off the inference path, we add perturbations to the path π_s corresponding to the maximum feature value, so the original inference result (h_s, r_s, t_s) no longer holds. Note that $\pi_s = (r_1, \dots r_i \dots r_n)$, where r_i is the one-hop path, and the path of triple (h_s, r_s, t_s) can be interpreted as $(h_s, r_1, e_1), \dots (e_{i-1}, r_i, e_i) \dots, (e_{n-1}, r_n, t_s)$. This indicates that we can remove any r_i to cut off this inference path.

To ensure more information utility and less privacy risk, it is necessary to selectively remove paths that contribute more to privacy inference rather than all paths. We assign a weight w_i to each path r_i, which denotes the proportion of path involving r_i among all inference paths. A larger weight indicates a greater contribution to the privacy set S, as it is involved in more inference paths. Thus, the principle of cutting off is to remove r_i when w_i is note less than δ, where δ is the threshold. A larger threshold setting means that a small number of paths with higher weight will be removed. The appropriate threshold setting can cut off more privacy paths and reduce unnecessary information loss, balancing the privacy risk and utility of the message.

4.3.3 Performance Analysis

Datasets We conduct our method on two knowledge graph datasets: real-world network and synthetic network. For the former, we extract the maximum connected subgraph and remove unimportant and duplicate triples of FB13 dataset [Wang et al., 2014] for use. FB13 is a subset of the well-known knowledge graph Freebase and contains a rich set of kinships and attributes. Finally, three relationships (children, parents, and spouse) and seven attributes (nationality, religion, gender, institution, ethnicity, location, and profession) are reserved. For the latter, we create a denser and regularized network for use, which contains three personal relationships (colleague, friend, and spouse) and four attributes (hobby, occupation, workplace, and income).

Moreover, for both datasets, we add some ambiguous nodes with the same concept but different semantics to represent the diversity of understanding. Details about the two datasets are shown in the Table 4.1.

The preprocessed datasets are called "the prior graph G_S." Then, a part of the nodes and edges is extracted from G_S, after which both relations and attributes are processed to derive the limited knowledge graph G_A. For the relationships, we adopt the anonymization methods used in previous

Table 4.1 Details about the datasets.

Datasets	Nodes	Edges	Edge Relation
FB13	71,718	226,847	Kinship and personal identity
Synthetic	1224	70,027	Social life circle

work [Fang et al., 2019] to change the edge structure of the graph. For the attributes, nodes such as occupations, incomes, and workplaces are replaced with more generalized ones, such as "organist" and "guitarist," both generalized as "musician."

Subsequently, we will analyze the performance of the proposed KDPP in terms of knowledge mapping and disambiguation accuracy, as well as the effect of privacy protection.

Knowledge Mapping As mentioned, the factor a denotes the weight of semantic similarity in the mapping confidence. For both synthetic and FB13 networks, Figure 4.10 shows that the knowledge mapping, which considers semantic similarity ($a > 0$), achieves higher accuracy compared to that of traditional de-anonymization ($a = 0$). Moreover, this figure also indicates that the assistance of semantic similarity in the FB13 network is rather few compared to synthetic network. This is because the nodes of FB13 exhibit significant structural differences, relying more on structural similarity for mapping. Conversely, the denser synthetic network endows nodes with more average structural characteristics, thus accentuating the role of semantics in enhancing mapping accuracy. This suggests that our method is more suitable for structurally denser networks.

Figure 4.11 shows that selecting the top 10 ($K = 10$) candidates is sufficient for the accurate mapping, as the neighbor sets of matched nodes are similar. Moreover, for the same K, a suitable threshold contributes to a high level of accuracy.

Disambiguation Since disambiguation is proposed to infer the receiver's understanding of ambiguous entities based on semantic labels, we test the generated precision of 437 labels in the datasets to evaluate the effectiveness of disambiguation.

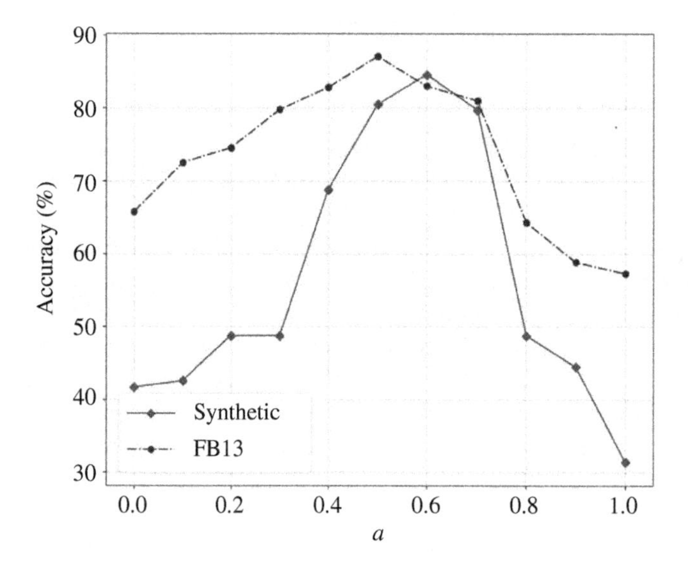

Figure 4.10 The effect of factor a on accuracy ($K = 10$, $\varepsilon = 0.5$).

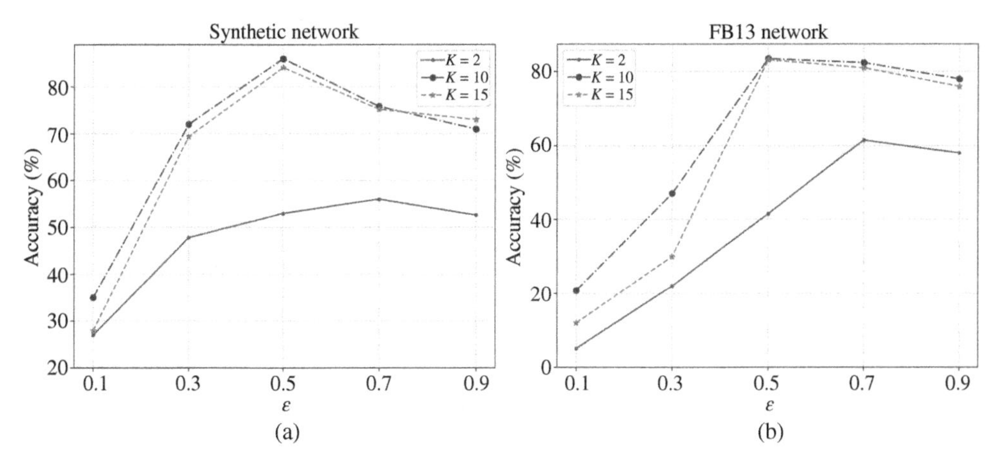

Figure 4.11 The effect of factor ε on accuracy ($a = 0.6$). (a) Synthetic network and (b) FB13 network.

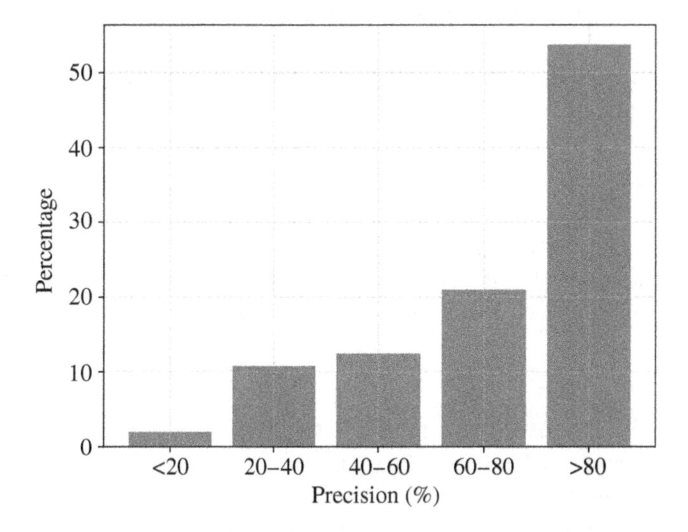

Figure 4.12 Percentage of different precision in semantic label generation results.

Figure 4.12 shows that most semantic labels of nodes are generated with precision above 80%. Some labels exhibit lower precision because the input neighbor nodes also appear in the training dataset of other labels. For example, label generation will be confused by the overlapping neighbor set of nodes with similar labels "oboist" and "musician." In order to solve this problem, the feasible approach is to optimize the dataset and reduce the overlapping portion of the training data between the labels.

Privacy Protection The proposed KDPP focuses on two simultaneous goals: one is to convey more useful information through the message, and the other is to constrain the receiver to recover less sensitive information. In this way, we define metrics as follows. To perform the comparison with the benchmark, privacy protection is conducted based on four methods: message-only, traditional de-anonymization, knowledge mapping, and knowledge mapping combined with disambiguation (KDPP).

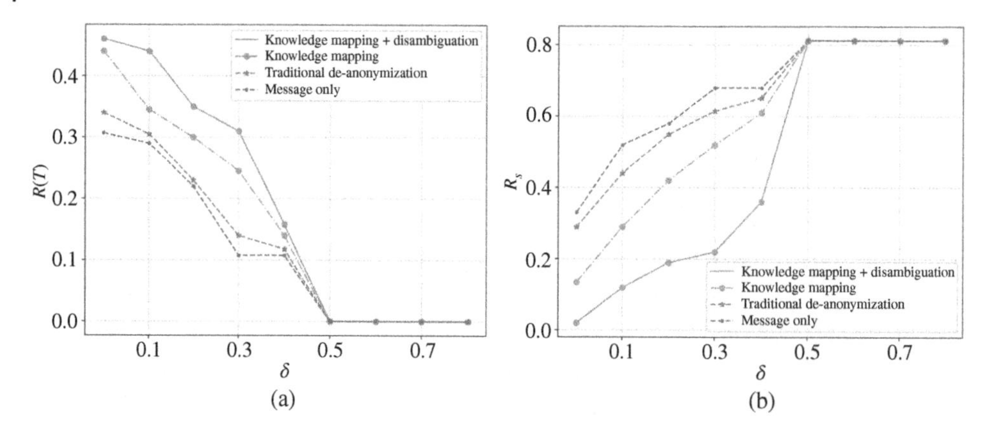

Figure 4.13 The information loss (a) and privacy risk (b) of four protection methods after removing varying amounts of edges.

Information Loss Since we transform the message into knowledge graphs, the information loss can be measured by the variation in edge structure. According to Zhang and Zhang [2009], we calculate the information loss $R(T)$ as follows, where $|E|$ is the total number of edges before processing, and $|E'|$ is the number of edges that remain after removing.

$$R(T) = \frac{|E| - |E'|}{|E|}. \tag{4.17}$$

Privacy Risk We measure the privacy risk, R_s, by the probability of privacy inference as mentioned here, where S' is the total inference results, and S is the results containing sensitive information. Note that the inferred results are based on the receiver's real knowledge graph, G_R.

$$R_s = \frac{|S|}{|S'|}. \tag{4.18}$$

For metric evaluation, we select FB13 network as the dataset (the results of the two networks are similar). Figure 4.13 illustrates $R(T)$ and R_s with different path thresholds, δ. The selection of the threshold δ is crucial, as it not only determines $R(T)$ but also relates to R_s. When $\delta > 0.5$, no edges satisfy the frequency of 0.5. Consequently, no edges are removed, resulting in the minimum $R(T)$ (a value of 0) and the maximum R_s. Moreover, there is an inverse relationship between $R(T)$ and R_s, indicating that as the privacy risk decreases, more edges are removed (resulting in higher information loss).

Furthermore, we observe that under all δ values, the privacy risk R_s after knowledge mapping and disambiguation is lower than that of the other three methods. Comparing the four methods, KDPP can narrow the knowledge discrepancy and reduce the privacy risk R_s, but it can also increase the information loss $R(T)$ due to more removal of edges. In a situation like $\delta = 0.4$, our method significantly reduces the risk of privacy breaches compared to message-based methods, with only around 6% increase in data utility loss.

4.4 Conclusion

Overall, this chapter proposes the robust semantic communication systems against semantic noise for image and speech transmission, as well as a privacy protection method to reduce the privacy risk arising from knowledge discrepancy. For the robust semantic communications, current researches focus on the image and text domains, while robust semantic communication in the video and speech domains has not yet received in-depth exploration. Therefore, future research should continue to explore robust semantic communication techniques in different modalities in depth to provide more effective and reliable semantic information transmission schemes. For the privacy protection, we consider the discrepancy between private knowledge bases. This can be seen as a pioneer in addressing privacy issues for semantic communication, where the proposed KDPP can be insightful for future design in privacy protection semantic communication system.

References

Naveed Akhtar, Ajmal Mian, Navid Kardan, and Mubarak Shah. Advances in adversarial attacks and defenses in computer vision: A survey. *IEEE Access*, 9:155161–155196, 2021. doi: 10.1109/ACCESS.2021.3127960.

E. Bourtsoulatze, D. Burth Kurka, and D. Gündüz. Deep joint source-channel coding for wireless image transmission. *IEEE Transactions on Cognitive Communications and Networking*, 5(3):567–579, 2019.

Abdelberi Chaabane, Gergely Acs, and Mohamed Ali Kaafar. You are what you like! information leakage through users' interests. In *Proceedings of the 19th Annual Network & Distributed System Security Symposium (NDSS)*. Citeseer, 2012.

Antonia Creswell, Tom White, Vincent Dumoulin, Kai Arulkumaran, Biswa Sengupta, and Anil A. Bharath. Generative adversarial networks: An overview. *IEEE Signal Processing Magazine*, 35(1):53–65, 2018. doi: 10.1109/MSP.2017.2765202.

Junbin Fang, Aiping Li, Qianyue Jiang, Shudong Li, and Weihong Han. A structure-based de-anonymization attack on graph data using weighted neighbor match. In *2019 IEEE Fourth International Conference on Data Science in Cyberspace (DSC)*, pages 480–486. IEEE, 2019.

Oana Goga, Howard Lei, Sree Hari Krishnan Parthasarathi, Gerald Friedland, Robin Sommer, and Renata Teixeira. Exploiting innocuous activity for correlating users across sites. In *Proceedings of the 22nd International Conference on World Wide Web*, pages 447–458, 2013.

Y. Gong and C. Poellabauer. Crafting adversarial examples for speech paralinguistics applications. In *Dynamic and Novel Advances in Machine Learning and Intelligent Cyber Security Workshop*, pages 1–8, 2018.

Ian J. Goodfellow, Jonathon Shlens, and Christian Szegedy. Explaining and harnessing adversarial examples, 2015.

Vinh Thinh Ho, Daria Stepanova, Mohamed H. Gad-Elrab, Evgeny Kharlamov, and Gerhard Weikum. Rule learning from knowledge graphs guided by embedding models. In *The Semantic Web–ISWC 2018: 17th International Semantic Web Conference*, Monterey, CA, USA, October 8–12, 2018, Proceedings, Part I 17, pages 72–90. Springer, 2018.

Q. Hu, G. Zhang, Z. Qin, Y. Cai, G. Yu, and G. Y. Li. Robust semantic communications against semantic noise. In *2022 IEEE 96th Vehicular Technology Conference (VTC2022-Fall)*, pp. 1–6, 2022.

Q. Hu, G. Zhang, Z. Qin, Y. Cai, G. Yu, and G. Y. Li. Robust semantic communications with masked VQ-VAE enabled codebook. *IEEE Transactions on Wireless Communications*, 22(12):8707–8722, 2023.

Armand Joulin, Edouard Grave, Piotr Bojanowski, and Tomas Mikolov. Bag of tricks for efficient text classification. *arXiv preprint arXiv:1607.01759*, 2016.

Michal Kosinski, David Stillwell, and Thore Graepel. Private traits and attributes are predictable from digital records of human behavior. *Proceedings of the National Academy of Sciences*, 110(15):5802–5805, 2013.

Felix Kreuk, Yossi Adi, Moustapha Cisse, and Joseph Keshet. Fooling end-to-end speaker verification with adversarial examples. In *2018 IEEE International Conference on Acoustics, Speech and Signal Processing (ICASSP)*, pages 1962–1966, 2018. doi: 10.1109/ICASSP.2018.8462693.

Markus Krötzsch, Maximilian Marx, Ana Ozaki, and Veronika Thost. Attributed description logics: Reasoning on knowledge graphs. In *IJCAI*, pages 5309–5313, 2018.

Ni Lao, Tom Mitchell, and William Cohen. Random walk inference and learning in a large scale knowledge base. In *Proceedings of the 2011 Conference on Empirical Methods in Natural Language Processing*, pages 529–539, 2011.

Jingming Liang, Yong Xiao, Yingyu Li, Guangming Shi, and Mehdi Bennis. Life-long learning for reasoning-based semantic communication. In *2022 IEEE International Conference on Communications Workshops (ICC Workshops)*, pages 271–276. IEEE, 2022.

Xuewen Luo, Hsiao-Hwa Chen, and Qing Guo. Semantic communications: Overview, open issues, and future research directions. *IEEE Wireless Communications*, 29(1):210–219, 2022. doi: 10.1109/MWC.101.2100269.

Tomas Mikolov, Kai Chen, Greg Corrado, and Jeffrey Dean. Efficient estimation of word representations in vector space. *arXiv preprint arXiv:1301.3781*, 2013.

S.-M. Moosavi-Dezfooli, A. Fawzi, O. Fawzi, and P. Frossard. Universal adversarial perturbations. In *Proceedings of the IEEE Conference on Computer Vision and Pattern Recognition (CVPR)*, 2017.

Paarth Neekhara, Shehzeen Hussain, Prakhar Pandey, Shlomo Dubnov, Julian McAuley, and Farinaz Koushanfar. Universal adversarial perturbations for speech recognition systems, 2019.

Danish Pruthi, Bhuwan Dhingra, and Zachary C. Lipton. Combating adversarial misspellings with robust word recognition, 2019.

Jianwei Qian, Xiang-Yang Li, Chunhong Zhang, Linlin Chen, Taeho Jung, and Junze Han. Social network de-anonymization and privacy inference with knowledge graph model. *IEEE Transactions on Dependable and Secure Computing*, 16(4):679–692, 2017.

Guangming Shi, Yong Xiao, Yingyu Li, and Xuemei Xie. From semantic communication to semantic-aware networking: Model, architecture, and open problems. *IEEE Communications Magazine*, 59(8):44–50, 2021.

Zhen Wang, Jianwen Zhang, Jianlin Feng, and Zheng Chen. Knowledge graph embedding by translating on hyperplanes. In *Proceedings of the AAAI Conference on Artificial Intelligence*, volume 28, 2014.

Yong Xiao, Yingyu Li, Guangming Shi, and H. Vincent Poor. Reasoning on the air: An implicit semantic communication architecture. In *2022 IEEE International Conference on Communications Workshops (ICC Workshops)*, pages 289–294. IEEE, 2022a.

Yong Xiao, Zijian Sun, Guangming Shi, and Dusit Niyato. Imitation learning-based implicit semantic-aware communication networks: Multi-layer representation and collaborative reasoning. *IEEE Journal on Selected Areas in Communications*, 41(3):639–658, 2022b.

Yulong Yang, Chenhao Lin, Xiang Ji, Qiwei Tian, Qian Li, Hongshan Yang, Zhibo Wang, and Chao Shen. Towards deep learning models resistant to transfer-based adversarial attacks via data-centric robust learning, 2023.

Xiufen Ye, Lin Wang, Huiming Xing, and Le Huang. Denoising hybrid noises in image with stacked autoencoder. In *2015 IEEE International Conference on Information and Automation*, pages 2720–2724, 2015. doi: 10.1109/ICInfA.2015.7279746.

Shengkang Yu, Xi Li, Xueyi Zhao, Zhongfei Zhang, Fei Wu, Jingdong Wang, Yueting Zhuang, and Xuelong Li. A bilinear ranking SVM for knowledge based relation prediction and classification. *IEEE Transactions on Big Data*, 5(4):588–600, 2018.

Lijie Zhang and Weining Zhang. Edge anonymity in social network graphs. In *2009 International Conference on Computational Science and Engineering*, volume 4, pages 1–8. IEEE, 2009.

5

Interplay of Semantic Communication and Knowledge Learning

Fei Ni[1], Bingyan Wang[1], Rongpeng Li[1], Zhifeng Zhao[1,2], and Honggang Zhang[1]

[1] *College of Information Science and Electronic Engineering, Zhejiang University, Hangzhou, China*
[2] *Zhejiang Lab, Hangzhou, China*

5.1 Introduction

Advancements in deep learning (DL) and natural language processing (NLP) have paved the way for the evolution of semantic communication (SemCom). As a communication paradigm that prioritizes the transmission of meaningful information over the precise delivery of bits or symbols, SemCom emerges as a feasible approach to achieve high-fidelity information delivery with lower communication consumption [Lu et al., 2023c]. To boost transmission efficiency and optimize adaptability in varying channel environments, several prior studies have employed DL-based joint source–channel coding (JSCC) in the context of the SemCom system. This approach leverages an end-to-end neural network (NN) architecture for both transmitter and receiver, effectively conceptualizing the entire communication system as an autoencoder (AE). Extending from this foundational structure, SemCom systems are eligible for the transmission of a wide array of data types, including but not limited to text [Farsad et al., 2018; Lu et al., 2021, 2023a, 2023b; Zhou et al., 2021, 2022b; Jiang et al., 2022a], image [Bourtsoulatze et al., 2019; Kurka and Gündüz, 2020; Ding et al., 2021; Lee et al., 2023; Tong et al., 2023; Wu et al., 2023; Yang et al., 2023], speech [Weng and Qin, 2021], and video [Jiang et al., 2022b]. These systems demonstrate superior performance over traditional communication systems, particularly in scenarios characterized by limited channel bandwidth and low signal-to-noise ratio (SNR) environments.

However, in the aforementioned frameworks, knowledge is implicitly encapsulated within the parameters of NNs. Nonetheless, it is noteworthy that such approaches exhibit deficiencies in their capacity to thoroughly comprehend and represent knowledge. In other words, it is feasible to enhance the efficacy of SemCom systems by investigating refined knowledge representations. A prevalent approach in this endeavor is the utilization of knowledge graphs (KGs), which conceptualize human knowledge in the form of a graph structure. A KG comprises entities and relationships, where entities symbolize objects or concepts in the tangible world and serve as vertices within the graph. Relationships, on the other hand, represent directed edges linking two entities. By leveraging a graph structure, KGs manifest enhanced computational compatibility, interpretability, and extensibility.

Notably, the breakthroughs in large language models (LLMs), exemplified by the GPT series [Radford et al., 2018, 2019; Brown et al., 2020] and LLaMA [Touvron et al., 2023a, 2023b], have

Wireless Semantic Communications: Concepts, Principles, and Challenges, First Edition.
Edited by Yao Sun, Lan Zhang, Dusit Niyato, and Muhammad Ali Imran.

demonstrated remarkable prowess in the fields of natural language understanding, generation, and reasoning. In conjunction with these advancements, prompt engineering (PE) emerges as a novel and versatile programming approach within LLMs, offering users a more adaptable operational framework. Going beyond traditional research boundaries, LLMs can skillfully construct and enrich KGs with their comprehension of knowledge under the instruction of appropriate prompts, thereby augmenting the usability and reliability of knowledge [Carta et al., 2023]. Hence, it sounds appealing to utilize LLMs as an alternative data augmentation solution, particularly for extracting and contextualizing knowledge within the SemCom process, so as to enrich the depth and breadth of SemCom.

In a nutshell, this chapter aims to investigate strategies for incorporating KGs into SemCom to improve the knowledge-learning capabilities (e.g., knowledge representation and reasoning) of the SemCom system. Compared to existing studies, the main contents and innovations of this study are as follows:

- A KG-enhanced SemCom system is proposed to facilitate the effective utilization of knowledge, in which a transformer-based knowledge extractor at the receiver side is designed to find semantically related factual triples from the received noisy signal for assisting the semantic decoding.
- On this basis, due to the potential variations in knowledge contents, a KG evolving-based approach, which attempts to acquire possible semantic representation from received signals in a unified space, is adopted with the aid of contrastive learning [Chen et al., 2020]. It can better capture the relationship between entities and the received signals, thus boosting the flexibility and robustness of the system.
- In addition, this study also explores a feasible solution for data augmentation by LLMs. In that regard, LLMs demonstrate the feasibility of extracting the knowledge of the dataset without manual annotation through appropriate prompts.

The remainder of the chapter is organized as follows: Section 5.2 introduces the fundamental concepts of KGs and reviews existing works on the integration of knowledge in SemCom. Section 5.3 describes a KG-enhanced SemCom system and presents corresponding numerical results. Furthermore, optimization strategies for the KG evolving-based SemCom system and accompanying experimental results are discussed in Section 5.4. Section 5.5 further explores a solution for data augmentation in LLMs. Finally, Section 5.6 concludes the chapter.

Beforehand, we summarize the mainly used notations in Table 5.1.

5.2 Basic Concepts and Related Works

In this section, we commence with fundamental concepts associated with KGs, accompanied by an exploration of pertinent techniques related to KGs. Subsequently, we conduct a comprehensive review of existing research regarding the integration of KGs in SemCom.

5.2.1 Introduction to the KG

The fundamental components of a KG encompass entities and relationships. Specifically, entities represent objects or concepts in the real world, while relationships characterize the connections between entities. From a mathematical perspective, a KG is typically represented as a graph structure, where entities form the vertices of the graph, and relationships constitute the directed edges of

Table 5.1 Mainly used notations in this chapter.

Notations	Description
(e_p, r_{pq}, e_q)	Triple with head and tail entity e_p, e_q and their relationship r_{pq}
$S_\beta(\cdot)$	Semantic encoder with trainable parameters β
$S_\gamma^{-1}(\cdot)$	Semantic decoder with trainable parameters γ
$C_\alpha(\cdot)$	Channel encoder with trainable parameters α
$C_\delta^{-1}(\cdot)$	Channel decoder with trainable parameters δ
$K_\theta(\cdot)$	Knowledge extractor with trainable parameters θ
$\mathbf{s}, \hat{\mathbf{s}}$	Input and decoded sentences
$\mathbf{h}, \hat{\mathbf{h}}$	Semantically encoded vector and channel decoded vector
$\mathbf{x}, \mathbf{y}$	Transmitted and received signals
$\mathbf{t}$	Indicator vector computes relevancy between triples in knowledge base and the embedding representation of $\hat{\mathbf{h}}$
n_t	Number of triples in the knowledge base
N	Length of sentence
w	Weight parameter for knowledge extraction
$\mathcal{F}_k(\cdot), \mathcal{F}_e(\cdot)$	Knowledge and entity embedding process
$\mathcal{U}$	Unified semantic representation space
$\mathbf{v}_h$	Mapping vector of $\hat{\mathbf{h}}$ in the space $\mathcal{U}$
$\mathcal{F}_{\hat{h}}(\cdot)$	Mapping function of $\hat{\mathbf{h}}$
$\mathcal{D}(\cdot)$	Generalized distance function
λ	Pre-defined distance threshold
τ	The temperature parameter of InfoNCE
$\{e\}$	The set of entities from the knowledge base
$\mathbf{v}_{e_i}$	Embedding vector of e_i in $\mathcal{U}$
$\hat{r}_{pq}$	Relationship computed by the relationship prediction module
$\{m\}$	Extraction triples from $K_\theta(\cdot)$ or $\mathcal{U}$
$\mathbf{k}$	Knowledge vector obtained by $\mathcal{F}_k(\{m\})$
$\{e_s\}$	Semantically related entity set with $\mathbf{s}$
e_+	Related entity from $\{e_s\}$
K	Number of the negative samples
e_-	Irrelevant entity chosen from the knowledge base
$\mathbf{v}_{e_+}, \mathbf{v}_{e_-}$	Embedding vector of e_+, e_- as positive and negative samples

the graph. Each factual statement can be expressed as a triple (e_p, r_{pq}, e_q), where e_p and e_q signify the head and tail entities, respectively, and r_{pq} represents the relationship connecting the two entities.

As a computational and analytical approach, knowledge representation learning (KRL), commonly referred to as "knowledge embedding," endeavors to acquire low-dimensional distributional embeddings for entities and relationships within a KG. Notably, prevalent KRL methodologies can be broadly classified into two distinct categories. The first category revolves around translational distance, as exemplified by the TransE model [Bordes et al., 2013]. This approach represents entities and relationships as vectors within a predetermined mathematical space. Contrastingly, the other category leverages deep neural networks (DNNs), encompassing alternative architectures such as multilayer perceptron (MLP), convolutional neural networks (CNNs), and graph neural networks (GNNs) [Bordes et al., 2014; Dettmers et al., 2018; Schlichtkrull et al., 2018]. These DL frameworks serve well to transform a knowledge embedding problem into a sophisticated task within the domain of NN-based approaches [Ji et al., 2021].

5.2.2 Knowledge Representation and Reasoning in SemCom

Many existing studies on SemCom typically regard knowledge as a statistical variable underlying data acquisition procedures. In these methodologies, instead of being confined to an explicitly represented knowledge base, knowledge can be implicitly encoded as parameters of DNNs at both the transmitter and the receiver sides. Furthermore, recent literature has begun to integrate KGs into SemCom systems for a more tangible representation of knowledge. For instance, Lan et al. [2021] explore the viability of incorporating KGs into SemCom systems, proposing a system where KGs are adeptly merged. Targeted at both human-to-human and human-to-machine communication, the authors suggest employing KGs at the transmitter side to achieve semantic representation while utilizing KGs at the receiver side for error correction and inference. Wang et al. [2022] focus on enhancing communication between base stations and users, leveraging KGs as a tool for depicting semantic information. Upon obtaining the transmitted KG, receivers employ a graph-to-text model to reconstruct the information. Furthermore, the authors introduce an attention policy gradient algorithm to assess the significance of individual triples within the semantic information [Wang et al., 2022]. Similarly, Zhou et al. [2022a] incorporate KGs into the process of semantic encoding and decoding. This involves transforming sentences into factual triples, then encoding these into bitstreams compatible with traditional communication methods. Together with recovery and error correction procedures, the retrieved triples are then restored into meaningful semantic information utilizing a finely tuned T5 model (a type of LLM) [Raffel et al., 2020]. In line with this framework, the proposed scheme is further extended to be implemented in multiuser scenarios [Zhou et al., 2023]. Furthermore, Jiang et al. [2022c] employ factual triples from KGs in their semantic encoding scheme by measuring the semantic importance and selectively transmitting knowledge with higher semantic significance.

Besides, effective knowledge reasoning has been investigated as a by-product of SemCom. Liang et al. [2022] propose a reasoning-based SemCom system that leverages knowledge reasoning to fill in missing or implicit entities and relationships in information. Meanwhile, in order to maintain lifelong learning, new knowledge is continuously acquired during the communication process. Xiao et al. [2022a] introduce a novel, implicit SemCom system, which harnesses KGs for the identification and representation of latent knowledge embedded within communication content. The authors further demonstrate that the integration with federated learning contributes to collaborative reasoning across multiple users as well [Xiao et al., 2022b]. Li et al. [2022] introduce the concept of the cross-modal knowledge graph (CKG), which establishes connections between knowledge extracted from various data sources such as video, audio, and haptic signals. Subsequently, the authors investigate the viability of incorporating a CKG into a cross-modal SemCom system. Recently, Jiang et al. [2023] explored the possibility to use large-scale artificial intelligence models (i.e., LLMs) to construct the knowledge base. The authors show that the introduction of LLMs provides more precise representations of knowledge, thus reducing the need for data training and lowering computational expenses.

In a nutshell, the incorporation of KGs into contemporary research has significantly enhanced the knowledge-learning capacities of SemCom systems. Nonetheless, such integration is still at a preliminary stage and introduces extra challenges. First, some studies regard factual triples as the semantic carriers [Zhou et al., 2022a], leading to these triples inevitably being subjected to the distortions of noisy channels during transmission. Moreover, a critical limitation arises from the inherent capacity constraints, resulting in the term "semantic" not necessarily aligning with the "knowledge" represented by KGs. Taking text as an illustrative example, KGs can solely depict semantics in straightforward declarative sentences but encounter difficulties with interrogative sentences or those with subjective sentiments. Furthermore, in practice, the limited content from knowledge bases inevitably results in artificial semantic loss. Moreover, it remains under-explored to simultaneously utilize and improve the decoding in SemCom by available KGs and apply received information from SemCom to evolve KGs. In other words, current research predominantly concentrates on the unidirectional application of KG for semantic encoding, with a lack of emphasis on the bidirectional interaction between KGs and SemCom at the receiver side. Consequently, there emerges a strong incentive for further exploration of techniques that interplay KGs with SemCom.

5.3 A KG-enhanced SemCom System

In this section, a KG-enhanced SemCom system is proposed with a particular focus on the receiver side. Notably, our approach entails no requisite modifications to the DNN structure of the transmitter. Meanwhile, at the receiver side, a transformer-based knowledge extractor is devised to extract semantically relevant factual triples from received signals and assist subsequent semantic decoding. Unlike the aforementioned studies [Zhou et al., 2022a], our approach positions the KG as an enhancement strategy rather than a complete carrier for semantic encoding [Wang et al., 2023]. Therefore, it not only enables effective knowledge harnessing but also substantially reduces the risk of semantic loss that might arise from limitations in representational capacity or challenges in the integration of knowledge. Additionally, extensive experiments are conducted to validate the performance, effectiveness, and superiority.

5.3.1 System Model

The SemCom system employed in this section comprises a semantic encoder and a semantic decoder, as depicted in Figure 5.1. Without loss of generality, the input sentence is denoted as $\mathbf{s} = [s_1, s_2, \dots, s_N] \in \mathbb{N}^N$, where s_i represents the ith word (i.e., token) in the sentence. In particular, the transmitter consists of two modules: the semantic encoder and the channel encoder. The semantic encoder $S_\beta(\cdot)$ with trainable parameters β extracts the semantic information in the content and represents it as a vector $\mathbf{h} \in \mathbb{R}^{N \times d_s}$, where d_s is the dimension of each semantic symbol. Mathematically,

$$\mathbf{h} = S_\beta(\mathbf{s}). \tag{5.1}$$

Afterward, the channel encoder $C_\alpha(\cdot)$ with trainable parameters α encodes $\mathbf{h}$ into symbols that can be transmitted over the physical channel, that is,

$$\mathbf{x} = C_\alpha(\mathbf{h}), \tag{5.2}$$

where $\mathbf{x} \in \mathbb{C}^{N \times c}$ is the channel vector for transmission, and c is the number of symbols for each token.

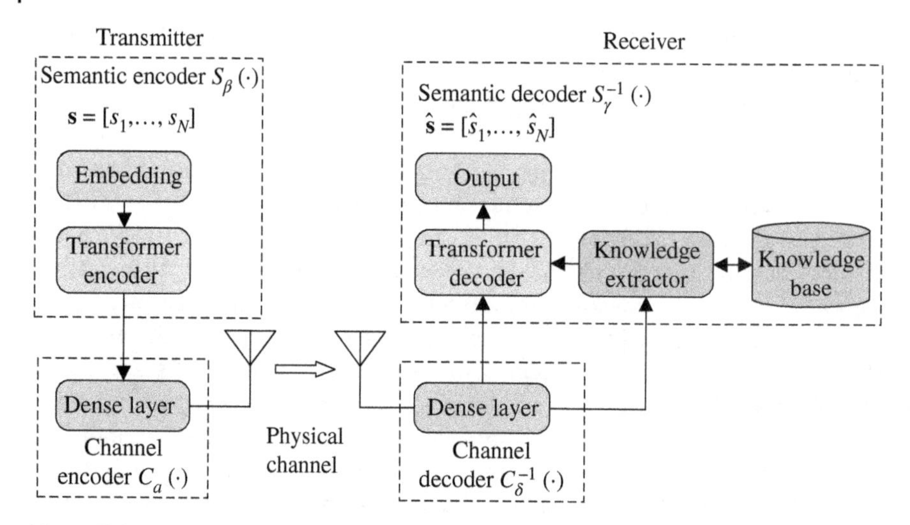

Figure 5.1 The framework of the SemCom system.

Given that $\mathbf{y} \in \mathbb{C}^{N \times c}$ is the received vector transmitted over the physical channel, it can be formulated as

$$\mathbf{y} = H\mathbf{x} + \mathbf{n}, \tag{5.3}$$

where H denotes the channel coefficient and $\mathbf{n} \sim \mathcal{N}(0, \sigma^2\mathbf{I})$ is the additive white Gaussian noise (AWGN).

After receiving $\mathbf{y}$, the receiver first decodes the transmitted symbols with the channel decoder $C_\delta^{-1}(\cdot)$. Hence, the decoded vector $\hat{\mathbf{h}} \in \mathbb{R}^{N \times d_s}$ can be given by

$$\hat{\mathbf{h}} = C_\delta^{-1}(\mathbf{y}), \tag{5.4}$$

where δ represents the trainable parameters of the channel decoder.

Notably, SemCom implicitly relies on some prior knowledge between the transmitter and receiver for the joint training process. However, different from such prior knowledge, the knowledge base in our model refers to some factual triples and can be located at the receiver side. To exploit the knowledge base, a knowledge extractor is further applied on the receiver side to extract and integrate relevant knowledge from the received signal to yield the aggregated knowledge $\mathbf{k}$. In particular, the knowledge extraction and aggregation process can be formulated as

$$\mathbf{k} = K_\theta(\hat{\mathbf{h}}), \tag{5.5}$$

where $K_\theta(\cdot)$ represents the knowledge extractor with trainable parameters θ.

Eventually, the knowledge-enhanced semantic decoder $S_\gamma^{-1}(\cdot)$ with trainable parameters γ leverages the channel-decoded vector $\hat{\mathbf{h}}$ and the extracted knowledge vector $\mathbf{k}$ to obtain the restored message $\hat{\mathbf{s}} = [\hat{s}_1, \hat{s}_2, \ldots, \hat{s}_N]$, which can be expressed as

$$\hat{\mathbf{s}} = S_\gamma^{-1}(\hat{\mathbf{h}} \,||\, \mathbf{k}), \tag{5.6}$$

where $||$ indicates a concatenation operator.

Generally, the accuracy of SemCom is determined by the semantic similarity between the transmitted and received contents. In order to minimize the semantic errors between $\mathbf{s}$ and $\hat{\mathbf{s}}$, the loss function $\mathcal{L}_{\text{model}}$ taking into account the cross entropy (CE) of the two vectors can be formulated as

$$\mathcal{L}_{model} = -\sum_{i=1}^{N} \left(q(s_i) \log p(\hat{s}_i)\right), \tag{5.7}$$

where $q(s_i)$ is the one-hot representation of s_i with $s_i \in \mathbf{s}$, and $p(\hat{s}_i)$ is the predicted probability of the ith word.

Instead of drawing upon traditional communication modules for physical-layer transmission, most existing studies have chosen to utilize end-to-end DNNs to accomplish the whole communication process. Typically, the semantic encoders and decoders are based on transformers [Vaswani et al., 2017]. Meanwhile, the channel encoding and decoding part can be viewed as an AE implemented by fully connected layers. The whole SemCom process is then reformulated as a sequence-to-sequence problem. On this basis, we primarily focus on developing appropriate implementation means of the KG-enhanced SemCom receiver, i.e., the knowledge extractor in Eq. (5.5) and the knowledge-enhanced decoder in Eq. (5.6), thereby minimizing $\mathcal{L}_{model}$ in Eq. (5.7).

5.3.2 The Design of the KG-Enhanced SemCom Receiver

In this part, we begin with the implementation details of the knowledge extractor. The whole knowledge extraction process, as illustrated in Figure 5.2, can be divided into two phases. The first phase initially involves an embedding task, utilizing transformer encoders to derive a representation of the decoded vector. Subsequently, in the second phase, it turns to identifying all triples corresponding with the representation via a multi-label classifier. Following the identification, these triples are then compacted into a condensed format, which plays a crucial role in facilitating the final decoding.

In particular, in order to extract the semantic representation, we adopt a model composed of a stack of L identical transformer encoder layers, each of them consisting of a multi-head attention mechanism, as well as some feed-forward and normalization sublayers [Vaswani et al., 2017]. Without loss of generality, assuming that $\mathbf{z}^{(l-1)}$ is the output of the $(l-1)$th encoder layer, where $\mathbf{z}^{(0)}$ is equivalent to $\hat{\mathbf{h}}$ for the input layer, the self-attention mechanism of the lth layer can be represented as

$$\text{Attention}(\mathbf{z}^{(l-1)}) = \text{softmax}\left(\frac{\mathbf{Q}^{(l)}\left(\mathbf{K}^{(l)}\right)^{\mathsf{T}}}{\sqrt{d_k}}\right)\mathbf{V}^{(l)}, \tag{5.8}$$

where $\mathbf{Q}^{(l)} = \mathbf{z}^{(l-1)}\mathbf{W}_Q^{(l)}$, $\mathbf{K}^{(l)} = \mathbf{z}^{(l-1)}\mathbf{W}_K^{(l)}$, $\mathbf{V}^{(l)} = \mathbf{z}^{(l-1)}\mathbf{W}_V^{(l)}$. $\mathbf{W}_Q^{(l)}$, $\mathbf{W}_K^{(l)}$ and $\mathbf{W}_V^{(l)}$ are the projection matrices of the lth layer, and d_k is the model dimension. Furthermore, $\mathbf{z}^{(l-1)}$ is added to the calculated $\text{Attention}(\mathbf{z}^{(l-1)})$ via a normalized residual connection to obtain the intermediate output

$$\mathbf{a}^{(l)} = \text{LayerNorm}(\text{Attention}(\mathbf{z}^{(l-1)}) + \mathbf{z}^{(l-1)}), \tag{5.9}$$

where $\text{LayerNorm}(\cdot)$ denotes a layer normalization operation. Afterward, a feed-forward network is involved as $\text{FFN}(\mathbf{a}^{(l)}) = \max(0, \mathbf{a}^{(l)}\mathbf{W}_{\text{F1}}^{(l)} + \mathbf{b}_{\text{F1}}^{(l)})\mathbf{W}_{\text{F2}}^{(l)} + \mathbf{b}_{\text{F2}}^{(l)}$, where $\mathbf{W}_{\text{F1}}^{(l)}$, $\mathbf{W}_{\text{F2}}^{(l)}$, $\mathbf{b}_{\text{F1}}^{(l)}$ and $\mathbf{b}_{\text{F2}}^{(l)}$ are parameters of the feed-forward layer in the l-th encoder block. Next, we adopt a residual connection and a layer normalization that can be formulated as

$$\mathbf{z}^{(l)} = \text{LayerNorm}(\text{FFN}(\mathbf{a}^{(l)}) + \mathbf{a}^{(l)}). \tag{5.10}$$

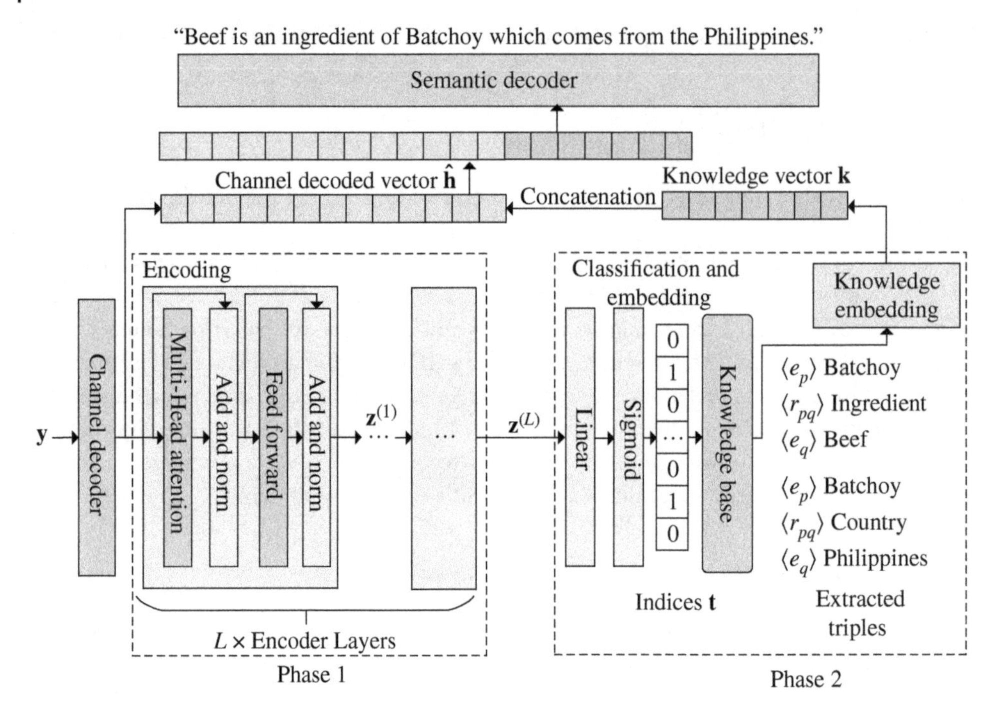

Figure 5.2 The KG-enhanced semantic decoder.

After L recursive encoding layers, the embedding representation $\mathbf{z}^{(L)}$ of the channel-decoded vector $\hat{\mathbf{h}}$ is obtained. Then, an indicator vector $\mathbf{t}$ is derived contingent upon a multi-label classifier, which computes the relevancy between triples and the representation, that is,

$$\mathbf{t} = \text{sigmoid}(\mathbf{z}^{(L)}\mathbf{W}_t + \mathbf{b}_t), \tag{5.11}$$

where $\mathbf{t} = [\hat{t}_1, \dots, \hat{t}_{n_t}] \in \mathbb{R}^{n_t}, \hat{t}_i \in [0, 1]$ for $i \in \{1, \dots, n_t\}$. Besides, n_t denotes the number of triples in the knowledge base, while $\mathbf{W}_t$ and $\mathbf{b}_t$ are parameters of the classifier. If $\hat{t}_i \geq 0.5$, the triple m_i corresponding to index i is predicted to be relevant to the received content.

Ultimately, the set of relevant factual triples $\{m\}$ predicted by the model is embedded into the knowledge vector $\mathbf{k} = \mathcal{F}_k(\{m\})$, where the embedding process is abstractly represented as $\mathcal{F}_k(\cdot)$. In particular, rather than computing the embedding of the entity and relationship separately, we choose to integrate the triples into the same compressed format. Subsequently, as in Eq. (5.6), the knowledge vector is concatenated with the decoding vector and fed into the semantic decoder. Therefore, the extracted knowledge $\mathbf{k}$ from $\hat{\mathbf{h}}$ with the assistance of KGs is capable of ameliorating the semantic decoding process. The procedure of the KG-enhanced SemCom system is illustrated in Algorithm 5.1.

5.3.3 The Training Methodology

The training of the knowledge extractor, $K_\theta(\cdot)$, heavily relies on an end-to-end SemCom model. On this basis, the sentences are first transmitted by the transmitter via the channel, and then the channel-decoded vectors are fed into the knowledge extractor. Afterward, the knowledge extractor is trained by gradient descent while the parameters of the transmitter are frozen. Since the number of negative labels is much more than that of positive labels, the loss function adopts the weighted

Algorithm 5.1 The SemCom Process with the KG-Enhanced Receiver.

Initialization: The parameters of the semantic encoder $S_\beta(\cdot)$, channel encoder $C_\alpha(\cdot)$, semantic decoder $S_\gamma^{-1}(\cdot)$, channel decoder $C_\delta^{-1}(\cdot)$ and knowledge extractor $K_\theta(\cdot)$.

Input: The tokenized sentence **s**.

Output: The restored sentence $\hat{\mathbf{s}}$.

Transmitter:

1: Semantic encoding: $\mathbf{h} \leftarrow (S_\beta(\mathbf{s}))$.
2: Channel encoding: $\mathbf{x} \leftarrow (C_\alpha(\mathbf{h}))$.
3: Transmit **x** over the physical channel: $\mathbf{y} \leftarrow \mathbf{Hx} + \mathbf{n}$.

Receiver:

4: Channel decoding: $\hat{\mathbf{h}} \leftarrow C_\delta^{-1}(\mathbf{y})$.
5: Knowledge extraction $K_\theta(\cdot)$:
6: Compute the embedding representation $\mathbf{z}^{(L)}$.
7: $\mathbf{t} \leftarrow \text{sigmoid}(\mathbf{z}^{(L)}\mathbf{W}_t + \mathbf{b}_t)$.
8: Find the triples $\{m\}$ where $\hat{t}_i \geq 0.5$.
9: Knowledge embedding: $\mathbf{k} \leftarrow \mathcal{F}_k(\{m\})$.
10: Semantic decoding: $\hat{\mathbf{s}} \leftarrow S_\gamma^{-1}(\hat{\mathbf{h}} \mid\mid \mathbf{k})$.

binary cross entropy (BCE), which can be represented as

$$\mathcal{L}_{knowledge} = \sum_{i=1}^{n_t} - w_i[t_i \cdot \log \hat{t}_i + (1 - t_i) \cdot \log(1 - \hat{t}_i)]. \tag{5.12}$$

Here $t_i \in \{0, 1\}$ represents the training label, and $\hat{t}_i$ is the prediction output in Eq. (5.11). w_i is the weight of the ith position, related to the hyperparameter w. Specifically, $w_i = w$ when $t_i = 0$; otherwise, $w_i = 1 - w$. Increasing w can result in a more sensitive extractor, but it also brings an increase in the false positive rate.

Same as the transformer encoder, the training complexity of a knowledge extractor is $O(LN^2 \cdot d_k)$. Notably, the knowledge extractor is not limited to the conventional transformer structure but can also be applied to different transformer variants such as universal transformer (UT) [Dehghani et al., 2018]. With the self-attention mechanism, the extracted factual triples can provide additional prior knowledge to the semantic decoder and therefore improve the performance of semantic decoding. Typically, the knowledge vector is concatenated to the received message, rather than being merged into the encoded vector of the input sentence, as done by previous works [Zhou et al., 2022a]. Therefore, our work ensures that when the extractor is of little avail (e.g., no relevant knowledge found by the knowledge extractor), the system can function normally by using a standard encoder–decoder transformer structure, while avoiding possible semantic losses introduced by the knowledge extraction procedure.

5.3.4 Simulation Results

5.3.4.1 Dataset and Parameter Settings

The dataset used in the numerical experiment is based on WebNLG v3.0 [Gardent et al., 2017], which consists of data–text pairs. For each data–text pair, the data is a set of triples extracted from DBpedia while the text corresponds to the verbalization of these triples. In this numerical experiment, the weight parameter w is set to 0.02, while the learning rate is set to 10^{-4}. Moreover, we set the dimension of the dense layer as 128×16 and adopt eight-head attention in transformer

Table 5.2 Experimental settings.

Parameter	Value
Train dataset size	24,467
Test dataset size	2,734
Weight parameter w	0.02
DNN optimizer	Adam
Batch size	32
Model dimension	128
Learning rate	10^{-4}
Channel vector dimension	16
Number of multi-heads	8

layers. The detailed settings of the proposed system are shown in Table 5.2. We train the models based on both the classical transformer and UT [Zhou et al., 2021]. Besides, we adopt two metrics to evaluate their performance, that is, the 1-gram bilingual evaluation understudy (BLEU) [Papineni et al., 2002] score for measuring word-level accuracy and sentence-BERT [Reimers and Gurevych, 2019] score for measuring semantic similarity. Notably, as a Siamese Bert-network model that generates fixed-length vector representations for sentences, sentence-BERT produces a score in terms of the cosine similarity of embedded vectors.

5.3.4.2 Numerical Results

Figure 5.3 shows the BLEU and sentence-BERT scores versus SNR for the KG-enhanced SemCom system based on transformer, respectively. It can be observed that the assistance of the knowledge extractor could significantly contribute to improving the performance. In particular, regardless of the channel type, the knowledge extractor can always bring more than 5% improvement in BLEU under low SNRs. For the sentence-BERT score, the KG-enhanced receiver also shows a similar performance improvement. The results demonstrate that the proposed scheme can improve the comprehension of semantics on the receiver side. On the other hand, Figure 5.4 manifests the

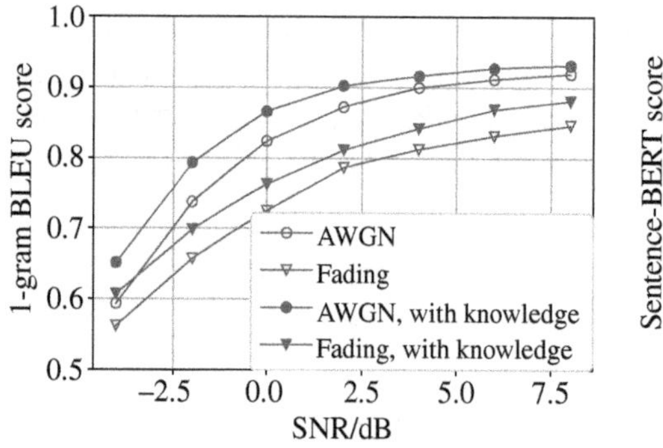

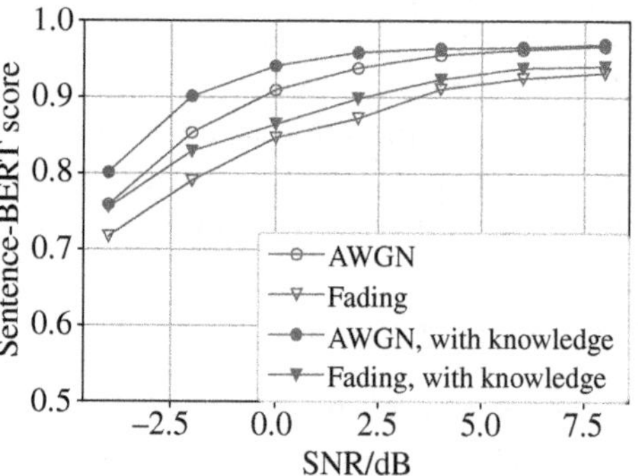

Figure 5.3 The BLEU and sentence-BERT score versus SNR for the KG-enhanced SemCom system based on transformer.

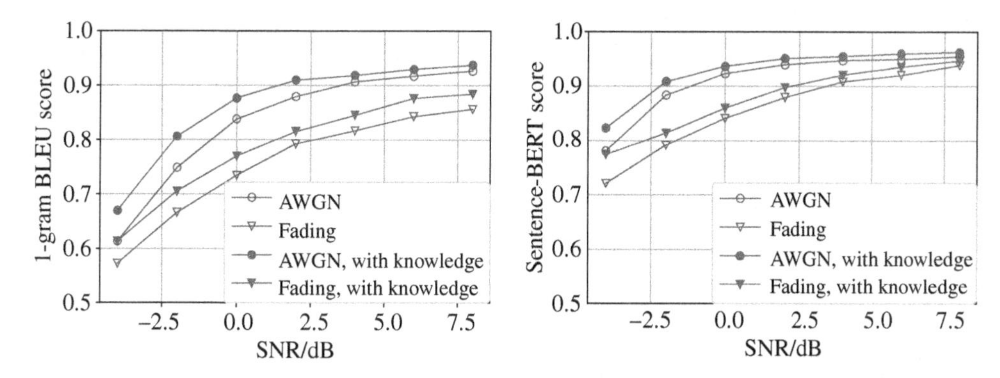

Figure 5.4 The BLEU and sentence-BERT score versus SNR for the KG-enhanced SemCom system based on UT.

Table 5.3 Example of semantic decoding process based on knowledge enhancement.

Enter sentence	Ayam penyet contains fried chicken and originates from Singapore. It can also be found in Java where the banyumasan people are one of the ethnic groups.
Related knowledge in the knowledge base	(Ayam_penyet, ingredient, Fried_chicken) (Java, ethnicGroup, Banyumasan_people) (Ayam_penyet, region, Singapore) (Ayam_penyet, country, Java)
Decoded results without knowledge enhancement	Ayam penyet contains fried **tomatoes** and originates from Singapore. It can also be found in **Japan** where the banyumasan people are one of the ethnic groups.
Knowledge-enhanced decoding results	Ayam penyet contains fried chicken and originates from Singapore. It can also be found in Java where the banyumasan people are one of the ethnic groups.

performance of the system based on UT under both the BLEU and sentence-BERT score, and a similar performance superiority could also be expected. In order to demonstrate the effect of knowledge enhancement more intuitively, Table 5.3 shows a case in the semantic decoding process. In particular, the original text characterizes a food item, encompassing related details such as its ingredients and the country of origin. Due to the annoying channel noise, the received signal may encounter some semantic errors. Although traditional semantic decoders are able to correct grammatical errors, other errors in semantics and knowledge may be ignored. Instead, the knowledge extractor can locate related knowledge from the knowledge base, so as to assist the semantic decoder for the output of the correct result.

On the other hand, we test the precision and recall rates versus SNR for the KG-enhanced SemCom system based on transformer. As shown in Figure 5.5, the KG-enhanced SemCom system can obtain a recall rate of over 90%. However, the received content may be polluted by noise, resulting in an increase in false positives and leading to a significant gap between precision and recall.

Intuitively, the number of encoder layers in the knowledge extractor may also affect the performance of the system. Therefore, we additionally implement the knowledge extractor with a different number of transformer encoder layers and present the BLEU performance comparison

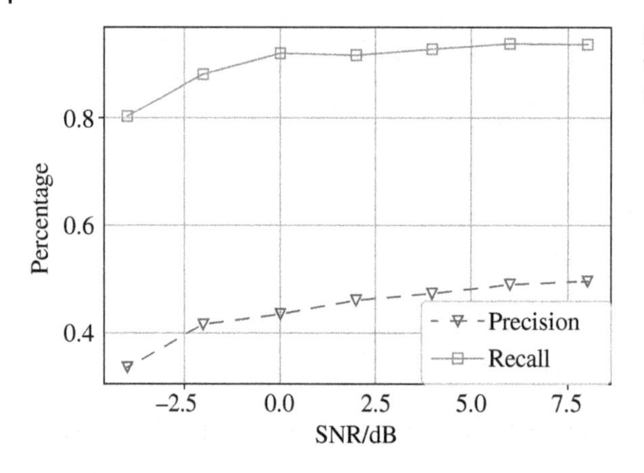

Figure 5.5 The precision and recall rate versus SNR for the KG-enhanced SemCom system based on transformer.

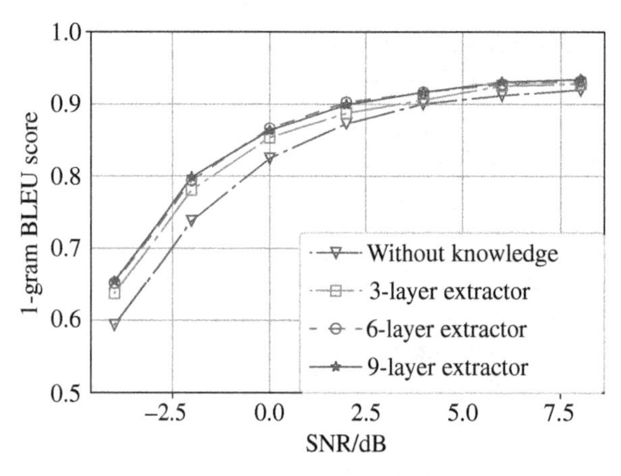

Figure 5.6 The comparison of the BLEU score for knowledge extractor with different numbers of transformer encoder layers.

in Figure 5.6. It can be observed that the six-layer model performs slightly better than the three-layer model. However, the performance remains almost unchanged when it increases to nine layers. Furthermore, in addition to utilizing the fixed model trained at a certain SNR, it is also possible to leverage several SNR-specific models, each corresponding to a specific SNR. Table 5.4 demonstrates the BLEU performance comparison between the fixed model trained at an SNR of 0 dB and different SNR-specific models. It can be observed that, compared to the fixed model, the SNR-specific model could yield superior performance improvements at the corresponding SNR.

Table 5.4 The comparison of the BLEU score between a fixed extractor model and SNR-specific models.

SNR/dB	Fixed	SNR-specific
−4	0.6514	0.6718
−2	0.7936	0.8126
0	0.8661	0.8661
2	0.9025	0.9134
4	0.9164	0.9201

5.4 A KG Evolving-based SemCom System

After exploring a KG-enhanced SemCom system in Section 5.3, in this section, we further investigate potential ways to empower the system in order to operate more effectively in evolving knowledge scenarios.

5.4.1 System Overview

The approach in Section 5.3 manifests the performance of a static KG-enhanced means that significantly depends on the knowledge base with a limited capacity. Thus, it mandates the need for continually retraining the classifier to capture potential updates in the knowledge base. In practical applications, complex knowledge-processing activities are often accompanied by frequent and subtle variations in knowledge content over time. Hence, in order to maintain the essential flexibility and adaptability of the knowledge extractor, this section proposes a KG evolving-based SemCom system on top of a dynamic knowledge base at the receiver side by incorporating the concept of a unified semantic representation space. Notably, the framework of KG evolving-based SemCom is the same as KG-enhanced SemCom depicted in Section 5.3.1 with a particular focus on the receiver's design due to a dynamic knowledge base.

5.4.2 The Design of the KG Evolving-Based SemCom Receiver

The structure of the KG evolving-based SemCom receiver is depicted in Figure 5.7. Similar to the framework proposed in Section 5.3, the receiver contains a knowledge base in which the knowledge is organized and stored in the form of factual triples. However, different from Section 5.3, a unified semantic representation space $\mathcal{U} \in \mathbb{R}^d$ is added here, where d denotes the space dimension. Notably, the unified semantic representation space implies that each entity e_i belonging to the entity set $\{e\}$ from the knowledge base has a corresponding embedding vector in the space $\mathcal{U}$, which can be formulated as

$$\mathbf{v}_{e_i} = \mathcal{F}_e(e_i), \tag{5.13}$$

where $\mathcal{F}_e(\cdot)$ is the abstraction of the entire embedding process and can be implemented by DNNs.

After obtaining the channel decoded vector $\hat{\mathbf{h}}$ as in Eq. (5.4), the receiver maps $\hat{\mathbf{h}}$ to the unified semantic representation vector $\mathbf{v}_h$ according to a DNN-induced mapping function $\mathcal{F}_{\hat{h}}(\cdot)$; that is,

$$\mathbf{v}_h = \mathcal{F}_{\hat{h}}(\hat{\mathbf{h}}). \tag{5.14}$$

Taking the linear mapping as an example of $\mathcal{F}_{\hat{h}}(\cdot)$, Eq. (5.14) can be rewritten as $\mathbf{v}_h = \mathbf{W}_h \cdot \hat{\mathbf{h}} + \mathbf{b}_h$, where $\mathbf{W}_h$ and $\mathbf{b}_h$ are trainable parameters of the DNN. In addition to linear mapping, more complex DNN structures such as transformer can also be used to fully exploit the semantic features of the signal.

By leveraging the mapping, the receiver tries to find all suitable embedding vectors (and accompanied entities) $\mathbf{v}_{e_i}, e_i \in \{e\}$ whose distances from $\mathbf{v}_h$ are within a preset threshold λ, that is,

$$\mathcal{D}(\mathbf{v}_h, \mathbf{v}_{e_i}) \leq \lambda. \tag{5.15}$$

Here, $\mathcal{D}(\cdot)$ represents a generalized distance function and can be calibrated according to the training method of the unified semantic space. Common forms of the distance function include Euclidean

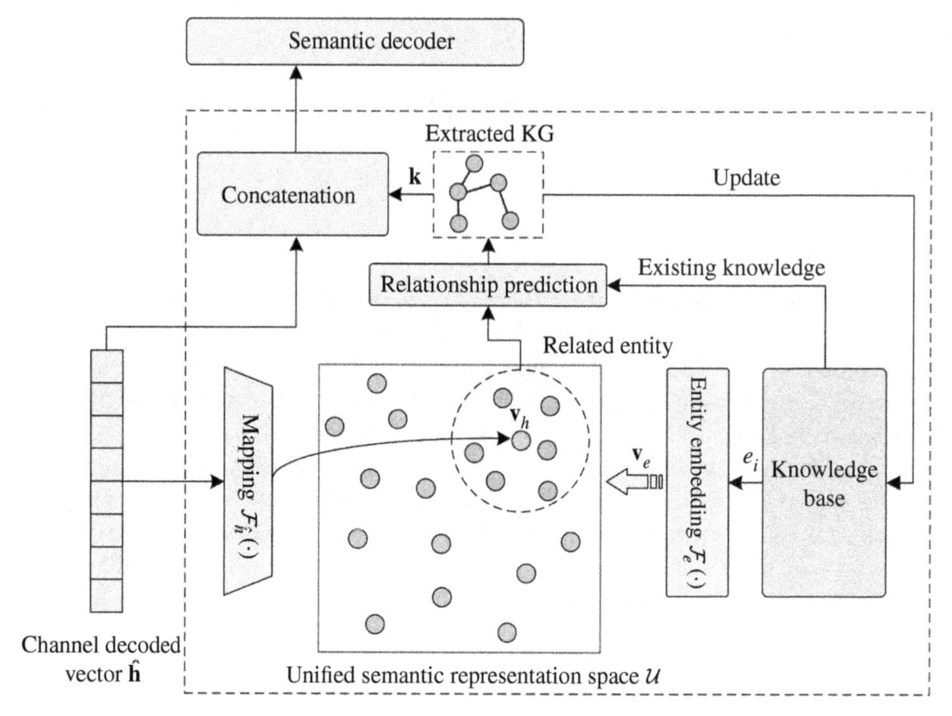

Figure 5.7 The structure of the KG evolving-based SemCom receiver.

distance $\mathcal{D}_{euclid}(\mathbf{x}, \mathbf{y}) = \sqrt{\| \mathbf{x} - \mathbf{y} \|^2}$, cosine distance $\mathcal{D}_{cosine}(\mathbf{x}, \mathbf{y}) = 1 - \frac{\mathbf{x} \cdot \mathbf{y}}{\|\mathbf{x}\| \cdot \|\mathbf{y}\|}$, etc. All entities satisfying Eq. (5.15) are added to a set $\{e_s\}$.

The identification of the entity set $\{e_s\}$ from the knowledge base lays the very foundation for extracting factual triples $\{m\}$. To elaborate, for each entity pair e_p and e_q within the set $\{e_s\}$, the presence of a preexisting triple (e_p, r_{pq}, e_q) in the knowledge base leads to its incorporation into the set of extraction results $\{m\}$. On the other hand, for instances where such a triple is absent, a relationship prediction module, which can be independently trained as a classification model with the CE loss function, is invoked to ascertain the potential existence of a relational $\hat{r}_{pq}$ between e_p and e_q. In cases where $\hat{r}_{pq}$ is predicted, this newly identified triple $(e_p, \hat{r}_{pq}, e_q)$ is then appended to the set of extraction results. Moreover, it can be updated in the existing knowledge base, thus leading to evolving KGs. Subsequently, subject to a knowledge embedding procedure, all extracted triples $\{m\}$ contribute to generating the knowledge vector $\mathbf{k}$. Consistent with the description in Section 5.3, this vector $\mathbf{k}$ is concatenated with the channel decoded vector $\hat{\mathbf{h}}$. Meanwhile, the missing factual triples are further incorporated into the knowledge base so as to facilitate subsequent decoding. The procedure of the KG evolving-based SemCom system is summarized in Algorithm 5.2.

5.4.3 The Training Methodology

As illustrated in Figure 5.8, we introduce contrastive learning [Chen et al., 2020] to train $\mathcal{F}_e(\cdot)$ and $\mathcal{F}_{\hat{h}}(\cdot)$ in a unified semantic representation space. Beforehand, we assume that for a sentence $\mathbf{s}$, there exists a set of semantically related entities $\{e_s\}$. Afterward, the semantic decoding vector $\hat{\mathbf{h}}$ can be

Algorithm 5.2 The SemCom Process with the KG Evolving-based Receiver.

Initialization: The parameters of the semantic encoder $S_\beta(\cdot)$, channel encoder $C_\alpha(\cdot)$, semantic decoder $S_\gamma^{-1}(\cdot)$, channel decoder $C_\delta^{-1}(\cdot)$ and the unified semantic representation space $\mathcal{U}$.

Input: The tokenized sentence $\mathbf{s}$.

Output: The restored sentence $\hat{\mathbf{s}}$.

Transmitter:

1: Semantic encoding: $\mathbf{h} \leftarrow (S_\beta(\mathbf{s}))$.

2: Channel encoding: $\mathbf{x} \leftarrow (C_\alpha(\mathbf{h}))$.

3: Transmit $\mathbf{x}$ over the physical channel: $\mathbf{y} \leftarrow \mathbf{Hx} + \mathbf{n}$.

Receiver:

4: Channel decoding: $\hat{\mathbf{h}} \leftarrow C_\delta^{-1}(\mathbf{y})$.

5: Compute unified semantic representation vector $\mathbf{v}_h$ with Eq. (5.14).

6: Find all suitable embedding vectors $\mathbf{v}_{e_i}, e_i \in \{e\}$ with Eq. (5.15) and add the corresponding entities into the set $\{e_s\}$.

7: **for** each entity pair e_p and e_q in $\{e_s\}$ **do**

8:   **if** (e_p, r_{pq}, e_q) exists in the knowledge base **then**

9:     Integrate them into the extraction triples $\{m\}$.

10:   **else**

11:     Obtain the relationship $\hat{r}_{pq}$ between e_p and e_q by the relationship prediction module.

12:     Add the corresponding triple $(e_p, \hat{r}_{pq}, e_q)$ to $\{m\}$.

13:     Update $(e_p, \hat{r}_{pq}, e_q)$ to the existing knowledge base.

14:   **end if**

15: **end for**

16: Knowledge embedding: $\mathbf{k} \leftarrow \mathcal{F}_k(\{m\})$.

17: Semantic decoding: $\hat{\mathbf{s}} \leftarrow S_\gamma^{-1}(\hat{\mathbf{h}} \,||\, \mathbf{k})$.

obtained after passing $\mathbf{s}$ through the channel decoder on the receiver side. Accordingly, a unified semantic representation vector $\mathbf{v}_h$ can be determined.

Furthermore, some entity e_+ can be found among the semantically related entities and mapped to $\mathbf{v}_{e_+}$ as a positive sample. Meanwhile, K irrelevant entities are randomly selected from the entity set $\{e\}$ in the knowledge base, and the corresponding unified semantic representation vectors $\{\mathbf{v}_{e_-}\}$ will be regarded as negative samples, where K is for controlling the ratio of positive and negative samples. On this basis, we take an InfoNCE loss function [He et al., 2020], which can be expressed as

$$\mathcal{L}_{\text{InfoNCE}} = -\log\left(\frac{\exp(\mathbf{v}_h \cdot \mathbf{v}_{e_+}/\tau)}{\exp(\mathbf{v}_h \cdot \mathbf{v}_{e_+}/\tau) + \sum_{e_k \in \{e_-\}} \exp(\mathbf{v}_h \cdot \mathbf{v}_{e_k}/\tau)}\right). \tag{5.16}$$

Here, τ is the temperature hyperparameter. Finally, through backpropagation, the parameters of each layer in NNs will be updated. The overall process is shown in Algorithm 5.3.

5.4.4 Simulation Results

5.4.4.1 Dataset and Parameter Settings

Consistent with the settings in Section 5.3.4, this section utilizes the preprocessed WebNLG v3.0 dataset [Gardent et al., 2017] and leverages the transformer-based model as the underlying

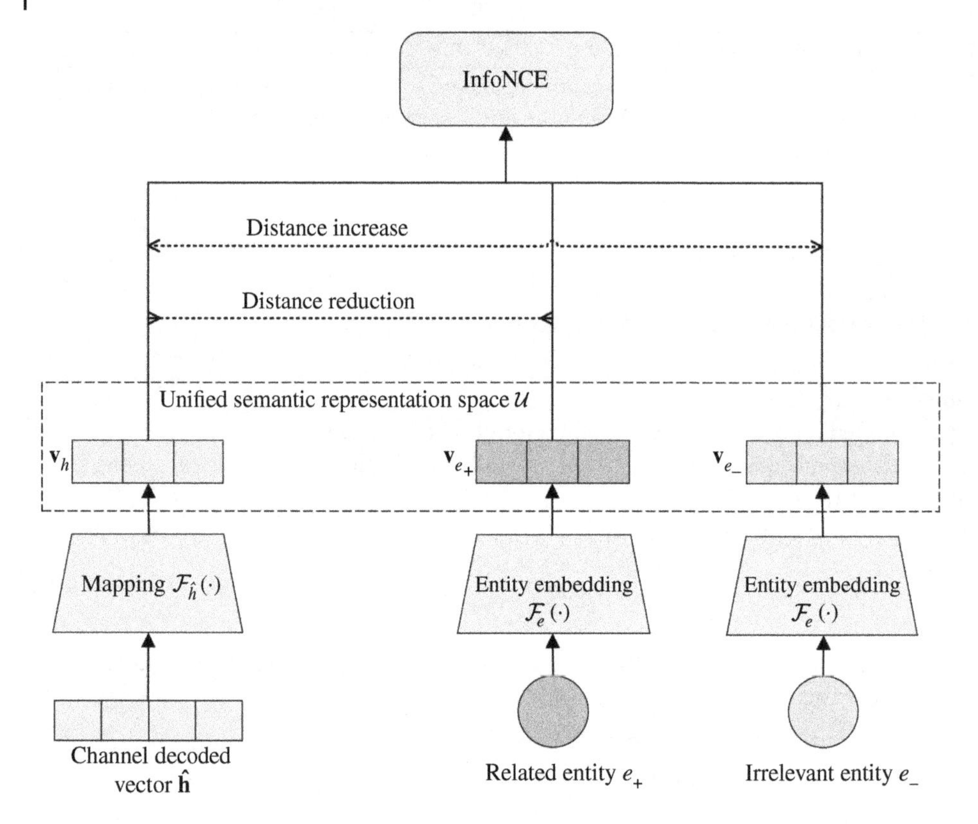

Figure 5.8 The schematic diagram of the training process of unified semantic representation.

Algorithm 5.3 Training Process of the Unified Semantic Representation Based On Contrastive Learning.

Initialization: The parameters of the unified semantic representation space $\mathcal{U}$; the number of negative samples K.

Input: The input sentence $\mathbf{s}$ and the entity set $\{e_s\}$.

Output: Embedding process of the entity $\mathcal{F}_e(\cdot)$ and mapping function of the channel decoded vector $\mathcal{F}_{\hat{h}}(\cdot)$.

1: Transmit the sentence $\mathbf{s}$ through the SemCom system, and obtain the channel decoded vector $\hat{\mathbf{h}}$ at the receiver side.
2: Take an entity e_+ from the entity set $\{e_s\}$ and calculate its embedding vector $\mathbf{v}_{e_+}$ by Eq. (5.13).
3: Calculate the unified semantic representation vector of the channel decoding vector $\mathbf{v}_h$ by Eq. (5.14).
4: Generate embedding vectors $\mathbf{v}_{e_-}$ as negative samples based on K randomly selected irrelevant entities $\{e_-\}$ belonging to the entity set $\{e\}$ in the knowledge base.
5: Calculate the InfoNCE loss function by Eq. (5.16).
6: Use stochastic gradient descent to optimize $\mathcal{F}_e(\cdot)$ and $\mathcal{F}_{\hat{h}}(\cdot)$.

SemCom system. In particular, the architecture comprises a three-layer codec with a model dimension of 128. The simulation is conducted on the AWGN channel.

During the simulation, the transformer-based JSCC is first trained on plain text without the assistance of any knowledge. Based on the pretrained transformer, the proposed knowledge enhancement module with an evolving KG is subsequently trained. During the training process, the hyperparameters are carefully calibrated for better performance. Specifically, the number of negative samples, denoted as K, is fixed at 63, and the temperature parameter τ is set to 0.2. Moreover, the distance threshold λ is fixed at 1.16.

5.4.4.2 Numerical Results

The performance of the newly developed KG evolving-based module versus SNR is experimentally validated, with the corresponding precision and recall rates depicted in Figure 5.9. Notably, the static KG-enhanced approach discussed in Section 5.3.4.2 is also compared. Figure 5.9 manifests that the recall rate of the proposed KG evolving-based approach remains consistently stable at around 80%, while demonstrating a marginal decrease compared to the method detailed in Section 5.3.4.2. On the contrary, the precision of the KG evolving-based approach exhibits a discernible superiority (i.e., 70% at 0 dB) and achieves about 80% at higher SNRs. This outcome holds substantial positive implications, as a notable improvement in precision serves to mitigate the potentially misleading effects on the decoder stemming from missing or inaccurate knowledge.

To visualize the impact of unified semantic representations, a t-SNE dimensionality reduction algorithm Van der Maaten and Hinton [2008] is leveraged to project the vectors onto a two dimensional space, with the corresponding results presented in Figure 5.10. Notably, the square markers indicate the embedding results of related entities, while the dot markers represent those of irrelevant entities. Furthermore, the embedding of the semantic decoding vector is marked by a star marker. Therefore, it validates that the contrastive learning-based training method successfully

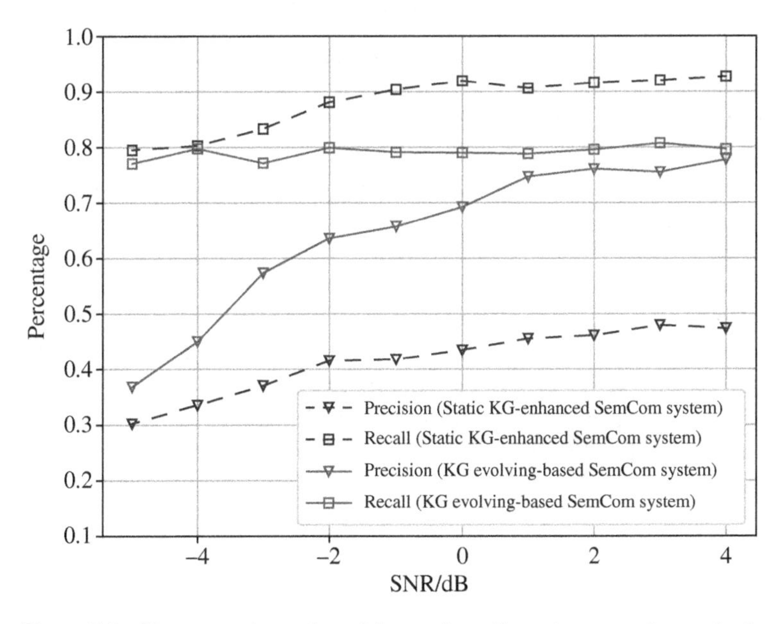

Figure 5.9 The comparison of precision and recall rate between the newly developed KG evolving-based approach and the static KG-based one in Section 5.3.

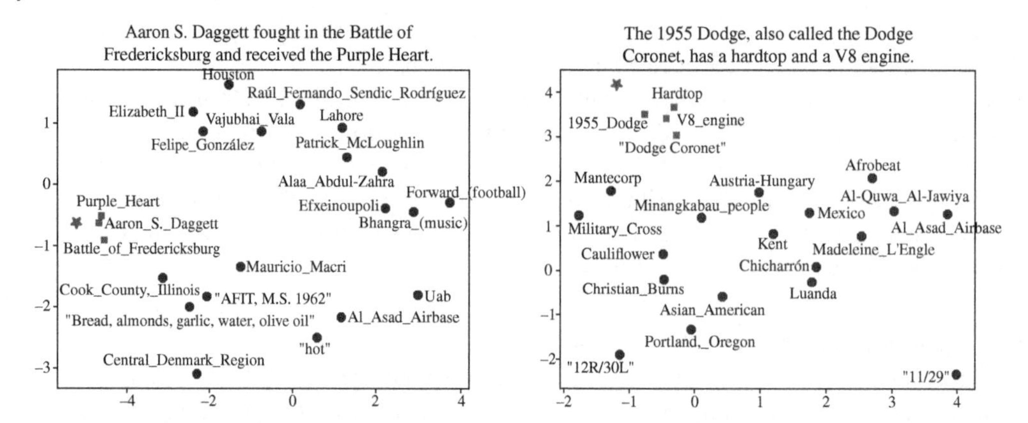

Figure 5.10 Examples of spatial visualization of unified semantic representation based on t-SNE dimensionality reduction algorithm. Source: Adapted from Van der Maaten and Hinton [2008].

maps semantically related entities to the vicinity of the semantic decoding vector, thereby enabling the feasibility of a distance-based extraction algorithm.

5.5 LLM-assisted Data Augmentation for the KG Evolving-Based SemCom System

After discussion on the KG evolving-based SemCom system, whose semantic representation is unified by contrastive learning, we shed some light on LLM-assisted data augmentation solutions for knowledge acquisition.

5.5.1 Description of the LLM-Assisted Data Augmentation Solution

The KG evolving-based approach proposed in Section 5.4 can adaptively update the recorded factual statements in the knowledge bases. However, within SemCom, the unannotated transmitted content proves challenging to directly correlate with existing knowledge bases. As a world model [Radford et al., 2018], LLMs can competently fill the deficiency and be utilized for data augmentation by extracting the knowledge without the apriority of the transmitted content. In other words, by designing appropriate prompts, it is feasible to instruct LLMs to incorporate relevant prior knowledge or constraint conditions. More importantly, by embracing a zero-shot learning paradigm, LLMs can extract knowledge without manual annotation of relevant datasets or retraining of models, thereby enhancing system flexibility.

Figure 5.11 illustrates the envisioned integration of LLMs to augment SemCom in Section 5.4. It can be observed in Figure 5.11 that by using appropriate prompts, LLMs can identify named entities and predict relationships so as to more swiftly formulate factual triples. Hence, this LLM-assisted data augmentation further fortifies the system's capability to manage dynamic knowledge effectively.

5.5.2 Simulation Results

5.5.2.1 Dataset

The feasibility of the LLM-assisted SemCom scheme discussed in this section is also validated through simulation experiments. Specifically, the experiment simulates scenarios with incomplete

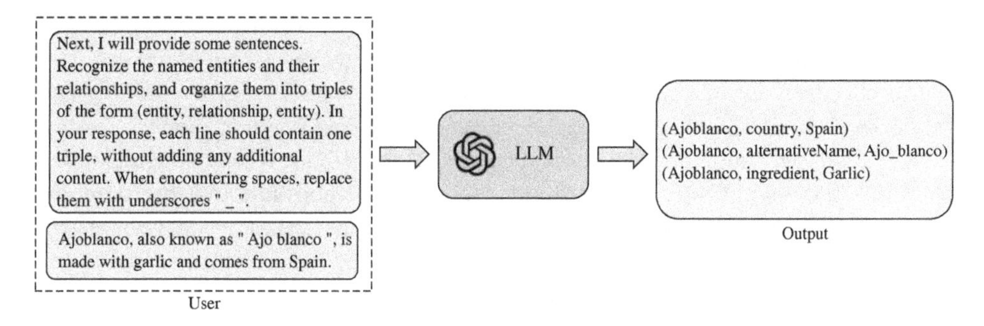

Figure 5.11 An example of LLM-assisted data augmentation solution for knowledge acquisition.

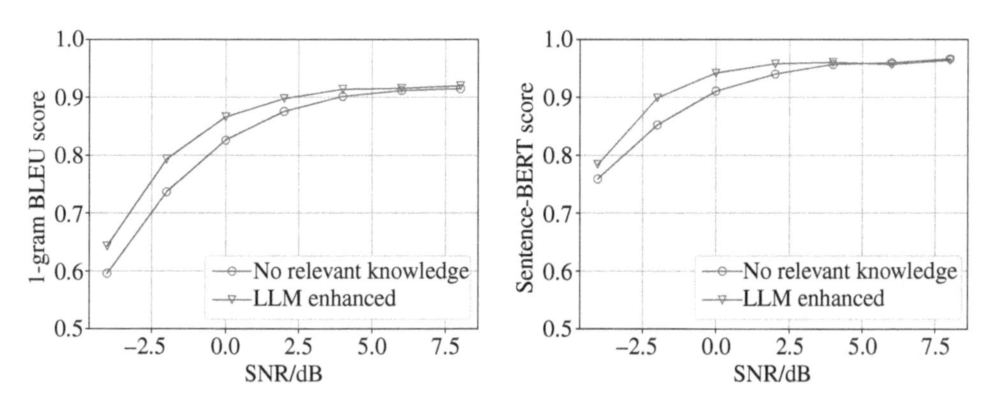

Figure 5.12 The performance evaluation of the proposed LLM-enhanced system.

knowledge matching by intentionally omitting a subset of the original knowledge dataset. To address this, a solution based on PE is employed to leverage the LLM for generating relevant knowledge, which is then integrated into the existing knowledge base. In detail, the experiment utilizes the GPT-3.5 Turbo model [Peng et al., 2023], incorporating 1000 samples from the original dataset. The data generated by the expansive model are utilized for an overall system performance simulation, evaluating its efficacy in SemCom tasks. Among the 1000 samples produced by the LLM, 800 of them are divided into the training dataset for instructional purposes, while the remaining 200 samples constitute the test dataset.

5.5.2.2 Numerical Results

Figure 5.12 illustrates the system's performance on the test dataset generated by the LLM. It can be observed from Figure 5.12 that the LLM possesses the ability to compensate for missing knowledge to a considerable degree, thus promising to improve the overall performance of the system. Particularly in scenarios with lower SNRs, the knowledge generated by the LLM contributes to an approximate increase of 0.05 in the BLEU score, while a similar phenomenon is observed for the sentence-BERT score. Therefore, it validates the efficacy of the LLM-empowered scheme and could further bolster SemCom's capability in processing and comprehending knowledge.

5.6 Conclusion

This chapter aims to investigate feasible and effective approaches for interplaying KGs and SemCom. It begins with an in-depth introduction to knowledge-processing challenges within

SemCom and proceeds to thoroughly investigate the limitations of pertinent existing studies. Subsequently, a KG-enhanced SemCom system is proposed, in which a knowledge extractor is leveraged to effectively exploit knowledge from KGs and facilitate semantic reasoning and decoding at the receiver side. On this basis, a contrastive learning-based optimization strategy for evolving KGs is described. Additionally, this chapter discusses methods for data augmentation through the utilization of LLMs. Extensive simulation results demonstrate that the proposed KG-enhanced SemCom systems can benefit from the prior knowledge in the knowledge base with significant performance gains. Looking ahead, future research could focus on further harnessing the semantic understanding and generation capabilities of LLMs to manage complex contextual and semantic information.

References

Antoine Bordes, Nicolas Usunier, Alberto Garcia-Duran, et al. Translating embeddings for modeling multi-relational data. In *Advances in Neural Information Processing Systems 26* (NIPS 2013), Lake Tahoe, Nevada, USA, Dec. 2013.

Antoine Bordes, Xavier Glorot, Jason Weston, et al. A semantic matching energy function for learning with multi-relational data: Application to word-sense disambiguation. *Machine Learning*, 94:233–259, 2014.

Eirina Bourtsoulatze, David Burth Kurka, and Deniz Gündüz. Deep joint source-channel coding for wireless image transmission. *IEEE Transactions on Cognitive Communications and Networking*, 5(3):567–579, 2019.

Tom Brown, Benjamin Mann, Nick Ryder, et al. Language models are few-shot learners. In *Advances in Neural Information Processing Systems 33 (NeurIPS 2020)*, Vancouver, BC, Canada, Dec. 2020.

Salvatore Carta, Alessandro Giuliani, Leonardo Piano, et al. Iterative zero-shot LLM prompting for knowledge graph construction. *arXiv preprint arXiv:2307.01128*, 2023.

Ting Chen, Simon Kornblith, Mohammad Norouzi, et al. A simple framework for contrastive learning of visual representations. In *International Conference on Machine Learning*, Virtual Edition, Jul. 2020.

Mostafa Dehghani, Stephan Gouws, Oriol Vinyals, et al. Universal transformers. *arXiv preprint arXiv:1807.03819*, 2018.

Tim Dettmers, Pasquale Minervini, Pontus Stenetorp, et al. Convolutional 2D knowledge graph embeddings. In *Proceedings of the Association for the Advancement of Artifical Intelligence Conference on Artificial Intelligence*, New Orleans, Louisiana, USA, Feb. 2018.

Mingze Ding, Jiahui Li, Mengyao Ma, et al. SNR-adaptive deep joint source-channel coding for wireless image transmission. In *IEEE International Conference on Acoustics, Speech and Signal Processing*, Toronto, Canada, Jun. 2021.

Nariman Farsad, Milind Rao, and Andrea Goldsmith. Deep learning for joint source-channel coding of text. In *IEEE International Conference on Acoustics, Speech and Signal Processing*, Calgary, Alberta, Canada, Apr. 2018.

Claire Gardent, Anastasia Shimorina, Shashi Narayan, et al. Creating training corpora for NLG micro-planning. In *Annual Meeting of the Association for Computational Linguistics*, Vancouver, BC, Canada, Jul. / Aug. 2017.

Kaiming He, Haoqi Fan, Yuxin Wu, et al. Momentum contrast for unsupervised visual representation learning. In *Proceedings of the IEEE/CVF Conference on Computer Vision and Pattern Recognition*, Virtual Edition, Jun. 2020.

Shaoxiong Ji, Shirui Pan, Erik Cambria, et al. A survey on knowledge graphs: Representation, acquisition, and applications. *IEEE Transactions on Neural Networks and Learning Systems*, 33(2):494–514, 2021.

Peiwen Jiang, Chao-Kai Wen, Shi Jin, et al. Deep source-channel coding for sentence semantic transmission with HARQ. *IEEE Transactions on Communications*, 70(8):5225–5240, 2022a.

Peiwen Jiang, Chao-Kai Wen, Shi Jin, et al. Wireless semantic communications for video conferencing. *IEEE Journal on Selected Areas in Communications*, 41(1):230–244, 2022b.

Shengteng Jiang, Yueling Liu, Yichi Zhang, et al. Reliable semantic communication system enabled by knowledge graph. *Entropy*, 24(6):846, 2022c.

Feibo Jiang, Yubo Peng, Li Dong, et al. Large AI model-based semantic communications. *arXiv preprint arXiv:2307.03492*, 2023.

David Burth Kurka and Deniz Gündüz. DeepJSCC-*f*: Deep joint source-channel coding of images with feedback. *IEEE Journal on Selected Areas in Information Theory*, 1(1):178–193, 2020.

Qiao Lan, Dingzhu Wen, Zezhong Zhang, et al. What is semantic communication? A view on conveying meaning in the era of machine intelligence. *Journal of Communications and Information Networks*, 6(4):336–371, 2021.

Changwoo Lee, Xiao Hu, and Hun-Seok Kim. Deep joint source-channel coding with iterative source error correction. In *IEEE International Conference on Acoustics, Speech and Signal Processing*, Rhodes island, Greece, Jun. 2023.

Ang Li, Xin Wei, Dan Wu, et al. Cross-modal semantic communications. *IEEE Wireless Communications*, 29(6):144–151, 2022.

Jingming Liang, Yong Xiao, Yingyu Li, et al. Life-long learning for reasoning-based semantic communication. In *IEEE International Conference on Communications Workshops*, Seoul, South Korea, May 2022.

Kun Lu, Rongpeng Li, Xianfu Chen, et al. Reinforcement learning-powered semantic communication via semantic similarity. *arXiv preprint arXiv:2108.12121*, 2021.

Kun Lu, Qingyang Zhou, Rongpeng Li, et al. Rethinking modern communication from semantic coding to semantic communication. *IEEE Wireless Communications*, 30(1):158–164, 2023a.

Zhilin Lu, Rongpeng Li, Ming Lei, et al. Self-critical alternate learning based semantic broadcast communication. *arXiv preprint arXiv:2312.01423*, 2023b.

Zhilin Lu, Rongpeng Li, Kun Lu, et al. Semantics-empowered communications: A tutorial-cum-survey. *IEEE Communications Surveys & Tutorials*, 26(1):41–79, 2023c.

Kishore Papineni, Salim Roukos, Todd Ward, et al. BLEU: A method for automatic evaluation of machine translation. In *Annual Meeting of the Association for Computational Linguistics*, Philadelphia, PA, USA, Jul. 2002.

Andrew Peng, Michael Wu, John Allard, et al. GPT-3.5 turbo fine-tuning and API updates. https://openai.com/blog/gpt-3-5-turbo-fine-tuning-and-api-updates, 2023.

Alec Radford, Karthik Narasimhan, Tim Salimans, et al. Improving language understanding by generative pre-training. https://openai.com/research/language-unsupervised, 2018.

Alec Radford, Jeffrey Wu, Rewon Child, et al. Language models are unsupervised multitask learners. *OpenAI Blog*, 1(8):9, 2019.

Colin Raffel, Noam Shazeer, Adam Roberts, et al. Exploring the limits of transfer learning with a unified text-to-text transformer. *The Journal of Machine Learning Research*, 21(1):5485–5551, 2020.

Nils Reimers and Iryna Gurevych. Sentence-BERT: Sentence embeddings using siamese BERT-networks. In *Conference on Empirical Methods in Natural Language Processing*, Hong Kong, China, Nov. 2019.

Michael Schlichtkrull, Thomas N. Kipf, Peter Bloem, et al. Modeling relational data with graph convolutional networks. In *European Semantic Web Conference*, Heraklion, Crete, Greece, Jun. 2018.

Siyu Tong, Xiaoxue Yu, Rongpeng Li, et al. Alternate learning based sparse semantic communications for visual transmission. In *IEEE International Symposium on Personal, Indoor and Mobile Radio Communications*, Toronto, ON, Canada, Sept. 2023.

Hugo Touvron, Thibaut Lavril, Gautier Izacard, et al. LLaMA: Open and efficient foundation language models. *arXiv preprint arXiv:2302.13971*, 2023a.

Hugo Touvron, Louis Martin, Kevin Stone, et al. LLaMA 2: Open foundation and fine-tuned chat models. *arXiv preprint arXiv:2307.09288*, 2023b.

Laurens Van der Maaten and Geoffrey Hinton. Visualizing data using t-SNE. *Journal of Machine Learning Research*, 9(11):2579–2605, 2008.

Ashish Vaswani, Noam Shazeer, Niki Parmar, et al. Attention is all you need. In *Advances in Neural Information Processing Systems 30 (NIPS 2017)*, Long Beach, CA, USA, Dec. 2017.

Yining Wang, Mingzhe Chen, Tao Luo, et al. Performance optimization for semantic communications: An attention-based reinforcement learning approach. *IEEE Journal on Selected Areas in Communications*, 40(9):2598–2613, 2022.

Bingyan Wang, Rongpeng Li, Jianhang Zhu, et al. Knowledge enhanced semantic communication receiver. *IEEE Communications Letters*, 27(7):1794–1798, 2023.

Zhenzi Weng and Zhijin Qin. Semantic communication systems for speech transmission. *IEEE Journal on Selected Areas in Communications*, 39(8):2434–2444, 2021.

Haotian Wu, Yulin Shao, Emre Ozfatura, et al. Transformer-aided wireless image transmission with channel feedback. *arXiv preprint arXiv:2306.09101*, 2023.

Yong Xiao, Yingyu Li, Guangming Shi, and H. Vincent Poor. Reasoning on the air: An implicit semantic communication architecture. In *IEEE International Conference on Communications Workshops*, Seoul, South Korea, May 2022a.

Yong Xiao, Zijian Sun, Guangming Shi, and Dusit Niyato. Imitation learning-based implicit semantic-aware communication networks: Multi-layer representation and collaborative reasoning. *IEEE Journal on Selected Areas in Communications*, 41(3):639–658, 2022b.

Ke Yang, Sixian Wang, Jincheng Dai, et al. WITT: A wireless image transmission transformer for semantic communications. In *IEEE International Conference on Acoustics, Speech and Signal Processing*, Rhodes island, Greece, Jun. 2023.

Qingyang Zhou, Rongpeng Li, Zhifeng Zhao, et al. Semantic communication with adaptive universal transformer. *IEEE Wireless Communications Letters*, 11(3):453–457, 2021.

Fuhui Zhou, Yihao Li, Xinyuan Zhang, et al. Cognitive semantic communication systems driven by knowledge graph. In *IEEE International Conference on Communications*, Seoul, South Korea, May 2022a.

Qingyang Zhou, Rongpeng Li, Zhifeng Zhao, et al. Adaptive bit rate control in semantic communication with incremental knowledge-based HARQ. *IEEE Open Journal of the Communications Society*, 3:1076–1089, 2022b.

Fuhui Zhou, Yihao Li, Ming Xu, et al. Cognitive semantic communication systems driven by knowledge graph: Principle, implementation, and performance evaluation. *arXiv preprint arXiv:2303.08546*, 2023.

6

VISTA: A Semantic Communication Approach for Video Transmission

Chengsi Liang[1], Xiangyi Deng[2], Yao Sun[1], Runze Cheng[1], Le Xia[1], Dusit Niyato[3], and Muhammad Ali Imran[1]

[1] *James Watt School of Engineering, University of Glasgow, Glasgow, Glasgow, Lanarkshire, UK*
[2] *Glasgow College, University of Electronic Science and Technology of China, Chengdu, Sichuan, China*
[3] *School of Computer Science and Engineering, Nanyang Technological University, Singapore*

6.1 Introduction

The boom in multimedia services has led to video streaming dominating about 82% of all internet traffic in 2022 [Ahmad et al., 2021], encompassing a variety of applications like live streaming, virtual/augmented/mixed reality, and virtual meetings. To enhance the quality of service (QoS) for these services, particularly for real-time applications such as augmented reality (AR) and virtual reality (VR), the implementation of ultra-reliable and low-latency communication (URLLC) is essential. Traditional wireless video transmission methods, which primarily rely on compressing and recovering video through pixel encoding and decoding, however, use a significant amount of wireless spectrum and transmission time. This approach often falls short of delivering satisfactory visual quality due to the instability of wireless channels.

Fortunately, semantic communication (SemCom) [Xie et al., 2021; Weng and Qin, 2021; Lan et al., 2021; Xia et al., 2022b], which focuses on the meaning of the source information rather than the bits or symbols themselves, has emerged as a significant paradigm shift in communication systems. Typically, a SemCom system's transmitter first extracts and encodes semantic information from the source, tailored to the conditions of the wireless channel. This semantic information is then wirelessly transmitted, with the receiver reconstructing the original meaning of the source to minimize semantic errors. This approach in SemCom is poised to substantially reduce the volume of transmitted bits, thereby significantly conserving wireless resource consumption. Additionally, SemCom can offer enhanced robustness, particularly in suboptimal wireless conditions. This is achieved by leveraging the tight semantic coupling between adjacent frames, allowing the decoder to accurately reconstruct unclear video pixels based on their semantic content.

While SemCom-enabled video transmission offers numerous advantages, it also presents certain critical challenges. A primary issue is the coexistence of static and dynamic objects in successive video frames. Typically, the semantics of static objects remain consistent across different frames, while dynamic objects exhibit a regular pattern of change. Efficiently representing and reconstructing these semantics in consecutive frames is a significant challenge. Additionally, signal degradation and distortion in wireless channels can lead to semantic ambiguities in the

Wireless Semantic Communications: Concepts, Principles, and Challenges, First Edition.
Edited by Yao Sun, Lan Zhang, Dusit Niyato, and Muhammad Ali Imran.

transmitted videos, adversely impacting the quality of the final video output. Therefore, another challenge is accommodating various channel conditions in SemCom-enabled video transmission.

Recent pioneering works in deep learning (DL)-based video transmission within SemCom have begun addressing these issues [Wang et al., 2023; Jiang et al., 2023; Xia et al., 2022a]. A deep joint source–channel coding (JSCC) framework is designed by Wang et al. [2023] to transmit the semantics of entire videos over diverse wireless channels. Jiang et al. [2023] focus on transmitting key semantic points for video conferencing, introducing a framework to adjust to fluctuating channels. Furthermore, the potential of URLLC in VR delivery from mobile edge computing servers to VR users has also been explored by Xia et al. [2022a].

Compared with these works, we propose a novel framework called VISTA (VIdeo transmission over Semantic communicaTion Approach) to address the challenges previously mentioned. VISTA comprises three key modules: the semantic segmentation module and the frame interpolation module, responsible for semantic encoding and decoding respectively, and the JSCC module, designed for signal-to-noise ratio (SNR)-adaptive wireless transmission.

The principal contributions of this work are outlined as follows:

- A semantic segmentation module is developed at the transmitter, where it first detects and recognizes the dynamic objects and static background for each frame in the source video. Then, a semantic location graph (SLG) is built to describe the locations and relationships for all dynamic objects and accurately extract semantics.
- VISTA separates each video frame into environment (image of static background) and behavior segments (image of all the dynamic objects), which require only one-frame environment and several key behavior segments to be fed into the JSCC module for wireless transmission.
- A frame interpolation module is developed at the receiver end to accurately recover the video based on the segments and semantics with the help of SLG.
- We test the performance of VISTA using a real video dataset, and the results demonstrate its superiorities in terms of transmitted data volume, video processing time, and video quality (especially under low SNR scenario) compared with two other benchmarks.

6.2 Video Transmission Framework in VISTA

In this section, we delve into the video transmission framework underpinning VISTA. Typically, the semantic coding model's role is to extract (on the sender's side) and reconstruct (on the receiver's side) the semantic information of the transmitted videos. Concurrently, the channel coding model is responsible for considering varying physical channel conditions to ensure the precise delivery of semantic content. Additionally, a shared knowledge base (KB), as discussed in Strinati and Barbarossa [2021], is a fundamental component in the video delivery process, serving as a common reference for both the transmitter and receiver.

We consider a source video composed of T sequential frames: $\mathbf{s} = \{s^1, \ldots, s^T\} \in \mathbb{R}^{H\times W\times T}$, where H and W, respectively, denote the height and width of a frame. These frames are first fed into the convolutional semantic encoder to distill the textual semantic information $\mathbf{g}$. In addition, the semantic encoder divides the source video into two parts: environment (static background) $\mathbf{s}_e$ and behavior segments (dynamic objects) $\mathbf{s}_b$ individually. Thus, the encoded frames can be written as $\hat{\mathbf{s}} = \{\mathbf{s}_e, \mathbf{s}_b, \mathbf{g}\}$ under the semantic encoder network $\mathcal{S}(\cdot)$ with parameter set α_s, i.e.,

$$\hat{\mathbf{s}} = \{\mathbf{s}_e, \mathbf{s}_b, \mathbf{g}\} = \mathcal{S}(\mathbf{s}; \alpha_s). \tag{6.1}$$

The encoded frames $\hat{\mathbf{s}}$ then flow into the JSCC module for SNR-adaptive wireless transmission. In this module, source encoder $\mathcal{E}$ and channel encoder C with parameter sets α_ϵ and α_c generate the symbols $\mathbf{x}$ to be transmitted,

$$\mathbf{x} = C\left(\mathcal{E}\left(\hat{\mathbf{s}}; \alpha_\epsilon\right); \alpha_c\right). \tag{6.2}$$

At the receiver side, $\mathbf{y}$ is denoted as the received symbols for the input $\mathbf{x}$ over the wireless channel with additive noise w, i.e.,

$$\mathbf{y} = h * \mathbf{x} + w, \tag{6.3}$$

where h denotes the channel gain, $\mathbf{y}$ is then fed to the channel decoder C^{-1} and source decoder $\mathcal{E}^{-1}$ sequentially to reconstruct the environment $\tilde{\mathbf{s}}_e$ and behavior segments $\tilde{\mathbf{s}}_b$ with the help of the semantics. The decoded frames $\tilde{\mathbf{x}}$ are presented as

$$\tilde{\mathbf{x}} = \left\{\tilde{\mathbf{s}}_e, \tilde{\mathbf{s}}_b\right\} = \mathcal{E}^{-1}\left(C^{-1}\left(\mathbf{y}; \beta_c\right); \beta_\epsilon\right), \tag{6.4}$$

where β_c and β_ϵ denote the parameters of the channel decoder and source decoder networks, respectively.

Finally, the recovered video $\tilde{\mathbf{s}}$ should be constructed as per the two parts of segments $\tilde{\mathbf{s}}_e$ and $\tilde{\mathbf{s}}_b$. The semantic-decoder network and its parameters are given as S^{-1} and β_s. Thus, the final recovered video is expressed as

$$\tilde{\mathbf{s}} = S^{-1}\left(\tilde{\mathbf{x}}; \beta_s\right). \tag{6.5}$$

In this work, the ultimate goal is minimizing the semantic ambiguity of the recovered video. We use the average peak signal-to-noise ratio (PSNR) [Hore and Ziou, 2010], a popular video quality metric, to measure the differences between the recovered and original video frames. In detail, for the tth frame with the size $m \times n$, the mean squared error (MSE) between the original frame s^t and the recovered one $\tilde{s}^t$ is calculated as

$$MSE^t = \frac{1}{mn}\sum_{i=1}^{m}\sum_{j=1}^{n}\left[s^t(i,j) - \tilde{s}^t(i,j)\right]. \tag{6.6}$$

Thus, the average of PSNR of the original and recovered video is expressed as

$$PSNR = \frac{1}{T}\sum_{t=1}^{T} 10 \cdot log_{10}\left(\frac{{I_{max}}^2}{MSE^t}\right), \tag{6.7}$$

where ${I_{max}}^2$ represents the maximum pixel value of the frame.

6.3 SLG-Based Transceiver Design in VISTA

Building on the previously described video transmission framework, we now focus on the design of the transceiver in the VISTA system. As shown in Figure 6.1, VISTA consists of three key modules: semantic segmentation, JSCC, and frame interpolation. The semantic segmentation and JSCC encoder modules are located at the transmitter side, while the receiver side features the frame interpolation module along with the JSCC decoder. Importantly, SLGs are created during the semantic segmentation process and later utilized in frame interpolation. These modules are individually trained with distinct loss functions, aiming to minimize the PSNR for the restored video. Next, we will explore the construction and training of the DL networks in these three modules.

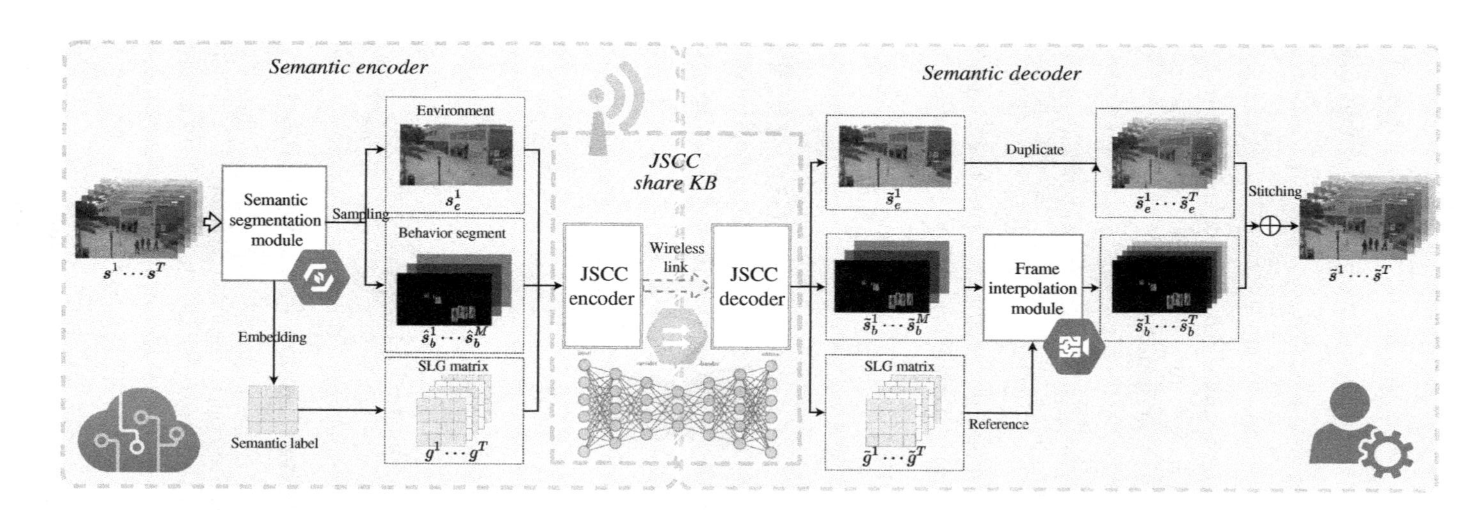

Figure 6.1 The diagram of transceiver in VISTA.

6.3.1 Semantic Segmentation Module

Semantic segmentation module is deployed at the transmitter to recognize and distill the dynamic objects from the video. Four tasks should be performed in this module: object detection, trajectory prediction, SLG construction, and frame sampling. Generally, we first bound all objects using rectangular boxes in each frame and differentiate the dynamic objects from the static background via velocity testing. The semantic information of each dynamic object can be extracted by the means of category recognition. However, the occlusion caused by overlapping objects in the video will affect the accuracy of position detection and semantics extraction. Thus, we predict the trajectory of each object for the continuous frames in the second task. After trajectory prediction, an SLG is designed to assist in reserving the estimated positions and semantics of all dynamic objects in each frame. Finally, we sample the frames and send them along with SLGs to JSCC module. Let us illustrate the design of the four tasks separately below.

6.3.1.1 Object Detection

Borrowing the idea from Redmon et al. [2016], we apply a conventional network to outline bounding boxes using features of the entire frame. Specifically, we initialize several bounding boxes, and they are projected to enlarge and shift dynamically until all the objects are bounded with the optimized confidence scores. In this way, each bounding box is associated with six predictions: 2D coordinates (u, v) of the center for the object, the width and height (w, h) of the box containing relative to the whole image of the object, object category l, and the associated confidence score c.

6.3.1.2 Trajectory Prediction

After getting these objective detections, we should guarantee that every dynamic object can be captured completely. Therefore, we deploy the trajectory prediction module to track the dynamic objects when they are occluded. The input of the network of trajectory prediction is the images of dynamic objects and the five predictions of the corresponding bounding boxes. We first define an "observation" of a bounding box as $z = [u, v, w, h, c]^\top$. Moreover, we employ the Kalman filter (KF) to generate the state $q = [u, v, a, r, \dot{u}, \dot{v}, \dot{a}]^\top$, where a is the bounding box scale (area), r is the width-to-height ratio of the bounding box, and the other three variables ($\dot{u}$, $\dot{v}$ and $\dot{a}$) are the corresponding time derivatives.

Next, we utilize an observation-centric tracker [Cao et al., 2022] with the object movement. Specifically, since a nonlinear motion can be regarded as a synthesis of many small-scale linear motions in a reasonably short time, we calculate the velocity consistency (momentum) to gain the accurate velocity value and direction. Then, for an untracked object, an observation-centric online smoothing strategy is performed through a virtual trajectory $\hat{\mathbf{z}}^t$ starting from its last occurrence and ending at the re-associated observation, which is denoted as

$$\hat{\mathbf{z}}^t = \mathcal{T}_v\left(\mathbf{z}^{t_1}, \mathbf{z}^{t_2}, t\right), t_1 < t < t_2, \tag{6.8}$$

where $\mathbf{z}^{t_1}$ is the last observation before being untracked, $\mathbf{z}^{t_2}$ is the observation triggering the re-association, and $\mathcal{T}_v()$ represents the network of virtual trajectory. Along this virtual trajectory, the status at t_1 is recalled to check the filter parameters. Thus, the refreshed state $\hat{q}^t$ is estimated as

$$\hat{q}^t = \mathbf{F}^t\hat{q}^{t-1} + \mathbf{K}^t\left(\hat{z}^t - \mathbf{H}^t\mathbf{F}^t\hat{q}^{t-1}\right), \tag{6.9}$$

where $\mathbf{K}^t$ denotes the KF matrix, $\mathbf{F}^t$ and $\mathbf{H}^t$ denote the state transition and observation model, respectively. With the instruction of $\hat{q}^t$, for the tth frame in the video, we update the bounding boxes of dynamic objects and use the behavior segments $\hat{s}_b^t$ to represent the images of all estimated boxes covering. Moreover, the rest of this frame is represented by the environment $\hat{s}_e^t$.

6.3.1.3 SLG Construction

Aiming at locating the dynamic objects and illustrating the association between their location and semantics, we deliver an SLG to concatenate the classes and locations from the refreshed states $\hat{q}$. With respect to a frame containing B boxes, the set of object categories is $\mathbf{l} = \{l_1, \dots, l_B\}$, and the 2D coordinate sets are $\hat{\mathbf{u}} = \{\hat{u}_1, \dots, \hat{u}_B\}$ and $\hat{\mathbf{v}} = \{\hat{v}_1, \dots, \hat{v}_B\}$. Thus, the SLG $g^t \in \{g^1, \dots, g^T\}$ of the tth frame can be represented as

$$g^t = \{\mathbf{l}, \hat{\mathbf{u}}, \hat{\mathbf{v}}\}. \tag{6.10}$$

6.3.1.4 Frame Sampling

According to the results of trajectory prediction, we split the whole video into environment and behavior segments and transmit them separately. Since the environment is fixed, it is supposed that only the environment of the first frame s_e^1 needs to be transmitted. It is also not thrifty for the encoder to process behavior segments throughout the entire video so that we sample them every T_s frames and denote $M = \lceil T/T_s \rceil$ samples as $\hat{\mathbf{s}}_b' = \left\{\hat{s}_b^1, \hat{s}_b^{T_s+1}, \dots, \hat{s}_b^T\right\}$. Could the term "thriftless" be replaced by "thrifty" in the sentence "It is also…". The output $\hat{s}^t$ of the t-th frame is illustrated as

$$\hat{s}^t = \begin{cases} \{s_e^1, \hat{s}_b^1, g^1\}, & t = 1, \\ \{\hat{s}_b^t, g^t\}, & t = nT_s + 1, n = \{1, \dots, M-1\}, \\ g^t, & otherwise. \end{cases} \tag{6.11}$$

Generally, the overall output for all frames after the semantic encoder is composed of the environment of the first frame, behavior segments from the sample frames, and SLGs of all frames.

6.3.2 JSCC Module

As illustrated, all the extracted semantic segments, along with an SLG, should be transmitted through a wireless channel. In VISTA, we employ an SNR-adaptive JSCC module, which can configure its parameters depending on the SNR of the channel [Yang and Kim, 2022]. Its overall structure can be described as source encoder, channel encoder, channel decoder, and source decoder. In more detail, the features $\mathbf{f} = \{f_e^1, \mathbf{f}_b\}$ are first extracted from the input of environment and behavior segments (s_e^1 and $\hat{\mathbf{s}}_b'$) via several conventional layers, and some of them are activated to be transmitted first. After getting $\mathbf{f}$, the channel encoder produces two groups of length-L features. The first group with the length G_s contains either active or inactive features selected by a policy network $\mathcal{P}$, while the following G_n groups are always active without selection. The selection for each input is conducted by a binary mask W_i, which can only be 0 or 1. The total number of active groups is demonstrated as $\tilde{G} = G_n + \sum_{i=1}^{G_s} W_i$. All the active features are passed through the power normalization network to generate complex-valued transmission symbols $\{x_e^0, \hat{\mathbf{x}}_b'\} \in \mathbb{C}^{G \times L/2}$ with unit average power using the first half of features as the real part and the other half as the imaginary part. Moreover, the textual SLGs $\mathbf{g} = \{g^1, \dots, g^T\}$ are encoded to bits $\mathbf{x}_g$ and transmitted directly. In a word, the total encoded symbols are represented by $\mathbf{x} = \{x_e^0, \hat{\mathbf{x}}_b', \mathbf{x}_g\}$.

Next, $\mathbf{y} = \{y_e^1, \hat{\mathbf{y}}_b', \mathbf{y}_g\}$ is received as $\mathbf{x}$ should be transmitted over the wireless channel model in (6.3), where y_e^1, $\hat{\mathbf{y}}_b'$ and $\mathbf{y}_g$ denote the transmitted symbols of the environment, behavior segments, and SLG, respectively. Then, $\mathbf{y}$ is fed to the channel decoder and source decoder sequentially to reconstruct the environment $\tilde{s}_e^1$, behavior segments $\tilde{\mathbf{s}}_b{}'$, and SLGs $\tilde{\mathbf{g}}$. Specifically, $\tilde{s}_e^1$ and $\tilde{\mathbf{s}}_b{}'$ are recovered through several convolutional layers while $\tilde{\mathbf{g}}$ is decoded to text directly.

It is worth noting that the SNR value is the part of the input fed to the policy network and the SNR adaptive network leveraged in the channel encoder and channel decoder [Xu et al., 2021].

Particularly, for the SNR adaptive network, the features in one frame are first pooled averagely across diverse feature channels (different from the wireless channels) of a neural network and then concatenated with the SNR value. Next, the results are received by two multilayer perceptrons to produce the factors for channel-wise scaling and addition. In this way, we adjust the network of the transceiver in the JSCC module depending on the SNR value.

6.3.3 Frame Interpolation Module

After receiving the environment and behavior segments of sample frames, we complement them and combine the results to rebuild the video with the help of SLGs $\tilde{\mathbf{g}}$ in the semantic decoder. In more detail, we make T copies of the one-frame environment and generate the sequence of environment as $\tilde{\mathbf{s}}_e = \{\tilde{s}_e^1, \dots, \tilde{s}_e^1\}$ at first. Then, according to the behavior segments $\tilde{\mathbf{s}}_b' = \{\tilde{s}_b^1, \dots, \tilde{s}_b^M\}$ of M sample frames, we utilize transformer for frame interpolation with the inspiration of video frame interpolation with transformer (VFI-former) [Lu et al., 2022], aiming at predicting the behavior segments for all the remaining frames. Consider the behavior segments $\tilde{s}_b^1$ and $\tilde{s}_b^2$ of the two adjacent sample frames, and the intermediate frame is denoted as $\tilde{s}_b^t$.

A convolutional network called "flow estimator" is utilized to obtain the optical flows $O^{t\to 1}$ and $O^{t\to 2}$. Additionally, the images w_b^1 and w_b^2 are restored as per the features $\tilde{f}_i^1$ and $\tilde{f}_i^2$, which are warped by $O^{t\to 1}$ and $O^{t\to 2}$ respectively. Further, the semantic decoder includes Transformer blocks (TFB), and each TFB consists of convolutional layers and several transformer layers (TFL) with cross-scale window-based attention (CSWA) network, which is a state-of-the-art attention mechanism. For the ith TFB, its output feature f_i^t is formulated as

$$f_i^t = TFB_i\left(f_{i-1}^t, \tilde{f}_i^1, \tilde{f}_i^2\right), \tag{6.12}$$

where f_{i-1}^t is the output of the $(i-1)$th TFB.

Then, the intermediate frame $\tilde{s}_b^t$ is generated by a soft mask H and an image residual $\Delta\tilde{s}_b^t$ (from flow errors and occlusion) in the decoder as follows:

$$\tilde{s}_b^t = H \odot w_b^1 + (1-H) \odot w_b^2 + \Delta\tilde{s}_b^t, \tag{6.13}$$

where $\odot$ signifies the Hadamard product. It is worth noting that the interpolation is under the guidance of SLG. In other words, the prediction of behavior segments in the intermediate frames is limited to the bounding boxes provided by SLG.

In terms of model training, the loss should be evaluated from three aspects. The first is reconstruction loss, which compares the recovered behavior segments s_{gt}^t and its ground truth $\tilde{s}_b^t$ in the tth frame.

$$\mathcal{L}_{rec} = \left\|s_{gt}^t - \tilde{s}_b^t\right\|_1. \tag{6.14}$$

Next, the census loss [Meister et al., 2018] is robust to illumination changes, which is defined as the soft Hamming distance between census-transformed [Zabih and Woodfill, 1994] image patches of s_{gt}^t and $\tilde{s}_b^t$. The last one is distillation loss for supervising the estimated flows explicitly,

$$\mathcal{L}_{dis} = \left\|\tilde{O}^{t\to 1} - O^{t\to 1}\right\|_1 + \left\|\tilde{O}^{t\to 2} - O^{t\to 2}\right\|_1, \tag{6.15}$$

where $\tilde{O}^{t\to 1}$ and $\tilde{O}^{t\to 2}$ are derived from a pretrained flow estimation network presented by Hui et al. [2018].

As a result, the total loss is presented as follows:

$$\mathcal{L} = \lambda_{rec}\mathcal{L}_{rec} + \lambda_{css}\mathcal{L}_{css} + \lambda_{dis}\mathcal{L}_{dis}, \tag{6.16}$$

where $\mathcal{L}_{rec}$, $\mathcal{L}_{css}$, and $\mathcal{L}_{dis}$ correspond to the reconstruction loss, census loss, and distillation loss with their weights λ_{rec}, λ_{css}, and λ_{dis}, respectively.

After frame interpolation, the behavior segments are estimated as the combination of the sample frames and intermediate frames. The recovered video $\tilde{\mathbf{s}}$ is the synthesis of the behavior segments $\tilde{\mathbf{s}}_b$ and the copies of the environment $\tilde{\mathbf{s}}_e$, which can be expressed by

$$\tilde{\mathbf{s}} = \tilde{\mathbf{s}}_b \oplus \tilde{\mathbf{s}}_e. \tag{6.17}$$

Herein, we stitch $\tilde{\mathbf{s}}_e$ and $\tilde{\mathbf{s}}_b$ via $\oplus$ and maintain $\tilde{\mathbf{s}}_b$ as their overlapping parts.

6.4 Simulation Results and Discussions

6.4.1 Simulation Setting

The pretraining dataset is DanceTrack [Sun et al., 2022], which includes 100 videos of group dances, 105k frames, and 877k high-quality bounding boxes. The test dataset is from an open dataset named VIRAT [Corona et al., 2021]. Our team conducts simulations to evaluate the performance of the proposed VISTA framework in comparison with two different benchmarks: (i) A JSCC integrated with VFIformer scheme (JSCC-VFI), which first employs a single deep neural network to transmit video frames over wireless channels without any awareness of semantics and then uses the powerful transformer model for behavior segment interpolation; (ii) A conventional bit-oriented communication scheme (LDPC codes proposed by Cai et al. [2006]), in which all pixels of each video frame should be encoded into bits based on the prescribed coding rule (low-density parity check in our simulations) for precise transmission.

For the simulation settings, the OC-SORT structure is first leveraged for object segmentation of video frames, which is consistent with the setup offered by Cao et al. [2022]. Besides, the parameters in JSCC-related channel encoding and decoding networks are proceeding as those presented by Yang and Kim [2022], where the wireless channel model is simulated as an additive white Gaussian noise channel with SNR values varying from -9 to 6 dB. Moreover, the architecture details of VFIformer-related networks can refer to Lu et al. [2022]. Note that the Adam optimizer is adopted to train the VISTA with an initial learning rate of 5×10^{-4}, and all subsequent simulations are implemented on a computer with six CPU cores and Inter Core i7 processor, where the main software environment is Python 3.9.

6.4.2 VISTA Framework Performance Evaluation

Here, Figure 6.2 shows four examples of SLGs with the labels of dynamic objects in the original frames. Guided by them, the videos are recovered by VISTA under varying SNRs from -9

Figure 6.2 The examples of SLGs in original video.

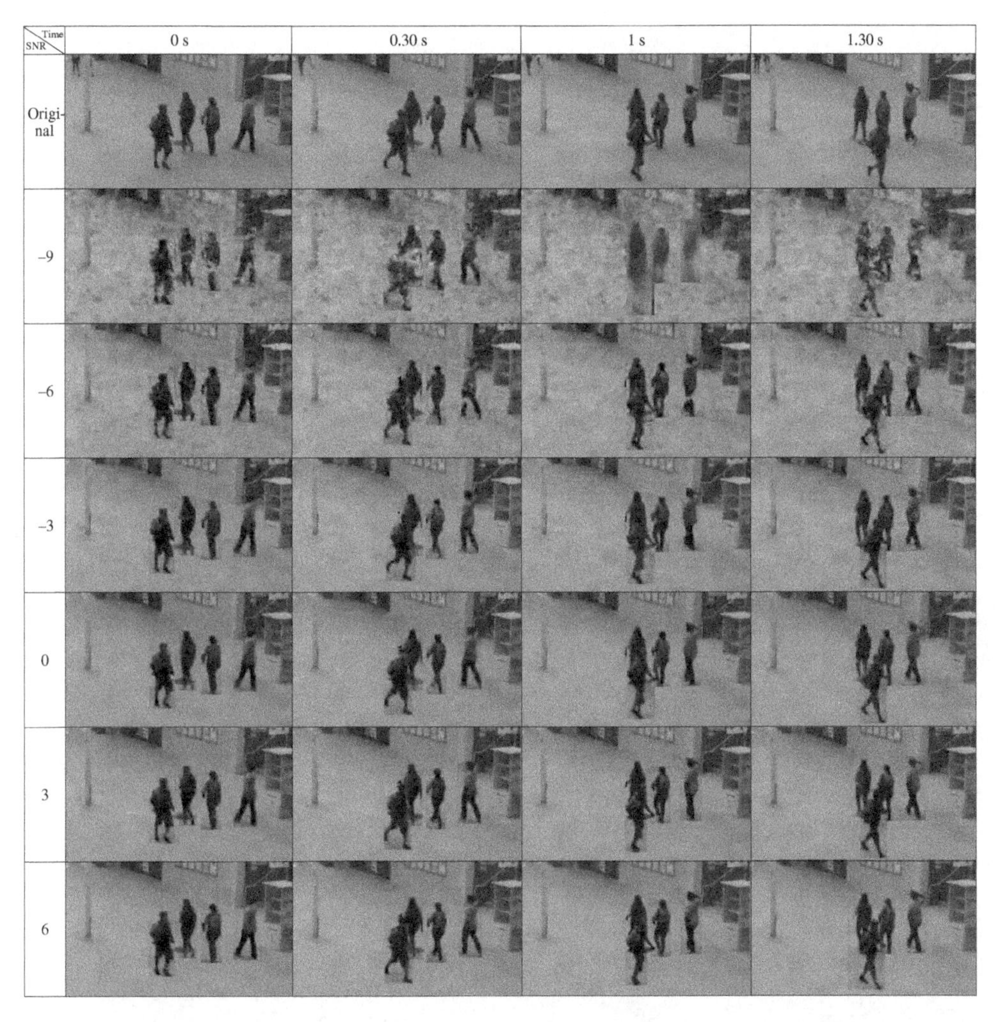

Figure 6.3 The frame samples recovered by VISTA under varying SNRs from −9 to 6 dB.

to 6 dB. Some examples are shown in Figure 6.3. Obviously, VISTA can restore relatively clear frames even at an SNR of −3 dB. Since it outperforms in low-SNR scenarios, we further exhibit some specific frames of the original video and the videos recovered by VISTA, JSCC-VFI, and low-density parity-check (LDPC) codes at an SNR of 0 dB, considering three differing interpolation proportions: 0%, 50%, and 75% (Figure 6.4). Importantly, it should be noted that a 50% interpolation corresponds to a 50% sampling ratio, while a 75% interpolation corresponds to a 25% sampling ratio. To clarify, a 50% sampling ratio indicates that only one out of every two sequential frames is used. All objects in the frames of VISTA and JSCC-VFI are well recovered in the contrast of the conventional scheme, and the recovered videos will be more blurred and inconsistent as the ratio of interpolation increases. Specifically, compared to JSCC-VFI, VISTA may not perfectly align moving objects with the environment. However, it offers superior frame quality and more precise detail in object rendering.

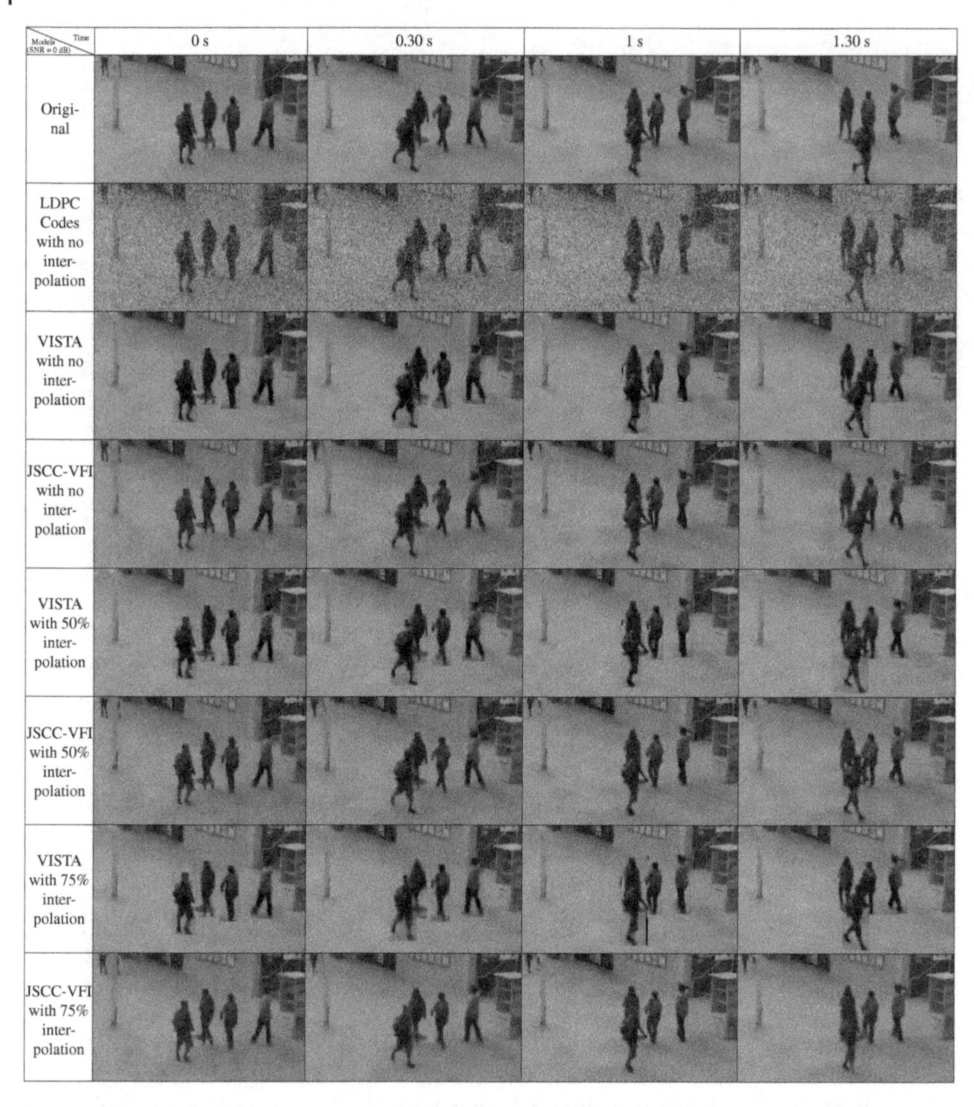

Figure 6.4 Visual comparison on frame samples for original video, recovered video by LDPC codes, VISTA, and JSCC-VFI with 0%, 50%, and 75% interpolation at an SNR of 0 dB.

To evidence it, we present the PSNR performance under varying SNRs from −9 to 6 dB with all situations in Figure 6.5. It can be seen that the PSNR of all schemes increases with SNR, which is because the higher SNR results in fewer impairment of transmitted semantic features so as to render a more accurate frame recovery. Additionally, there is a better PSNR of VISTA at a lower interpolation proportion. This trend is attributed to the fact that using fewer behavior frames for transmission means that more compressed features could be lost between consecutive dynamic objects, thereby resulting in a worse PSNR performance. Notably, VISTA with no interpolation consistently exhibits strong performance across a range of SNRs, starting with a high PSNR of 29.0 dB at −6 dB and culminating at 31.91 dB at 9 dB, which outperforms other models at SNRs from −6 to 6 dB. Such a performance gain of VISTA can be credited to its accurate semantic calibration function provided by SLG, which sufficiently guarantees high reliability of video transmission even in low SNR conditions.

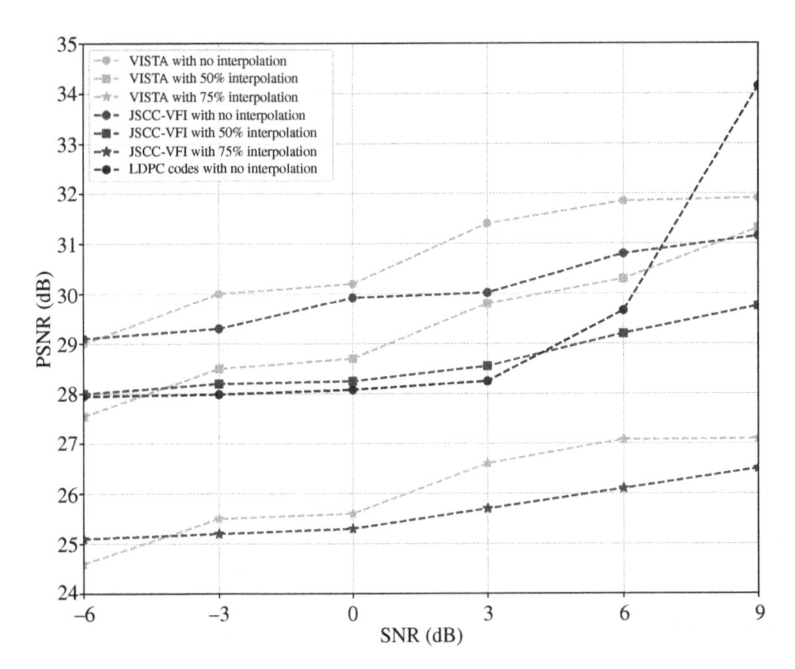

Figure 6.5 PSNR performance of recovered video frames versus varying SNRs from −9 to 6 dB.

However, LDPC codes exceed it, reaching 34.15 dB at an SNR of 9 dB because the semantics extraction inevitably causes the loss of information. Thereby, the recovered video via reasoning will be different from the original one, even in the high SNR scenario. In this regard, LDPC codes convey the entire information content for each frame in the original video, resulting in superior video quality when the SNR is high. However, such a conventional approach necessitates a substantial allocation of resources to transmit the full information of the video. Additionally, JSCC-VFI lacks the classification of the environment and dynamic objects, and the utilization of SLGs. As a result, its performance is worse than VISTA under the same environment at SNRs from −3 to 9 dB.

Next, we test the processing time (including encoding, transmission, and decoding time) for a total of 20 consecutive video frames with different interpolation proportions in Figure 6.6a, which is precisely obtained based on the aforementioned computer configuration. In particular, LDPC codes do not involve a sampling process, so interpolation is not applicable. Consequently, results for 50% and 75% interpolation are not available for LDPC-based approaches. It can be seen that without interpolation, the proposed VISTA only needs 0.25 seconds of processing time, which saves around 2.63 seconds/frame compared to the conventional scheme and saves nearly half the time of JSCC-VFI at any proportion of interpolation. This is because only behavior segments of a few video frames need to be processed in VISTA, thanks to the used SLG mechanism. Thereby fewer pixels are required to be encoded and decoded so as to save a significant processing time. In addition, a higher proportion of interpolation leads to a higher processing time as more intermediate frames need to be sampled and interpolated.

Finally, Figure 6.6b demonstrates the number of required bits for transmitting 20 consecutive video frames with the same three interpolation proportions. Like Figure 6.6a, there are no results for LDPC codes with 50% and 75% interpolation. It is implied that the proposed VISTA indeed shows its amazing superiority in communication resource saving, whose bit consumption is only 6.4% of the conventional scheme and 19.2% of the JSCC-VFI scheme when no interpolation is set. Moreover, it can be observed that the amount of transmitted bits of VISTA decreases as the

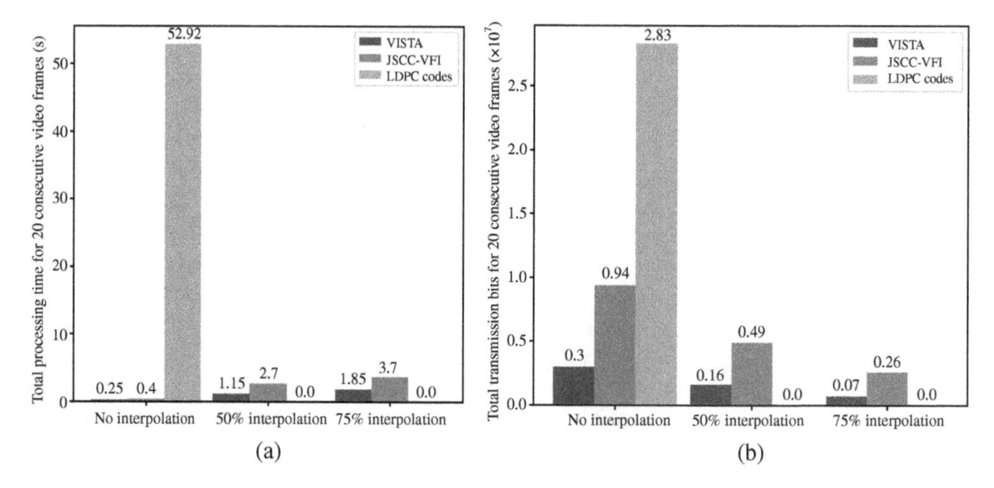

Figure 6.6 Total processing time (a) and transmission bits (b) for 20 consecutive video frames under different interpolation proportions.

interpolation proportion improves. Such a trend is easy to understand because the core semantics are more compressed at the high interpolation proportion, enabling these video frames to be sent with fewer bits.

6.5 Conclusions

In this chapter, we propose a SemCom-enabled wireless video transmission framework, named "VISTA." In VISTA, a unique transceiver is developed for semantic encoding and decoding, incorporating a SLG that operates alongside various neural networks for the extraction and restoration of video semantics. Simulations demonstrate a substantial decrease in the number of bits transmitted, maintaining (and even enhancing under SNR below 3 dB) video quality and transmission efficiency. This innovative approach is anticipated to be groundbreaking in utilizing SemCom for wireless video transmission, potentially easing bandwidth constraints in upcoming communication systems.

References

Arslan Ahmad, Atif Bin Mansoor, Alcardo Alex Barakabitze, Andrew Hines, Luigi Atzori, and Ray Walshe. Supervised-learning-based QoE prediction of video streaming in future networks: A tutorial with comparative study. *IEEE Communications Magazine*, 59(11):88–94, 2021.

Z. Cai, J. Hao, P. H. Tan, S. Sun, and P. S. Chin. Efficient encoding of IEEE 802.11n LDPC codes. *Electronics Letters*, 42(25):1, 2006.

Jinkun Cao, Xinshuo Weng, Rawal Khirodkar, Jiangmiao Pang, and Kris Kitani. Observation-centric SORT: Rethinking SORT for robust multi-object tracking. *arXiv preprint arXiv:2203.14360*, 2022.

Kellie Corona, Katie Osterdahl, Roderic Collins, and Anthony Hoogs. MEVA: A large-scale multiview, multimodal video dataset for activity detection. In *Proceedings of the IEEE/CVF Winter Conference on Applications of Computer Vision*, pages 1060–1068, 2021.

Alain Hore and Djemel Ziou. Image quality metrics: PSNR vs. SSIM. In *2010 20th international Conference on Pattern Recognition*, pages 2366–2369. IEEE, 2010.

Tak-Wai Hui, Xiaoou Tang, and Chen Change Loy. LiteFlowNet: A lightweight convolutional neural network for optical flow estimation. In *Proceedings of the IEEE Conference on Computer Vision and Pattern Recognition*, pages 8981–8989, 2018.

Peiwen Jiang, Chao-Kai Wen, Shi Jin, and Geoffrey Ye Li. Wireless semantic communications for video conferencing. *IEEE Journal on Selected Areas in Communications*, 41(1):230–244, 2023. doi: 10.1109/JSAC.2022.3221968.

Qiao Lan, Dingzhu Wen, Zezhong Zhang, Qunsong Zeng, Xu Chen, Petar Popovski, and Kaibin Huang. What is semantic communication? A view on conveying meaning in the era of machine intelligence. *Journal of Communications and Information Networks*, 6(4):336–371, 2021.

Liying Lu, Ruizheng Wu, Huaijia Lin, Jiangbo Lu, and Jiaya Jia. Video frame interpolation with transformer. In *Proceedings of the IEEE/CVF Conference on Computer Vision and Pattern Recognition*, pages 3532–3542, 2022.

Simon Meister, Junhwa Hur, and Stefan Roth. UnFlow: Unsupervised learning of optical flow with a bidirectional census loss. In *Proceedings of the AAAI Conference on Artificial Intelligence*, volume 32, 2018.

Joseph Redmon, Santosh Divvala, Ross Girshick, and Ali Farhadi. You only look once: Unified, real-time object detection. In *Proceedings of the IEEE Conference on Computer Vision and Pattern Recognition*, pages 779–788, 2016.

Emilio Calvanese Strinati and Sergio Barbarossa. 6G networks: Beyond Shannon towards semantic and goal-oriented communications. *Computer Networks*, 190:107930, 2021.

Peize Sun, Jinkun Cao, Yi Jiang, Zehuan Yuan, Song Bai, Kris Kitani, and Ping Luo. DanceTrack: Multi-object tracking in uniform appearance and diverse motion. In *Proceedings of the IEEE/CVF Conference on Computer Vision and Pattern Recognition*, pages 20993–21002, 2022.

Sixian Wang, Jincheng Dai, Zijian Liang, Kai Niu, Zhongwei Si, Chao Dong, Xiaoqi Qin, and Ping Zhang. Wireless deep video semantic transmission. *IEEE Journal on Selected Areas in Communications*, 41(1):214–229, 2023. doi: 10.1109/JSAC.2022.3221977.

Zhenzi Weng and Zhijin Qin. Semantic communication systems for speech transmission. *IEEE Journal on Selected Areas in Communications*, 39(8):2434–2444, 2021.

Le Xia, Yao Sun, Chengsi Liang, Daquan Feng, Runze Cheng, Yang Yang, and Muhammad Ali Imran. WiserVR: Semantic communication enabled wireless virtual reality delivery. *arXiv preprint arXiv:2211.01241*, 2022a.

Le Xia, Yao Sun, Dusit Niyato, Xiaoqian Li, and Muhammad Ali Imran. Wireless semantic communication: A networking perspective. *arXiv preprint arXiv:2212.14142*, 2022b.

Huiqiang Xie, Zhijin Qin, Geoffrey Ye Li, and Biing-Hwang Juang. Deep learning enabled semantic communication systems. *IEEE Transactions on Signal Processing*, 69:2663–2675, 2021.

Jialong Xu, Bo Ai, Wei Chen, Ang Yang, Peng Sun, and Miguel Rodrigues. Wireless image transmission using deep source channel coding with attention modules. *IEEE Transactions on Circuits and Systems for Video Technology*, 32(4):2315–2328, 2021.

Mingyu Yang and Hun-Seok Kim. Deep joint source-channel coding for wireless image transmission with adaptive rate control. In *ICASSP 2022-2022 IEEE International Conference on Acoustics, Speech and Signal Processing (ICASSP)*, pages 5193–5197. IEEE, 2022.

Ramin Zabih and John Woodfill. Non-parametric local transforms for computing visual correspondence. In *Computer Vision—ECCV'94: Third European Conference on Computer Vision Stockholm*, Sweden, May 2–6 1994 Proceedings, Volume II 3, pages 151–158. Springer, 1994.

7

Content-Aware Robust Semantic Transmission of Images over Wireless Channels with GANs

Xuyang Chen[1], Daquan Feng[1], Qi He[2], Yao Sun[3], and Xiang-Gen Xia[4]

[1] *College of Electronics and Information Engineering, Shenzhen University, Shenzhen, China*
[2] *National Key Laboratory of Science and Technology on Communications, University of Electronic Science and Technology of China, Chengdu, China*
[3] *James Watt School of Engineering, University of Glasgow, Glasgow, Lanarkshire, UK*
[4] *Department of Electrical and Computer Engineering, University of Delaware, Newark, DE, USA*

7.1 Introduction

SemCom is different from traditional communication because all the technologies use the semantics of transmitting data, rather than simple bitstreams, which can provide the same quality of service with lower data transmission requirements. Thanks to the rapid development of deep learning technology, SemCom has attracted a lot of attention in such fields as natural language processing (NLP), image transmission, and speech recognition, making SemCom promising in next-generation communications [Yang et al., 2023].

Existing SemCom works can be categorized into three types according to the type of source data. Regarding speech-based and text-based SemCom systems, a joint semantic-channel coding method called "DeepSC" was developed in Xie and Qin [2021], which encodes the input text into variable-length output symbols for complex channels. Moreover, the author of Xie et al. [2021] further proposed an environmentally friendly SemCom system called "L-DeepSC," for edge devices with limited computational and storage capabilities. In Wang et al. [2022], semantic information is extracted using a knowledge graph, where all texts are represented by a collection of triplets, and the authors utilized an attention network to obtain the importance of triplets for the user's resource allocation. A federated learning-based speech semantic transmission framework was proposed in Tong et al. [2021], using an automatic codec to recover the speech signal.

Apart from text-based and speech-based SemCom systems, there exist numerous works on image-based semantic transmission systems. A joint source–channel coding (JSCC) technique for image transmission using convolution neural networks was proposed in Bourtsoulatze et al. [2019]. Li et al. used a dual-path semantic encoder and decoder to enhance vessel details in retinal images [Li et al., 2023], helping doctors make better diagnoses. An attention deep learning-based JSCC (ADJSCC) was proposed in Xu et al. [2022], which can successfully transmit images with different signal-to-noise ratios (SNRs). Dong et al. proposed a semantic slice model (SeSM) [Dong et al., 2023] that can be flexibly adjusted under different model performances, channel conditions, and transmission targets. Gündüz et al. proposed a collaborative non-orthogonal multiple access (NOMA) and JSCC scheme [Lo et al., 2023] to improve the accuracy of multi-edge image retrieval.

Wireless Semantic Communications: Concepts, Principles, and Challenges, First Edition.
Edited by Yao Sun, Lan Zhang, Dusit Niyato, and Muhammad Ali Imran.

A progressive deep JSCC image transmission method was proposed in Kurka and Gündüz [2021], which considers the transmission of images in multiple parallel channels, and the receiver uses a hierarchical decoder to improve robustness against channel fading. Yang et al. designed an adaptive deep JSCC coding scheme [Yang and Kim, 2022], which can select parts of semantic features using a policy network based on channel conditions and image contents.

Meanwhile, video-based SemCom systems have also attracted a lot of attention. An end-to-end JSCC video transmission scheme was first proposed in Tung and Gündüz [2022], which allocates the optimal image coding rate for each group of pictures (GOP) to maximize the overall video quality. The authors of Wang et al. [2023] also proposed a JSCC video transmission scheme. Unlike the approach of Tung and Gündüz [2022], they allocate bandwidth to individual P-frames instead of the entire GOP and consider contextual information between P-frames, which can greatly reduce transmission bits. Regarding video conference scenarios in Galteri et al. [2020], more bits are allocated to the avatar area through a mask, and a generative adversarial network (GAN)-based method is utilized for compression and recovery to dramatically reduce bandwidth consumption. A semantic video conferencing (SVC) was proposed in Jiang et al. [2023], which maintains a high resolution by transmitting some key points to represent the motions. Two feedback mechanisms were designed: the hybrid automatic repeat request (HARQ) mechanism with acknowledgement (ACK) feedback can correct the error of key points information, and the channel state information (CSI) is utilized to learn to allocate more key points on the sub-channels with higher gain.

SemCom is task-oriented, semantic-aware, and content-aware, where only important information related to the task is extracted and transmitted. For image transmission, the image content that different tasks care about varies. For example, in monitoring and analysis scenarios, roads and street scenes generally do not change, and pedestrians and vehicles are more worthy of attention. In defect detection of components in the factory, the area where the components are located in the image is more important. Therefore, we can divide the image into two parts, regions of interest (ROI) and regions of non-interest (RONI), based on downstream tasks. During transmission, we preserve more image details in ROI and fewer image details in RONI, thereby saving a lot of bandwidth while maintaining downstream task performance.

Our model is conducted in Rayleigh channels. Experiments show almost no performance degradation when performing the downstream segmentation task on the receiver while saving up to 60.53% of bandwidth. In addition, our method can perform very well in different channel conditions. Compared to models trained for a certain SNR, our model significantly improves the multi-scale structural similarity index measure (MS-SSIM) and learned perceptual image patch similarity (LPIPS) in the [0, 10] dB testing environment.

The rest of this chapter is organized as follows. The proposed SemCom framework is described in Section 7.2. The system architecture and training details are introduced in Section 7.3. In Section 7.4, the experimental details and the performance are shown. Finally, conclusions are drawn in Section 7.5.

7.2 System Model

Figure 7.1 illustrates the proposed framework of the content-aware robust semantic transmission of images with GAN. The semantic encoder extracts the semantic information of the input image and attaches semantic labels to it. The quantizer selects the appropriate ROI based on downstream tasks and then adopts different quantization levels for ROI and RONI. The semantic decoder obtains a

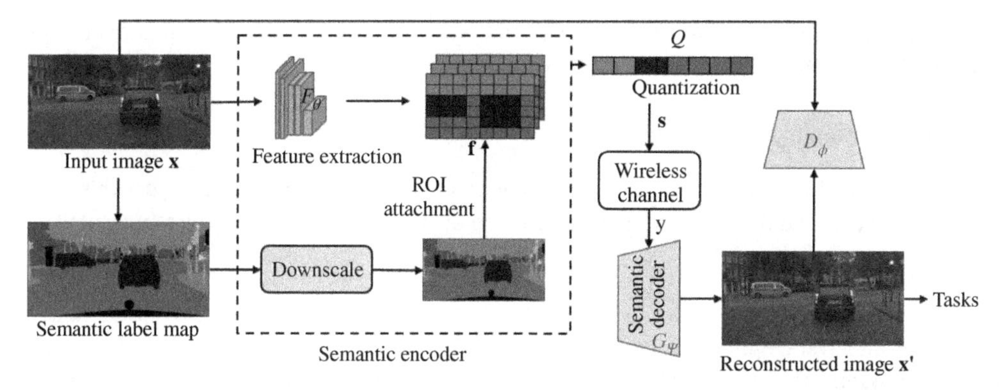

Figure 7.1 The proposed content-aware robust semantic transmission systems, G_ψ and D_ϕ, represent generator and discriminator modules respectively.

noisy semantic representation through wireless channels. Then, the original image is restored by the noisy semantic representations and performs downstream tasks.

7.2.1 Semantic Encoder

The semantic encoder is responsible for extracting feature maps with semantic descriptions. An input image is denoted as $\boldsymbol{x} \in \mathbb{R}^{3 \times H \times W}$, where "3" is the RGB channels, and H and W are the height and width of the image, respectively. The input image $\boldsymbol{x}$ is fed into F_θ to extract relevant feature representations.

$$\boldsymbol{f} = F_\theta(\boldsymbol{x}),\ \boldsymbol{f} \in \mathbb{R}^{C \times \frac{H}{n} \times \frac{W}{n}}. \tag{7.1}$$

$F_\theta(\cdot)$ is the semantic encoder network with the parameter set θ. C is the number of channels for $\boldsymbol{f}$, and n is the downsampling scale.

Feature maps will be annotated with semantic labels. Specifically, we downsample the semantic label map to the same size as $\boldsymbol{f}$ using nearest neighbor interpolation and attach different semantic labels to $\boldsymbol{f}$ to obtain the semantically annotated feature map. According to the downstream tasks, the feature map is divided into two parts: ROI and RONI. We choose humans, vehicles, traffic lights, and traffic signs on the road for ROI here. These instances are receiving more attention in the collaboration of the internet of vehicles.

7.2.2 Quantizer

The quantizer quantifies ROI and RONI at different levels to reduce the amount of transmission data without affecting the accuracy of downstream tasks. Q denotes a quantizer, and $\boldsymbol{s}$ denotes a quantized bit representation, which is calculated as $\boldsymbol{s} = Q(\boldsymbol{f})$. Each entry of $\boldsymbol{s}$ can be computed using the nearest neighbor assignment method. The process can be written as

$$\boldsymbol{s} = \arg\min_t \|\boldsymbol{f} - l_t\|, \tag{7.2}$$

where $l_t \in \{l_1, l_2, \dots, l_N\}$ is a quantization center. Note that $\boldsymbol{f}$ includes two parts: ROI and RONI, with different quantization levels, i.e., $N_{RONI} < N_{ROI}$. However, using the quantization method of (7.2) will result in the cutoff of gradients during the backpropagation process. To solve this problem, we conduct pretraining without the quantizer. This process aims to facilitate the semantic encoder in acquiring precise semantic feature representations. After that, we introduce the quantizer (7.2) and only update the semantic decoder and discriminator.

7.2.3 Wireless Channel

The semantic symbol $\boldsymbol{s}$ will be transmitted through wireless channels, which we consider the most widely used Rayleigh channels. The transfer function of the Rayleigh channel can be represented as

$$\boldsymbol{y} = \sqrt{P_s} h \boldsymbol{s} + \epsilon, \tag{7.3}$$

where P_s denotes the transmit power. $\epsilon \sim \mathcal{CN}\left(0, \sigma^2 \boldsymbol{I}\right)$ denotes complex Gaussian channel noise, with σ^2 being the average noise power and $\boldsymbol{I}$ being the identity matrix. $h \sim \mathcal{CN}(0,1)$ is a complex normal random variable sampled from a circularly symmetric complex Gaussian distribution. We set the noise power constraint $\sigma^2 = 1$ and vary the transmit power P_s to emulate the varying average SNR:

$$SNR = 10 \log_{10} \frac{P_s}{\sigma^2}. \tag{7.4}$$

7.2.4 Semantic Decoder

The semantic decoder restores corrupted semantic representations to the original image. Similar to the semantic encoder, the semantic decoder is composed of neural networks. GANs have shown great potential in generating high-quality images [Goodfellow et al., 2020]. GANs attempt to approximate the distribution of real images, rather than pursuing the minimum pixel-by-pixel difference of the image. This coincides with our pursuit of semantic transmission of images. Therefore, the semantic decoder is composed of a GAN, where the generator acts as the semantic decoder on the receiver. The decoder and the encoder share a semantic knowledge base, which facilitates a shared semantic understanding and inference capabilities between the transmitter and receiver. Let ψ be the parameter of the semantic decoder; then the decoded image $\boldsymbol{x}'$ can be represented as

$$\boldsymbol{x}' = G_\psi(\boldsymbol{y}), \tag{7.5}$$

where G_ψ indicates the semantic decoder parameterized with ψ. Generator G_ψ attempts to generate an image $\boldsymbol{x}'$ that is consistent with the distribution of the original image to deceive the discriminator D_ϕ. The discriminator is parameterized with $\boldsymbol{\phi}$. D_ϕ attempts to distinguish the recovered image $\boldsymbol{x}'$ from the original image $\boldsymbol{x}$, and in a zero-sum game, G_ψ and D_ϕ have trained alternatively until the min-max saddle point is found, which is described as

$$\begin{aligned} \min_{G_\psi} \max_{D_\phi} & \mathbb{E}[p(D_\phi(\boldsymbol{x}))] + \mathbb{E}[g(D_\phi(G_\psi(\boldsymbol{y})))] \\ & + \lambda \mathbb{E}[d(\boldsymbol{x}, G_\psi(\boldsymbol{y}))], \end{aligned} \tag{7.6}$$

where p and g are scalar functions, d denotes the mean square error (MSE) distortion term, and $\lambda > 0$ controls the balance between GAN loss and image distortion loss. We use the least squares GAN [Mao et al., 2017], and p and g are defined as

$$\begin{cases} p(z) = (z-1)^2, \\ g(z) = z^2, \end{cases} \tag{7.7}$$

which denotes the Pearson χ^2 divergence. It should be noted that (7.6) differs from traditional GAN. We hope to use representations with noise to restore the original image, rather than generating new images from random signals. Therefore, the image distortion loss term has been added to (7.6).

7.3 System Architecture

The network structure of our system is illustrated in Figure 7.2. The semantic encoder adopts a pretrained VGG-16 [Simonyan and Zisserman, 2015], which can be replaced by other excellent feature extractors, such as ResNet [He et al., 2016]. The dimension of $\boldsymbol{f}$ is $\mathbb{R}^{8\times\frac{H}{4}\times\frac{W}{4}}$, where the downsampling scale of an image is 4, and the output bottleneck is 8. All convolution blocks use instance normalization to make the distribution of the feature maps conform to the standard normal distribution, which means $z' = (z-\mu)/\sqrt{\delta^2+\varepsilon}$, where μ and δ are the means and the standard deviation of features, respectively. The quantizer quantifies the ROI and RONI of the feature maps with different levels.

We use multiple cascading residual blocks at the receiver to extract depth semantic information [He et al., 2022]. Then, the feature maps are upsampled to the size of $\boldsymbol{x}$ by using the transposed convolution, and finally, the restored image $\boldsymbol{x}'$ is obtained through the *tanh* activation function. We use a multi-scale discriminator [Wang et al., 2018] to measure the difference between $\boldsymbol{x}'$ and $\boldsymbol{x}$. A multi-scale discriminator can effectively evaluate global coarse similarity to fine local details, thus guiding G to generate images with better global similarity and local fine details. $\boldsymbol{x}$ and $\boldsymbol{x}'$ are respectively input into D with the same structure. Each discriminator is cascaded with four convolution blocks and finally outputs a one-channel small matrix. Compare it with the all-zero or all-one matrix to calculate discriminator loss. The final discriminator loss is the average of the discriminator losses across three scales.

7.4 Experimental Results

7.4.1 Datasets and Settings

Our model is trained with the Cityscapes [Cordts et al., 2016], which comprises a total of 5,000 images, collected from streetscapes in 50 different German cities, and consists of 30 classes of

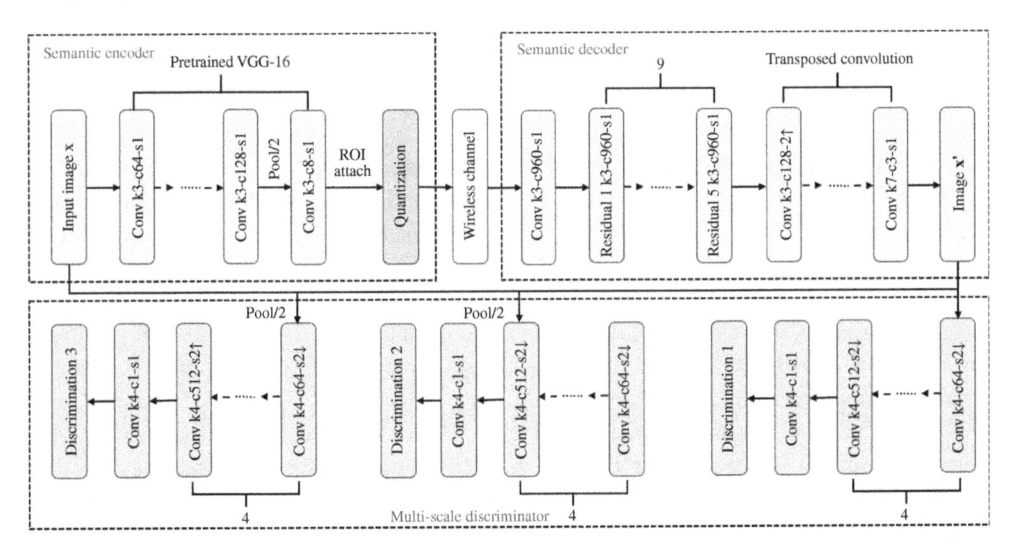

Figure 7.2 The network architecture of our proposed system. It is denoted that "k4-c64-s2↓" represents a convolution layer with 64 filters of size 4 × 4 and a downsampling stride of 2. Each residual block or convolution block is followed by an instance normalization and a ReLU.

objects. The dataset includes 2,975 training images, 1,225 testing images, and 500 validation images. Humans, vehicles, traffic lights, and traffic signs are considered as ROI. Our quantization center consists of eight discrete integer values: $\{0, 1, 2, 3, 4, 5, 6, 7\}$; thus each pixel can be represented by exactly 3 bits. The original image size of Cityscapes is 2048×1024. Following the settings in Huang et al. [2023], we downscale the size to 512×256, and we use the semantic label map provided by Cityscapes. The Adam solver is adopted here, where the momentum term is set to be $\beta_1 = 0.5$ and $\beta_2 = 0.999$. The learning rate of the generator and discriminator is initialized to 2×10^{-4}, decaying exponentially at a rate of 0.5. The hyperparameter λ in (7.6) is set to 150. It takes 50 epochs to complete the entire training.

7.4.2 Performance Metrics

We adopt a new performance metric, $\theta PSNR$, proposed in Wu et al. [2023]. $\theta PSNR$ is more sensitive to pixel loss in the ROI and can better reflect the performance of the algorithm, which is defined as

$$\theta PSNR = 10 \lg \frac{MAX^2}{MSE_{ROI} \cdot \theta + MSE_{RONI} \cdot (1 - \theta)}, \tag{7.8}$$

where MAX is the maximum possible pixel value of the image. MSE_{ROI} is the MSE of the ROI, and MSE_{RONI} is the MSE of the RONI. θ is set to 0.8. We also use MS-SSIM and LPIPS to measure the multi-scale structural similarity and perceptual loss between two images. For downstream segmentation tasks, we use intersection over union (IoU), which measures the ratio of the intersection area and the union area to evaluate the ROI segmentation accuracy, which is defined as

$$IoU\left(H_{seg}(\boldsymbol{x}), H_{seg}(\boldsymbol{x}')\right) = \frac{1}{M} \sum_{m=1}^{M} \frac{overlap\left(\boldsymbol{B}_g^m, \boldsymbol{B}_d^m\right)}{union\left(\boldsymbol{B}_g^m, \boldsymbol{B}_d^m\right)}, \tag{7.9}$$

where H_{seg} represents the pretrained semantic segmentation network PSPNet [Zhao et al., 2017]. $\boldsymbol{B}_g^m$ is the ground truth bounding set on $\boldsymbol{x}$ of class m, while $\boldsymbol{B}_d^m$ is the bounding set on $\boldsymbol{x}'$ of class m. mean-IoU is the average ROI segmentation accuracy over the total testing dataset and is a value between 0 and 1.

7.4.3 Results Analysis

A visual comparison between images with different quantization levels is shown in Figure 7.3. Our method exhibits superior recovery details when compared to the traditional JPEG and Polar coding methods while utilizing only 1/16 of the transmission bandwidth required by the traditional method. If we reduce the quantization levels of RONI, as shown in Figure 7.3g, RONI becomes blurry, but ROI remains the restoration quality. In downstream tasks such as object detection and semantic segmentation, the clarity of the RONI is not very important. Therefore, by using different quantization levels for the ROI and RONI, we can save a lot of bandwidth without losing the performance of downstream tasks.

As shown in Figure 7.4, we conduct training and testing under 10, 5, and 0 dB. The JPEG and Polar coding methods suffer from severe "cliff effects," making it almost impossible to decode correctly under 5 and 0 dB and unable to perform downstream tasks. Our method performs much better while saving a significant amount of bandwidth. For ROI, we uniformly use 3 bits for quantization, and for RONI, we use 1, 2, and 3 bits for quantization, respectively. When RONI uses 3-bit quantization, the system achieves optimal performance. As the quantization bits of RONI decrease, there is a slight decrease in image reconstruction quality. However, due to the performance loss mainly

Figure 7.3 The visual comparison between images with different quantization levels over Rayleigh channels. Cars are considered as ROI, while roadside houses and trees are regarded as RONI. BPP means bits per pixel. (c) and (d) represent JPEG and Polar coding and the ROI segmentation results. In (e), ROI and RONI are quantized with 3 bits. In (g), ROI and RONI are quantized with 3 bits and 1 bit, respectively. (a) Input image, (b) semantic label, (c) JPEG and Polar coding (24 bpp), (d) ROI segmentation result of (c), (e) The proposed (1.5 bpp), (f) ROI segmentation result of (e), (g) the proposed (0.59 bpp), (h) ROI segmentation result of (g), (i) ROI comparison, (j) ROI comparison, (k) RONI comparison, and (l) RONI comparison.

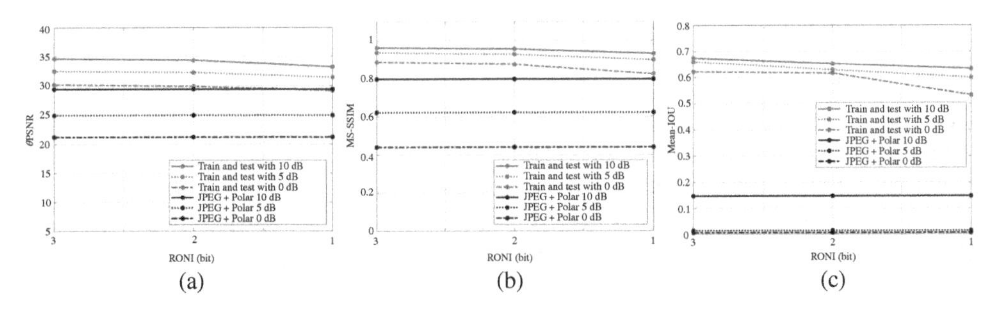

Figure 7.4 Performance under different quantization bits of RONI over Rayleigh channels. ROI is quantized with 3 bits, and the mean IoU of ROI is calculated by PSPNet [Zhao et al., 2017]. (a) θPSNR under different quantization levels. (b) MS-SSIM under different quantization levels. (c) Mean-IoU of ROI under different quantization levels.

coming from the RONI region, our downstream tasks still perform well, with almost no decrease in mean-IoU of ROI.

We find that when there is a difference in noise intensity between testing and training, the system performance will experience substantial attenuation. For example, we use 10 dB noise for training and 0 dB noise for testing, and θPSNR and MS-SSIM decrease by 23.1% and 22.9%, respectively. This is due to the significant difference in distribution between training and testing noise, resulting in a considerable decrease in the effectiveness of GAN. To solve this problem, we randomly initialize the SNRs from 0 to 10 dB. We conduct tests in the Rayleigh channel with SNRs ranging from 0 to 10 dB and find that both θPSNR and MS-SSIM are more robust. Figure 7.5 shows that by the method of randomly initializing the SNRs in training, compared to training with only a single SNR,

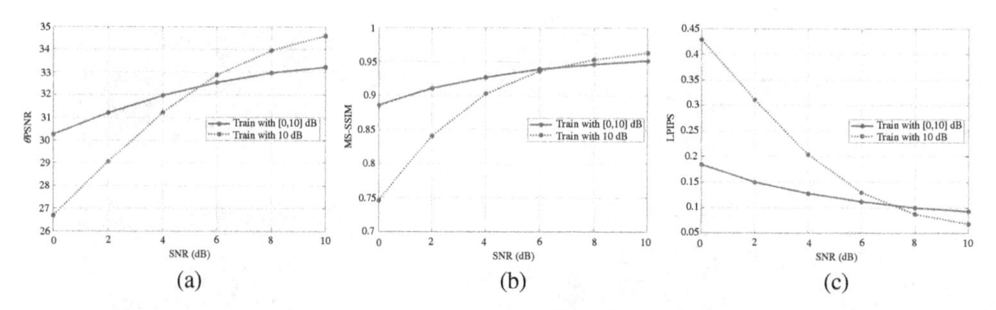

Figure 7.5 Robustness performance under Rayleigh channels. [0, 10] dB represents SNR randomly sampled from 0 to 10 dB. (a) θPSNR under Rayleigh channel. (b) MS-SSIM under Rayleigh channel. (c) LPIPS under Rayleigh channel.

MS-SSIM and θPSNR increase by 3.2% and 4.5% respectively, and LPIPS decreases by 35%. This is because the GAN has learned the comprehensive distribution across SNRs ranging from 0 to 10 dB, allowing it to achieve effective image restoration within a designated noise range.

7.5 Conclusion

This chapter proposed a novel, robust semantic transmission method for images based on GANs. Our method can extract deep semantic features from images and distinguish between ROI and RONI according to downstream tasks. ROI and RONI are quantized at different levels, aiming to save a large number of bits without affecting the performance of downstream tasks. Furthermore, to improve the robustness of the model under different SNRs, we proposed a training method with SNR randomly sampling from a certain range, which improves the overall performance of the model. We adopted Rayleigh fading channels and trained our model using random SNRs. The experimental results showed that our model can save up to 60.53% bandwidth compared with baselines while achieving almost identical performance in the downstream semantic segmentation task. Moreover, our model has better robustness, with testing SNR varying from 0 to 10 dB, obtaining the improvement of MS-SSIM, θPSNR, and LPIPS.

References

E. Bourtsoulatze, D. Burth Kurka, and D. Gündüz. Deep joint source-channel coding for wireless image transmission. *IEEE Transactions on Cognitive Communications and Networking*, 5(3):567–579, 2019. ISSN 2332-7731.

Marius Cordts, Mohamed Omran, Sebastian Ramos, Timo Rehfeld, Markus Enzweiler, Rodrigo Benenson, Uwe Franke, Stefan Roth, and Bernt Schiele. The cityscapes dataset for semantic urban scene understanding. In *2016 IEEE Conference on Computer Vision and Pattern Recognition (CVPR)*, pages 3213–3223, Las Vegas, NV, USA, 2016.

C. Dong, H. Liang, X. Xu, S. Han, B. Wang, and P. Zhang. Semantic communication system based on semantic slice models propagation. *IEEE Journal on Selected Areas in Communications*, 41(1):202–213, 2023. ISSN 1558-0008.

Leonardo Galteri, Marco Bertini, Lorenzo Seidenari, Tiberio Uricchio, and Alberto Del Bimbo. Increasing video perceptual quality with GANs and semantic coding. In *Proceedings of the 28th ACM*

International Conference on Multimedia, page 862–870, Seattle, WA, USA, 2020. Association for Computing Machinery. ISBN 9781450379885.

Ian Goodfellow, Jean Pouget-Abadie, Mehdi Mirza, Bing Xu, David Warde-Farley, Sherjil Ozair, Aaron Courville, and Yoshua Bengio. Generative adversarial networks. *Communications of the ACM*, 63(11):139–144, Oct 2020. ISSN 0001-0782.

Kaiming He, Xiangyu Zhang, Shaoqing Ren, and Jian Sun. Deep residual learning for image recognition. In *2016 IEEE Conference on Computer Vision and Pattern Recognition (CVPR)*, pages 770–778, Las Vegas, NV, USA, 2016.

Qi He, Haohan Yuan, Daquan Feng, Bo Che, Zhi Chen, and Xiang-Gen Xia. Robust semantic transmission of images with generative adversarial networks. In *2022 IEEE Global Communications Conference*, pages 3953–3958, Rio de Janeiro, Brazil, 2022.

D. Huang, F. Gao, X. Tao, Q. Du, and J. Lu. Toward semantic communications: Deep learning-based image semantic coding. *IEEE Journal on Selected Areas in Communications*, 41(1):55–71, 2023. ISSN 1558-0008.

P. Jiang, C. K. Wen, S. Jin, and G. Y. Li. Wireless semantic communications for video conferencing. *IEEE Journal on Selected Areas in Communications*, 41(1):230–244, 2023. ISSN 1558-0008.

D. B. Kurka and D. Gündüz. Bandwidth-agile image transmission with deep joint source-channel coding. *IEEE Transactions on Wireless Communications*, 20(12):8081–8095, 2021. ISSN 1558-2248.

Aini Li, Xiaohong Liu, Guangyu Wang, and Ping Zhang. Domain knowledge driven semantic communication for image transmission over wireless channels. *IEEE Wireless Communications Letters*, 12(1):55–59, 2023.

W. F. Lo, N. Mital, H. Wu, and D. Gündüz. Collaborative semantic communication for edge inference. *IEEE Wireless Communications Letters*, 12(7):1125–1129, 2023. ISSN 2162-2345.

Xudong Mao, Qing Li, Haoran Xie, Raymond Y. K. Lau, Zhen Wang, and Stephen Paul Smolley. Least squares generative adversarial networks. In *2017 IEEE International Conference on Computer Vision (ICCV)*, pages 2813–2821, Venice, Italy, 2017.

Karen Simonyan and Andrew Zisserman. Very deep convolutional networks for large-scale image recognition. In *3rd International Conference on Learning Representations, ICLR 2015*, San Diego, CA, USA, May 7–9, 2015, Conference Track Proceedings, 2015.

H. Tong, Z. Yang, S. Wang, Y. Hu, W. Saad, and C. Yin. Federated learning based audio semantic communication over wireless networks. In *2021 IEEE Global Communications Conference (GLOBECOM)*, pages 1–6, Madrid, Spain, 2021.

T. Y. Tung and D. Gündüz. DeepWiVe: Deep-learning-aided wireless video transmission. *IEEE Journal on Selected Areas in Communications*, 40(9):2570–2583, 2022. ISSN 1558-0008.

Ting-Chun Wang, Ming-Yu Liu, Jun-Yan Zhu, Andrew Tao, Jan Kautz, and Bryan Catanzaro. High-resolution image synthesis and semantic manipulation with conditional GANs. In *2018 IEEE/CVF Conference on Computer Vision and Pattern Recognition*, pages 8798–8807, Salt Lake City, UT, USA, 2018.

Y. N. Wang, M. Z. Chen, T. Luo, W. Saad, D. Niyato, H. V. Poor, and S. G. Cui. Performance optimization for semantic communications: An attention-based reinforcement learning approach. *IEEE Journal on Selected Areas in Communications*, 40(9):2598–2613, 2022. ISSN 0733-8716.

S. Wang, J. Dai, Z. Liang, K. Niu, Z. Si, C. Dong, X. Qin, and P. Zhang. Wireless deep video semantic transmission. *IEEE Journal on Selected Areas in Communications*, 41(1):214–229, 2023. ISSN 1558-0008.

Jiale Wu, Celimuge Wu, Yangfei Lin, Tsutomu Yoshinaga, Lei Zhong, Xianfu Chen, and Yusheng Ji. Semantic segmentation-based semantic communication system for image transmission. In *Digital Communications and Networks*, 2023. ISSN 2352-8648.

H. Xie and Z. Qin. A lite distributed semantic communication system for Internet of Things. *IEEE Journal on Selected Areas in Communications*, 39:142–153, 2021. ISSN 1558-0008.

H. Q. Xie, Z. J. Qin, G. Y. Li, and B. H. Juang. Deep learning enabled semantic communication systems. *IEEE Transactions on Signal Processing*, 69:2663–2675, 2021. ISSN 1053-587X.

J. Xu, B. Ai, W. Chen, A. Yang, P. Sun, and M. Rodrigues. Wireless image transmission using deep source channel coding with attention modules. *IEEE Transactions on Circuits and Systems for Video Technology*, 32(4):2315–2328, 2022. ISSN 1558-2205.

M. Yang and H. S. Kim. Deep joint source-channel coding for wireless image transmission with adaptive rate control. In *ICASSP 2022 - 2022 IEEE International Conference on Acoustics, Speech and Signal Processing (ICASSP)*, pages 5193–5197, Singapore, 2022. ISBN 2379-190X.

W. Yang, H. Du, Z. Q. Liew, W. Y. B. Lim, Z. Xiong, D. Niyato, X. Chi, X. Shen, and C. Miao. Semantic communications for future internet: Fundamentals, applications, and challenges. *IEEE Communications Surveys & Tutorials*, 25(1):213–250, 2023. ISSN 1553-877X.

Hengshuang Zhao, Jianping Shi, Xiaojuan Qi, Xiaogang Wang, and Jiaya Jia. Pyramid scene parsing network. In *2017 IEEE Conference on Computer Vision and Pattern Recognition (CVPR)*, pages 2881–2890, Honolulu, HI, USA, 2017.

8

Semantic Communication in the Metaverse

Yijing Lin[1]*, Zhipeng Gao*[1]*, Hongyang Du*[2]*, Jiacheng Wang*[3]*, and Jiakang Zheng*[4]

[1] *State Key Laboratory of Networking and Switching Technology, Beijing University of Posts and Telecommunications, Beijing, China*
[2] *Department of Electrical and Electronic Engineering, The University of Hong Kong, Hong Kong*
[3] *College of Computing and Data Science, Nanyang Technological University, Singapore*
[4] *School of Electronic and Information Engineering, Beijing Jiaotong University, Beijing, China*

8.1 Introduction

Metaverse was first elaborated in a novel *Snow Crash*, where it described a virtual world in which physical users are represented by digital avatars, allowing them to interact with one another [Stephenson, 2003]. To realize this virtual world, a combination of technologies such as augmented reality (AR), virtual reality (VR), artificial intelligence (AI), and blockchain is employed [Lin et al., 2023b]. These technologies aim to ensure an immersive experience and decentralized trust for participants. Physical users capture their surrounding environments using edge devices powered by AR or VR and then transmit this data to metaverse service providers (MSPs). Upon receiving the data, MSPs process it through AI models to generate customized content. This content is then fed back to the users, ensuring synchronization between the physical and virtual worlds [Huang et al., 2022].

Synchronization in the metaverse necessitates a significant communication overhead, especially when transmitting data from physical users to MSPs for creation of customized content. For example, sensing devices can produce as much as 3.072 megabytes of data every second [Wang et al., 2023]. Unfortunately, the capacities of the traditional communication paradigm are insufficient to handle such vast data streams, which cannot provide a seamless user experience. Semantic communication (SemCom) represents a novel communication paradigm. Instead of transmitting raw data directly, SemCom extracts and sends only meaningful information, e.g., semantic information, from the data collected by the transmitters to the receivers, thereby reducing communication overhead [Yang et al., 2022]. The semantic information can take various forms, such as part of raw data, tensors of a neural network, labels of an image, or prompts indicating user preferences. Within the metaverse, SemCom can extract semantic information from images, videos, and audio captured by AR/VR devices used by physical users. This semantic information is then transmitted to MSPs, effectively reducing communication overhead. Upon receiving semantic information about physical users, MSPs input it into AI models for training, subsequently generating customized content for users, such as digital avatars and unique stickers. This content can be used for creating immersive virtual scenarios exemplified by digital twins and virtual transportation, ensuring seamless synchronization between the real and virtual worlds.

Wireless Semantic Communications: Concepts, Principles, and Challenges, First Edition.
Edited by Yao Sun, Lan Zhang, Dusit Niyato, and Muhammad Ali Imran.

However, the integration of SemCom and the metaverse has not been extensively researched. There are several challenges to address. First, even though transmitting semantic information can significantly reduce communication overhead compared to raw data, verifying the authenticity of this semantic information is challenging. Second, it is necessary to consolidate and coordinate the allocation of communication resources. Given that synchronization operates as a disjointed structure between physical users and MSPs, it is challenging for individual modules to fully utilize communication resources and maximize utility for participants. Specially, the following issues for SemCom in metaverse need to be addressed:

- Before transmission, how to verify the authenticity of the semantic information intended for transmission?
- During transmission, what strategies can be employed to optimize the utilization of communication resources, especially during the synchronization process between physical users and MSPs?

To solve the aforementioned challenges, this chapter focuses on SemCom in the metaverse and establishes a unified framework to manage interactions between the physical and virtual worlds. We also adopt zero-knowledge proof (ZKP) [Song et al., 2022] to authenticate the validity of semantic information. The chapter then introduces a diffusion model [Wang et al., 2022b]-based resource allocation method, aiming to formulate a near-optimal strategy for maximizing utility. The primary contributions of this chapter are outlined as follows:

- We introduce a unified framework for SemCom in the metaverse. This framework encompasses modules for semantic extraction, AI-generated content (AIGC), and rendering, ensuring smooth interactions between the physical and virtual worlds.
- Within this framework, we introduce a verification mechanism based on ZKP. This mechanism employs cryptographic methods to track the transformations of semantic information, ensuring authenticity before transmission.
- We design a resource allocation mechanism based on diffusion models. By leveraging the add-and-remove noise process inherent to the diffusion model, the framework learns the distribution of available communication resources, enabling the formulation of near-optimal allocation strategies across multiple modules.

The rest of this chapter is illustrated as follows: Section 8.2 describes the related works of SemCom in metaverse. Section 8.3 designs a unified framework that integrates SemCom and metaverse. Section 8.4 illustrates a ZKP-based semantic verification mechanism to verify the authenticity of semantic information within metaverse. Section 8.5 constructs a diffusion model-based resource allocation mechanism to achieve dynamic strategies for SemCom in metaverse. Section 8.6 demonstrates the simulation results to show the performance of the proposed mechanisms. Section 8.7 points out the possible challenges and future directions for SemCom in metaverse. Section 8.8 concludes this chapter.

8.2 Related Work

In this section, we first review studies on SemCom in metaverse and then discuss previous works on the authenticity of semantic information and efficiency of SemCom in metaverse.

8.2.1 Semantic Communication in Metaverse

Semantic communication [Yang et al., 2022] offers a way to alleviate the communication load on the interactions between real and virtual networks. It processes raw data using AI technologies in edge devices and then forwards only the relevant semantic information to receivers. Du et al. [2023c] proposed reconfigurable intelligent surface-based encoding and self-supervised decoding methods to extract task-relevant hyper-source messages from raw data to reduce complexity. Weng and Qin [2021] employed deep learning to emphasize crucial speech data with higher weights, particularly in fluctuating channel conditions. To enable edge devices to carry out deep learning-based semantic tasks, Xie and Qin [2020] introduced a streamlined SemCom framework. This framework focuses on transmitting less computationally intensive text by optimizing the training procedures. Liew et al. [2023] introduced an auction-based data trading incentive mechanism that leverages SemCom to select relevant data. The mechanism takes into account factors such as data relatedness and freshness to incentivize users to update their data more frequently when communicating with MSPs. Ismail et al. [2022] employed SemCom to minimize the data transmission size and introduce a truthful reverse auction mechanism. This mechanism is designed to facilitate users in selling their semantic information to MSPs, thereby enhancing the overall quality of the metaverse. Semantic communication can significantly reduce the amount of data that needs to be transmitted. In the metaverse, with potentially billions of interactions per second, this efficiency is crucial for scalability.

Despite the advantages of SemCom in reducing data redundancy in the metaverse, it falls short of establishing trust between unidentified edge devices and MSPs. Lin et al. [2023b] introduced a comprehensive blockchain and semantic framework aimed at enabling Web 3.0 services to carry out both on-chain and off-chain activities. They included a proof-of-semantic mechanism to authenticate semantic information. Lin et al. [2023d] presented a blockchain-based framework for semantic data exchange, which allows the minting of non-fungible tokens (NFTs) for semantic information. While they incorporated game theory and ZKPs for more efficient and privacy-preserving exchanges, they did not account for the potential impact of semantic attacks on exchange efficiency.

However, the integration of SemCom in the metaverse is riddled with challenges that demand focused research and solutions. Primary among these is the issue of authenticating the semantic information that is being transmitted. As semantic information is often processed and abstracted form of raw data, verifying its accuracy and reliability becomes a critical concern. Another challenge lies in the efficient allocation and management of communication resources [Zheng et al., 2022]. Given that the metaverse is a complex ecosystem involving multiple actors, such as physical users and MSPs, the process of synchronizing these disparate elements can be cumbersome.

8.2.2 Authenticity of Semantic Information in Metaverse

The authenticity of semantic information can have a significant impact on the virtual world. For instance, to achieve synchronization with real-world conditions, the virtual world may rely on traffic information from physical users. However, malicious users could transmit irrelevant data, such as landscape images, to the virtual world. It serves no purpose for synchronization and could undermine the system's efficiency. Therefore, some studies focus on semantic attack and defense schemes to protect the authenticity of semantic information. Hu et al. [2022] proposed a weight perturbation-based adversarial training scheme that filters out noise-related and task-unrelated features to prevent semantic noise from causing semantic information to convey misleading meanings. Luo et al. [2022a] designed a symmetric encryption-based adversarial training scheme to encrypt and decrypt semantic information to preserve privacy when sharing background knowledge between the semantic transmitters and receivers. Du et al. [2023a] proposed a semantic attack-based adversarial training scheme to produce adversarial semantic information that minimizes semantic similarities while displaying different visual characteristics. Luo et al. [2023] integrated symmetric encryption with SemCom to protect privacy, employing both a semantic encryptor and decryptor. The paper also introduced an adversarial encryption training scheme to ensure the accuracy of semantic information in both encrypted and unencrypted states. Hu et al. [2023] proposed an adversarial training scheme that incorporates weight perturbation to mitigate semantic noise. The paper also designed a masked vector quantized-variational autoencoder to enhance the robustness of SemCom. Zhou et al. [2021] employed a universal transformer as opposed to a fixed transformer, along with an adaptive circulation mechanism within the universal transformer. This approach was used to establish SemCom through joint semantic source and channel coding, allowing for the flexible transmission of semantic information. Luo et al. [2022b] proposed an anti-noise autoencoder to serve as the channel encoder and decoder, designed to extract sentences from the semantic dimension. Additionally, the paper introduced a semantic forward mode as a relay mechanism for transmitting semantic information. This approach aimed to address the issue of a lack of common knowledge between transmitters and receivers.

However, there are few studies that focus on preventing the aforementioned types of attacks. In this chapter, we use ZKP to record the transformations from raw data to semantic information and detect adversarial semantic information, ensuring its authenticity. ZKP is a technique that has been widely adopted in AI and blockchain technologies for similar authentication objectives. Fan et al. [2023] proposed a ZKP-based verification mechanism that allows for the verification of model predictions without exposing the private parameters of models. Song et al. [2022] utilized ZKP to record transformation and achieve traceable data exchange for NFTs transactions. Lin et al. [2023d] utilized ZKP to facilitate the NFTs circulation. With ZKP, the users could initially share only the semantic information. Once payment is received, the raw data are then revealed. This approach allowed users to verify the authenticity of the semantic information without disclosing the raw data prior to payment. Wan et al. [2022] designed a ZKP-based oracle scheme that can securely transfer off-chain sensitive data to smart contracts, thereby enhancing privacy. Due to the transparency of blockchain, Galal and Youssef [2022] integrated ZKP into the implementation logic of smart contracts to protect the details of NFTs and transact them to ensure fairness.

8.2.3 Efficiency of SemCom in Metaverse

The integration of SemCom is a critical factor for the efficient operation of the metaverse. Consider scenarios where various types of data, such as environmental sensory data, user-generated content, and AIGC, need to be communicated simultaneously. Inefficient allocation and management of resources in these cases could lead to bottlenecks, reducing the overall effectiveness of the virtual environment. Wang et al. [2023] proposed a semantic transmission framework for metaverse that uses sensing data processed through semantic encoding. This framework significantly reduces communication overhead by converting the data into semantic information. Additionally, the paper employs contest theory to incentivize users to contribute data more frequently, thereby improving the service quality of the metaverse. Zhang et al. [2023a] introduced a framework based on SemCom for the extended reality that leverages semantic encoders to compress unimportant data, thereby minimizing communication overhead. The study further proposed a unique variable-length coding method for semantic channels and a rate allocation network to optimize code length, aiming to enhance transmission efficiency. Xu et al. [2023b] introduced the first mathematical model for SemCom, which could frame the goal-oriented source entropy as an optimization problem. This model leveraged the interplay between distributed edge learning techniques and advanced communication optimization designs for the fifth-generation technologies. Lin et al. [2023b] integrated blockchain with semantic information to facilitate both on-chain and off-chain operations for Web 3.0, one of the scenarios of metaverse, thereby achieving semantic verification to ensure the authenticity of the semantic information. Wang et al. [2022a] employed denoising diffusion probabilistic models for semantic image synthesis. The approach incorporated a noisy image into the encoder and applied a semantic layout to the decoder through the use of multilayer spatially adaptive normalization operators. Additionally, a classifier-free guidance sampling strategy was introduced to enhance both the quality of generated images and their semantic interoperability. Du et al. [2023b] introduced a SemCom-based information-sharing system tailored for mixed reality. This system not only extracted semantic details from visual images but also transmitted AIGC to enrich the user experience. Furthermore, the paper outlined an incentive mechanism built on contract theory to promote frequent information sharing among users. Han et al. [2023] proposed a SemCom approach based on generative models, as opposed to traditional deep learning-based methods to enhance the efficiency of image transmission. In this approach, the transmitter utilized a generative adversarial networks (GAN) inversion-based generative model to extract an interpretable latent representation from the raw data. Zhang et al. [2022b] proposed a task-unaware SemCom system that employs a data adaptation network for transfer learning. This system converted data into a form that resembles empirical data, eliminating the need for shared background knowledge between the transmitter and receiver. Lokumarambage et al. [2023] employed a pretrained GAN network along with a common knowledge base at the receiver end. This setup was used to reconstruct realistic images from semantic information, such as segmented images, that had been extracted from the original images. Zhang et al. [2023b] employed deep deterministic policy gradients to address the wireless resource allocation model as a joint optimization problem within a task-oriented SemCom network. The model took into account factors such as semantic compression ratio, transmit power, and bandwidth, aiming to maximize transmission efficiency. Wang et al. [2021] introduced a metric for evaluating the semantic accuracy and completeness of recovered semantic information. In addition, the paper formulated an optimization problem aimed at identifying which portions of semantic information should be transmitted. To solve this problem, the study employed a policy gradient-based reinforcement learning approach. Su et al. [2023] proposed a resource allocation mechanism that takes into account user satisfaction, queue stability, and communication delay. The paper employed Bernstein approximation and Lyapunov optimization to establish deterministic constraints. It then uses successive convex approximation and Karush–Kuhn–Tucker conditions to solve the optimization problem effectively.

8.3 Unified Framework for SemCom in the Metaverse

In this section, we initially introduce a unified framework for SemCom in the metaverse. This framework incorporates modules for semantic extraction, AIGC inference, and rendering, enabling efficient interactions between physical users and MSPs. Following this, we delve into the rationale behind the integration of these components. We highlight the existing research gaps in the field of SemCom within the metaverse context.

8.3.1 Framework Overview

Figure 8.1 presents an integrated framework for bridging SemCom within the metaverse, organized into three distinct layers. These layers correspond to three key modules: the semantic module, the AIGC module, and the rendering module. The structure aims to encapsulate the advantages of integrating SemCom and the metaverse, offering a comprehensive view of their interplay.

8.3.1.1 Semantic Module

To reduce communication overhead, the semantic module is carried out on the edge devices of physical users. For example, physical users can take images via edge devices like sensors and cameras to capture conditions of physical environments and transmit semantic information to

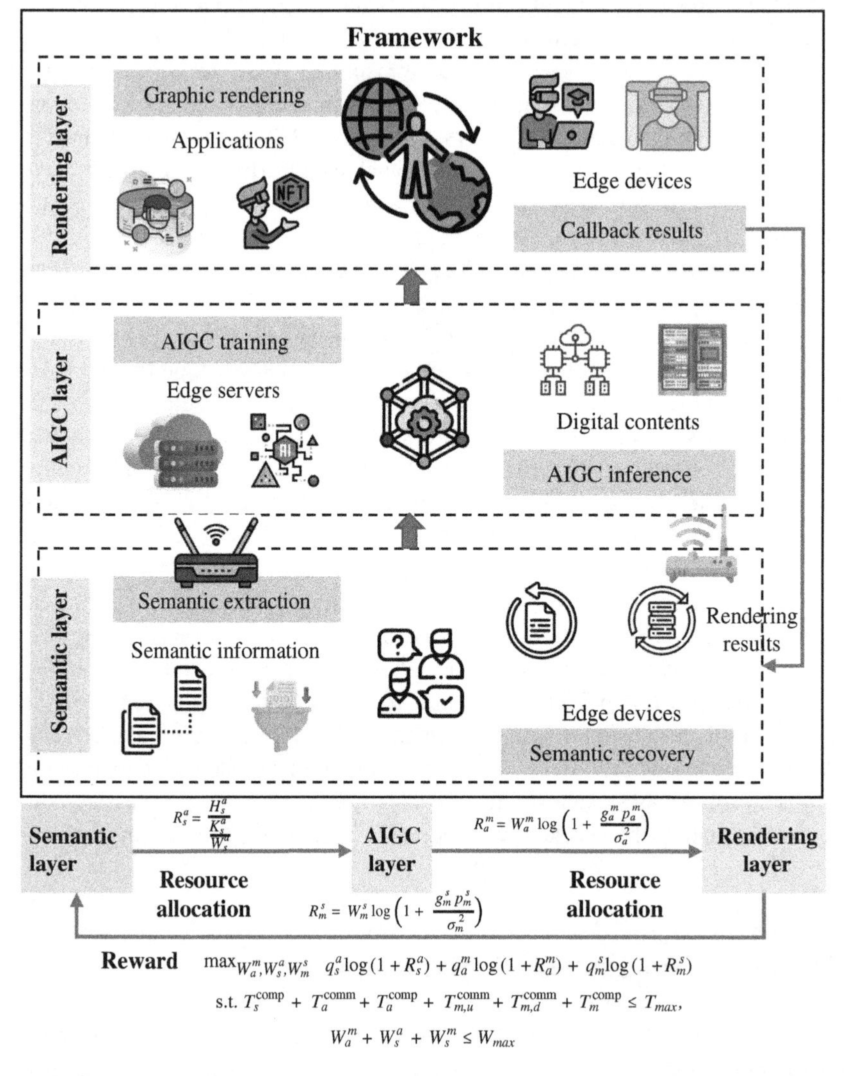

$$R_s^a = \frac{H_s^a}{\frac{K_s^a}{W_s^m}}$$

$$R_a^m = W_a^m \log\left(1 + \frac{g_a^m p_a^m}{\sigma_a^2}\right)$$

$$R_m^s = W_m^s \log\left(1 + \frac{g_m^s p_m^s}{\sigma_m^2}\right)$$

$$\max_{W_a^m, W_s^a, W_m^s} \; q_s^a \log(1 + R_s^a) + q_a^m \log(1 + R_a^m) + q_m^s \log(1 + R_m^s)$$

$$\text{s.t. } T_s^{\text{comp}} + T_a^{\text{comm}} + T_a^{\text{comp}} + T_{m,u}^{\text{comm}} + T_{m,d}^{\text{comm}} + T_m^{\text{comp}} \le T_{max},$$

$$W_a^m + W_s^a + W_s^m \le W_{max}$$

Figure 8.1 A unified framework bridging meaning in the metaverse. Source: [Lin et al., 2023c]/ arXivLabs/CC-BY 4.0.

MSPs instead of raw images. This allows for the transmission of meaningful information rather than raw bits, making it more efficient given the limited communication resources for frequent interactions between the real and virtual worlds. When physical users aim to send data to MSPs via edge devices, the data are first channeled through the semantic module. The semantic module employs a semantic encoder (a type of neural network) to extract meaningful content or knowledge from the raw data, which are depicted as images in this chapter. This process serves to decrease the data size for transmission while maintaining the desired meanings. Once the semantic information reaches the MSPs, they utilize both the AIGC and rendering modules to construct a customized virtual environment.

8.3.1.2 AIGC Module

The transmission of semantic information is susceptible to channel disturbances and potential inaccuracies from the semantic encoder, which could result in low-quality or incomplete data reaching the MSPs. To rectify this, the AIGC module is strategically implemented. It acts as a sophisticated layer of data refinement, generating high-quality, customized content based on the partially compromised or incomplete semantic information received. Within the scope of this chapter, the AIGC module leverages a pretrained diffusion model for this purpose. The model employs a forward and reverse diffusion process, adding and removing noise from the received semantic information to reconstruct it. This process ensures the generation of tailored, high-quality content that significantly enhances the overall user experience in the metaverse.

8.3.1.3 Rendering Module

With the support of both the semantic and AIGC modules, a high level of quality and customization is achieved in transmitting meaningful information from physical users to MSPs. This rich, contextual data serves as the basis for rendering intricate and immersive virtual environments. Beyond merely presenting a static or monolithic world, the rendering module leverages the semantic details and AIGC inputs to dynamically adapt the virtual space according to user interactions and real-world conditions. The rendering result is fed back to the physical users, not only enhancing their sensory experiences but also enabling more authentic and interactive engagements within the metaverse.

8.3.2 SemCom: Bridging Meaning in the Metaverse

Figure 8.2 showcases a range of research activities gathered from IEEE Xplore and arXiv as of April 2023. The illustration makes it evident that each module within the proposed unified framework has garnered significant scholarly attention in the context of the metaverse.

8.3.2.1 Integration of SemCom and AIGC

The synergistic integration of SemCom and AIGC serves to enhance digital experiences by achieving the efficient, meaning-rich data transmission of SemCom with the personalized, high-quality content creation of AIGC. The role of SemCom in transmitting only semantically essential data reduces bandwidth usage and computational load, while AIGC leverages this streamlined data to generate highly tailored and contextually appropriate content. The integration of these two technologies enables real-time adaptation to user behaviors and needs, with the additional benefit of improved resource allocation and resilience to out-of-distribution issues.

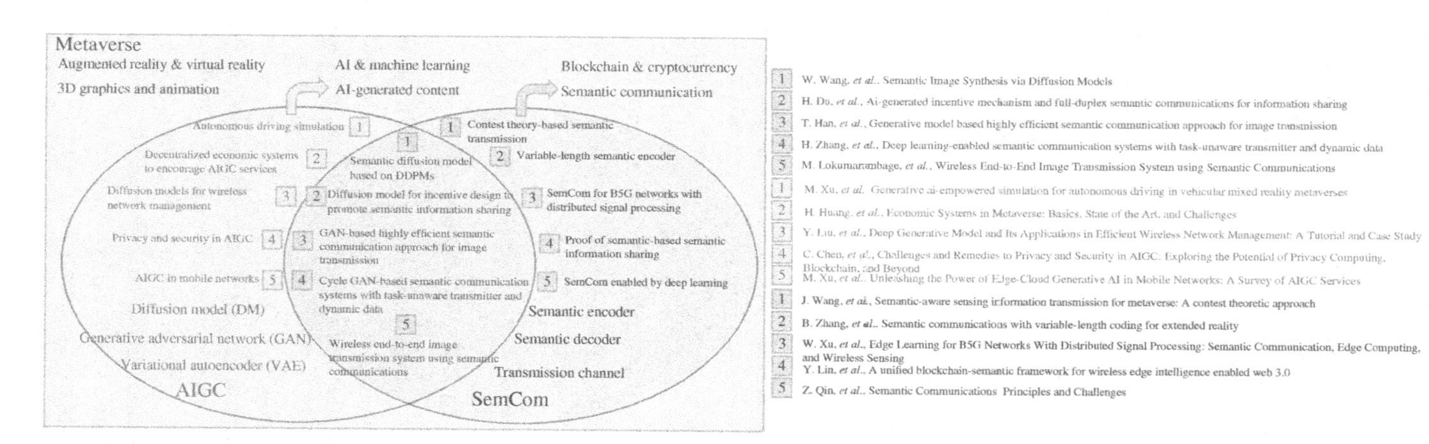

Figure 8.2 SemCom in metaverse. Source: [Lin et al., 2023c]/arXivLabs/CC-BY 4.0.

8.3.2.2 SemCom and Metaverse

The integration of SemCom into the metaverse framework stands as a transformative strategy to efficiently manage both computational resources and user experience. By focusing on the transmission of semantically relevant information rather than raw data, SemCom facilitates a streamlined interaction between users and MSPs, thereby reducing latency and bandwidth consumption. This approach allows for a more interactive and immersive experience within the virtual world, accommodating real-time updates and complex simulations. Furthermore, by limiting the transmitted data to what is semantically important, SemCom can help mitigate the risk of data overload and system slowdowns, ensuring a smoother, more dynamic virtual environment [Zheng et al., 2023b].

8.3.2.3 AIGC and Metaverse

The incorporation of AIGC into the metaverse amplifies its versatility, adaptability, and user engagement. AIGC technologies enable the creation of responsive and evolving virtual landscapes, characters, and narratives. These elements can adapt in real time to user behavior, making each individual's experience within the metaverse genuinely personalized. Furthermore, AIGC allows for the rapid prototyping and deployment of new content, helping to keep the virtual world fresh and captivating. By generating intricate, user-specific scenarios, AIGC also contributes to resolving the limitations associated with predesigned or static content. This creates an ever-changing and dynamic virtual universe that can continuously engage its inhabitants.

8.3.2.4 SemCom, AIGC, and Metaverse

The integration of SemCom and AIGC within the metaverse elevates the virtual experience to new dimensions of personalization, efficiency, and realism. SemCom enables targeted and intelligent exchange of information between users and the MSPs, ensuring that only relevant and context-aware data are transmitted. AIGC, on the other hand, provides the creative engine that populates these virtual worlds with adaptable and dynamic content. When integrated, SemCom can inform AIGC algorithms about user preferences and contextual needs, leading to the real-time creation of highly personalized and relevant virtual experiences. In return, AIGC can provide feedback loops that enrich the semantic layers of communication. This symbiotic relationship creates a virtual ecosystem that is not only immersive and interactive but also efficient and tailored to individual user experiences.

8.3.3 Research Gaps

While multiple research gaps are discussed in Section 8.7, this section narrows its focus to two primary areas: the dishonesty of semantic information and the inefficient utilization of communication resources for SemCom within the metaverse.

8.3.3.1 Dishonest Semantic Transformation

The transmission of semantic information, as opposed to raw data, between users and MSPs introduces the risk of deceptive transformations. Malicious users could employ training-based adversarial techniques to manipulate semantic information in such a way that, while appearing similar to authentic information, it conveys entirely different or misleading meanings. This form of adversarial semantic information compromises the integrity of the virtual environment [Du et al., 2023a]. As identified in Figure 8.3, semantic information can be sorted into three types: the target, which is the authentic semantic information originally extracted from raw data; the carrier, serving

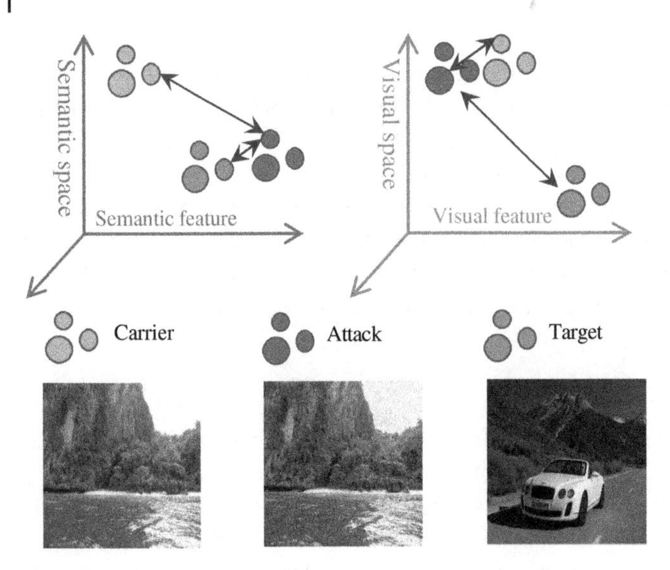

Figure 8.3 Dishonest semantic transformation. Source: [Lin et al., 2023a]/IEEE/CC BY 4.0.

as an auxiliary set of data used to facilitate the attack; and the falsified information, generated using targeted semantic attacks to manipulate pixel values in the data [Tolias et al., 2019]. Given that edge devices have limitations on their computational resources, it becomes crucial to record transformations from raw data to semantic information. This record-keeping helps in auditing and verification processes designed to prevent malicious manipulations and maintain the integrity of the virtual world.

8.3.3.2 Inefficient Resource Utilization

While various resource allocation mechanisms have been developed for individual modules such as semantic, AIGC, and rendering modules, these mechanisms often operate in isolation. As a result, they fail to provide a cohesive, a unified framework that can dynamically allocate resources based on varying environmental conditions and fluctuating user demands [Du et al., 2023d], [Zheng et al., 2021]. Such standalone strategies can lead to inefficient utilization of resources and may not fulfill the goals of maximizing utility for MSPs. It becomes imperative, therefore, to integrate the capabilities of these individual modules into a comprehensive framework. Resource allocation can become more strategic and adaptive, effectively taking into account real-time conditions and diverse requirements across semantic, AIGC, and rendering modules. This integrated approach aims to enhance the operational efficiency of the entire metaverse experience.

8.4 Zero-Knowledge Proof-Based Semantic Verification

8.4.1 Training-Based Semantic Attack Mechanism

While the SemCom framework for the metaverse helps minimize data redundancy between edge devices and MSPs, there exists the risk of malicious edge devices deploying training-based targeted semantic attacks. These attacks aim to compromise the integrity of semantic information in the form of images that are extracted from raw data circulating within metaverse services.

Specifically, targeted semantic attacks involve the transmission of adversarial semantic information that, while having nearly identical semantic descriptors, are visually distinct from the genuine information [Du et al., 2023a]. This poses a challenge for MSPs, who find it difficult to differentiate between authentic and adversarial information using conventional methods such as evaluating semantic similarities through the inner product of descriptors among images [Tolias et al., 2019; Du et al., 2023a].

8.4.1.1 Extraction

The challenge of accurately assessing semantic information calls for advanced techniques. To this end, edge devices frequently make use of convolutional neural networks (CNNs) for converting semantic information into high-dimensional descriptor vectors. These vectors become invaluable tools for MSPs in discerning and evaluating the semantic links across diverse pieces of information. As illustrated in Figure 8.3, the semantic information landscape comprises three primary categories when dealing with untrustworthy edge devices: adversarial ($\mathbf{x_a}$), authentic ($\mathbf{x_t}$), and carrier ($\mathbf{x_c}$) semantic information. The authentic information is harvested from original imagery via edge-located semantic extraction modules and tagged with a label $y_t = f_\mathbf{c}(x_t)$. On the other side, the carrier semantic information $\mathbf{x_c}$ functions as an enabler, assisting edge devices in crafting the more malicious adversarial information, $\mathbf{x_a}$. It is categorized under a different label, $y_c = f_\mathbf{c}(x_c) \neq y_t$ yet bears a striking visual resemblance to the adversarial information. Through sophisticated machine learning techniques, adversarial entities are designed to subtly alter pixel arrays. This process results in adversarial descriptors that are nearly identical to authentic ones, although they differ visually. This complexity is further highlighted in Figure 8.3, where $\mathbf{x_a}$ appears similar to $\mathbf{x_c}$ but gets misclassified as y_t. This complexity underscores the importance and intricacy of the feature extraction process, particularly for malicious edge devices aiming to create misleading adversarial samples.

To ensure uniformity in resolution, the incoming image, denoted by $\mathbf{x}$, undergoes a resampling process to yield a new image, $\mathbf{x}^s$, with a standard dimension s. This adjustment guarantees that the adversarial and carrier semantic information, $\mathbf{x_t}$ and $\mathbf{x_c}$, respectively, share identical resolution parameters. The resampled image $\mathbf{x}^s$ then becomes the input to a fully convolutional neural network. This network, represented mathematically by $\mathbf{g}_{\mathbf{x}^s} = g(\mathbf{x}^s)$, performs a dimensional mapping from $\mathbb{R}^{W \times H \times 3}$ to $\mathbb{R}^{w \times h \times d}$. These dimensions correspond to the image's width, height, and channel attributes. Within the fully CNN, a specialized pooling layer operates on the input tensor $\mathbf{g}_{\mathbf{x}^s}$, transforming it into a descriptor $\mathbf{h}_{\mathbf{x}^s} = h(\mathbf{g}_{\mathbf{x}^s})$. This transformation is guided by the adjustable parameters of the network, collectively represented by $\boldsymbol{\theta}$. Mathematically, this transformation process can be captured by the function $h : \mathbb{R}^{w \times h \times d} \rightarrow \mathbb{R}^d$. The resultant descriptor $\mathbf{h}_{\mathbf{x}^s}$ not only becomes an output of the pooling layer but also serves as a tool for evaluating semantic similarities between various descriptors. To facilitate easier and more accurate comparisons of these semantic attributes, the descriptor undergoes l_2 normalization. This additional step aids in the standardization of the descriptor, making it more compatible for subsequent operations, including the critical task of distinguishing between authentic and adversarial semantic information.

8.4.1.2 Loss

In the context of malicious objectives, edge devices aim to generate adversarial information, $\mathbf{x_a}$, that is semantically close to authentic information, $\mathbf{x_t}$, while maintaining visual similarities with carrier information, $\mathbf{x_c}$. To achieve this, the corresponding loss function comprises two distinct components: a performance loss term $l_{ts}(\mathbf{x}, \mathbf{x}_t)$ that quantifies the semantic differences between $\mathbf{x}$

and $\mathbf{x}_t$ and a distortion loss term that gauges the visual dissimilarities between $\mathbf{x}$ and $\mathbf{x}_c$. Formally, the loss function can be articulated as

$$L_{ts}(\mathbf{x}_c, \mathbf{x}_t; \mathbf{x}) = l_{ts}(\mathbf{x}, \mathbf{x}_t) + \lambda ||\mathbf{x} - \mathbf{x}_c||^2, \tag{8.1}$$

where λ serves as a hyperparameter that indicates the significance of the distortion loss within the overall loss function.

Operating under the assumption that malicious edge devices are privy to the architecture used for descriptor extraction, as explored in Tolias et al. [2019], these devices have a critical advantage. This knowledge enables them to more effectively adapt to the evaluation metrics applied by MSPs. To address this, we introduce three specialized forms of performance loss, inspired by the work cited. These loss functions are tailored to different aspects of the descriptor and can be summarized as follows: For the global descriptor, the performance loss is the inner product between descriptors and is calculated as $l_{global}(\mathbf{x}, \mathbf{x}_t) = 1 - \mathbf{h}_{\mathbf{x}}^\top \mathbf{h}_{\mathbf{x}_t}$. In the case of activation tensor, the performance loss captures the mean squared difference across the tensor dimensions and is expressed as $l_{tensor}(\mathbf{x}, \mathbf{x}_t) = \frac{||\mathbf{g}_{\mathbf{x}} - \mathbf{g}_{\mathbf{x}_t}||^2}{w \cdot h \cdot d}$. Lastly, for the activation histogram, the performance loss considers the first-order statistics of activations. It is defined as $l_{hist}(\mathbf{x}, \mathbf{x}_t) = \frac{1}{d} \sum_{i=1}^{d} ||u(\mathbf{g}_x, \mathbf{b})_i - u(\mathbf{g}_{x_t}, \mathbf{b})_i||$. Each of these performance loss forms is specifically designed to capture distinct aspects of the descriptor, thereby providing a better understanding of how well the adversarial information mimics the authentic data. This approach to calculating loss ensures a more robust and accurate evaluation, especially vital when edge devices have prior knowledge of the descriptor extraction architecture.

8.4.1.3 Optimization Strategy

The overarching goal is to pinpoint an optimal loss function that minimizes the semantic resemblance between the authentic $\mathbf{x_t}$ and the candidate $\mathbf{x}$, thereby improving the model's parameters, denoted as θ. To fulfill this objective, we propose using the Adam optimization algorithm, a widely accepted method for adaptive learning rate optimization. This allows for a dynamic and iterative update of the model parameters θ as $\theta_{t+1} = \theta_t - \eta \frac{\rho_t}{\sqrt{v_t + \epsilon}}$, where η stands for the learning rate, ρ_t and v_t denote the first and second momenta of the gradients, respectively. The term ϵ is included to avert division by zero in $\sqrt{v_t + \epsilon}$. As a result, the adversarial semantic information $\mathbf{x}_a$ can be formulated as

$$\mathbf{x}_a = \arg\min_{\mathbf{x}} L_{ts}(\mathbf{x}_c, \mathbf{x}_t; \mathbf{x}). \tag{8.2}$$

Here, L_{ts} could be substituted with l_{global}, l_{tensor}, or l_{hist} depending on the specific circumstances being considered.

8.4.2 ZKP-Based Semantic Verification Mechanism

To prevent attacks where malicious semantic information has almost the same semantic similarity but differs in visual similarity, this chapter presents a ZKP-based verification scheme [Lin et al., 2023a] that records transformations of semantic information before transmitting it to the virtual world, as shown in Figure 8.4. In this chapter, we concentrate on semantic information. These extracted images constitute a portion of the original raw data, an approach that aligns with the work presented in Ng et al. [2022]. The verification scheme comprises six steps before transmitting semantic information to MSPs: semantic extraction, transformation, ZKP extraction, ZKP proof, and ZKP verification. The transformation logic is captured in a circuit, generated via a zero-knowledge algorithm. Edge devices leverage this extracted semantic information as inputs to formulate proofs of transformations, subsequently producing transformed semantic information

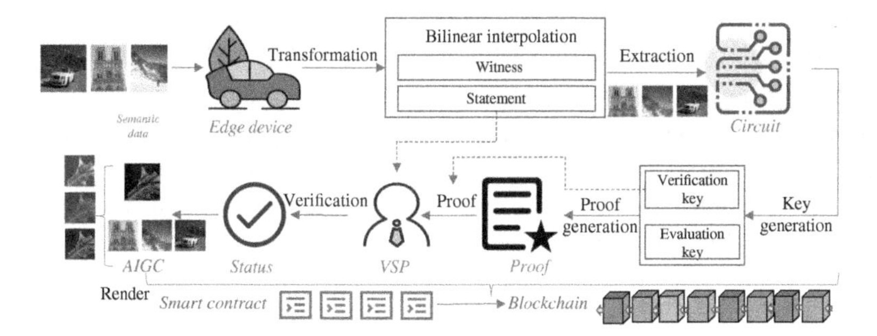

Figure 8.4 Zero-knowledge proof-based verification. Source: [Lin et al., 2023a]/IEEE/CC BY 4.0.

as outputs. These proofs and transformed semantic information are then forwarded to MSPs for validation. Considering that both MSPs and edge devices operate within unpredictable metaverse settings, it becomes essential to ensure that transformations are accurately recorded and verified, thereby mitigating the risk of unauthorized data alterations.

8.4.2.1 Semantic Extraction

The process of semantic extraction involves using a semantic encoder to transform raw data into semantic information, often represented as images. Given the potential for adversarial semantic attacks by malicious users, a layer of additional transformations is applied to the extracted semantic information to secure its authenticity. These transformations act as a security measure, minimizing the chances that the semantic information could be compromised or altered in a manner that would be damaging or misleading.

8.4.2.2 Transformation for Security

Malicious actors may attempt semantic attacks aimed at subtly altering the pixels in semantic information. Such alterations could yield semantic information that is misleading yet appears highly similar to authentic semantic data. To guard against this form of semantic deception, this chapter employs the bilinear interpolation algorithm [Castleman, 1996]. This technique not only enhances the data but also allows us to detect any unusual changes in the pixel distribution that might indicate an attack. We consider four key points in the original semantic information: (x_1, y_1), (x_1, y_2), (x_2, y_1), and (x_2, y_2). Bilinear interpolation is applied across both the x and y axes to compute the transformed points (x, y). The function $g(\cdot,\cdot)$ serves as a spatial transformation in this process. For the x-axis, the interpolation function can be expressed as $g(x, y_1) \approx \frac{x_2-x}{x_2-x_1} g(x_1, y_1) + \frac{x-x_1}{x_2-x_1} g(x_2, y_1)$. Similarly, for the y-axis: $g(x, y_2) \approx \frac{x_2-x}{x_2-x_1} g(x_1, y_2) + \frac{x-x_1}{x_2-x_1} g(x_2, y_2)$. Finally, the transformed points (x, y) can be calculated as follows: $g(x, y) \approx \frac{y_2-y}{y_2-y_1} g(x, y_1) + \frac{y-y_1}{y_2-y_1} g(x, y_2)$.

8.4.2.3 ZKP Extraction for Verifiable Transformation

The incorporation of ZKPs brings a layer of security and verifiability to the transformation process of semantic information. While transformations like bilinear interpolation can robustly distinguish authentic data from adversarial attacks, the challenge lies in convincing the MSPs that these transformations have genuinely been executed by the users. ZKPs provide a method for verification without revealing the actual data transformed, thereby maintaining the integrity of both the semantic information and the virtual world. In this scheme, the transformation function $g(\cdot,\cdot)$, which is based on bilinear interpolation, is represented as a computational logic in the circuit C. This circuit is implemented using a ZKP circuit compiler known as circom [Iden3, 2022].

Utilizing a security parameter 1^λ and the transformation circuit C, each physical user generates a common shared reference string crs, comprised of an evaluation key $\mathsf{crs.ek}$ and a verification key $\mathsf{crs.vk}$. Mathematically, this can be defined as $\mathsf{KeyGen}(1^\lambda, \mathsf{C}) \xrightarrow{\mathsf{C}} \mathsf{crs(ek, vk)}$. Subsequently, each user applies the transformation $g(\cdot, \cdot)$ and uses the circuit C to generate a public record s and a confidential witness $\mathbf{w}$ [Fan et al., 2023]. The ZKP extraction process, symbolically expressed as $\mathsf{ZKPExtract}(g) \rightarrow (\mathsf{s}, \mathsf{w})$, ensures a verifiable linkage between the initial and the final semantic information while keeping the original data confidential.

8.4.2.4 Generation of Zero-Knowledge Proof

In the process of recording the transformation, each user employs the evaluation key $\mathsf{crs.ek}$, a public statement s, and a confidential witness w to generate a ZKP π. The proof π serves as a mathematical validation that the user has indeed applied the legitimate transformations on the semantic information without revealing the specifics of what those transformations are. For a user's statement s and witness w to be valid, they must satisfy the equation $\mathsf{C}(\mathsf{s}, \mathsf{w}) = 1$, indicating compliance with the established computational logic in circuit C. Mathematically, the ZKP π can be generated through the $\mathsf{CreateProof}$ function as follows: $\mathsf{CreateProof}(\mathsf{crs.ek}, \mathsf{s}, \mathsf{w}) \xrightarrow{\mathsf{C}} \pi$. The generated π then becomes a part of the transmission to the MSP, allowing them to verify the legitimacy of the transformation without needing to access the raw semantic information or the specific transformations applied.

8.4.2.5 Zero-Knowledge Proof Verification

Once the ZKP π is received by the MSPs, it undergoes a verification process to ensure the legitimacy of the transmitted semantic information. The MSPs utilize the verification key $\mathsf{crs.vk}$, the public statement s, and the received proof π to perform this verification. The verification is considered successful if the output status is 1, indicating that the semantic information is valid and can be accepted. Conversely, if the output status is 0, the semantic information is deemed invalid and is subsequently rejected. This process effectively filters out malicious or tampered information, ensuring the integrity and reliability of the semantic content in the metaverse. Symbolically, the ZKP verification function can be expressed as follows: $\mathsf{Verify}(\mathsf{crs.vk}, \mathsf{s}, \pi) \rightarrow \{0, 1\}$. By employing this verification mechanism, MSPs can confirm the authenticity of the user-generated semantic information without violating user privacy.

8.4.3 Security Analysis

The ZKP-based verification mechanism fulfills several essential properties, namely completeness, soundness, and zero knowledge, as discussed in prior works [Galal and Youssef, 2022; Song et al., 2022; Wan et al., 2022; Fan et al., 2023].

8.4.3.1 Completeness

The completeness property ensures that if a physical user correctly executes the cryptographic protocols, the MSPs should be able to verify the authenticity of the semantic information with high probability. Specifically, suppose a physical user transforms the semantic information using bilinear interpolation and generates a cryptographic proof, denoted as π, based on the evaluation key $\mathsf{crs.ek}$ and a specific proof circuit C. In that case, the MSPs can employ the associated verification key $\mathsf{crs.vk}$ and the public record s to validate the transmitted information. If the proof circuit $\mathsf{C}(\mathsf{s}, \mathsf{w})$ evaluates to 1, indicating a successful authentication, an honest physical user's proof π will be accepted by the MSPs with a probability approaching 1. This ensures that authentic information is accepted without error, bolstering the reliability and integrity of the SemCom process.

8.4.3.2 Soundness

The soundness is a critical property that ensures malicious users cannot trick the system into accepting incorrect proofs. In the context of our SemCom model, this property guarantees that only valid proofs corresponding to valid witnesses will be accepted. If the proof circuit evaluates to zero, $\mathsf{C}(\mathsf{s},\mathsf{w}) = 0$, this implies that (s,w) is not a valid input-witness pair. In such cases, a polynomial-time extractor ZKPExt is constructed to make the probability that the ZKP verification $\mathsf{ZKPVerify}(\mathsf{crs.vk},\mathsf{s},\pi^*) = 1$ is negligibly small, denoted by $\mathsf{negl}(\lambda)$. This ensures that a malicious user, $\mathcal{A}$, would find it nearly impossible to produce a proof π that would incorrectly convince an honest MSP $\mathcal{V}$ to accept an invalid transformation g. Therefore, the system maintains its integrity by minimizing the malicious user's adversarial advantage in soundness, formally represented as $\mathsf{Adv}^{sd}_{\mathcal{A}}(\lambda)$, which can be mathematically expressed as

$$\mathsf{Adv}^{sd}_{\mathcal{A}}(\lambda) = \Pr\begin{bmatrix} \mathsf{KeyGen}(1^\lambda,\mathsf{C}) \to \mathsf{crs}(\mathsf{ek},\mathsf{vk}) \\ \mathcal{A}(\mathsf{crs.ek},\mathsf{s}) \to \pi^* \\ \mathsf{ZKPExt}(\mathsf{crs.ek},\mathsf{s},\pi^*) \to \mathsf{w} \\ \mathsf{ZKPVerify}(\mathsf{crs.vk},\mathsf{s},\pi^*) = 1 \end{bmatrix} \le \mathsf{negl}(\lambda). \tag{8.3}$$

8.4.3.3 Zero Knowledge

The zero-knowledge property ensures that the MSPs cannot access the original semantic information unless explicitly shared by the physical users. This attribute is crucial for privacy concerns, guaranteeing that information leakage is minimized. Although the MSPs can view public statements related to the semantic information, these statements are structured in such a way that prevents MSPs from extracting detailed information from them. Thus, even with access to these public elements, the MSPs find it difficult to uncover the underlying semantic content. This formalization underpins the privacy safeguards built into the SemCom system, ensuring that users can confidently interact in a secure and private manner. To validate this property, we introduce probabilistic polynomial-time simulators, S_1 and S_2, and malicious physical users, $\mathcal{A}_1$ and $\mathcal{A}_2$. The simulator S_1 creates a common reference string, which is then used by the second simulator S_2, to generate a simulated proof. This establishes the zero-knowledge characteristic of the system, denoted mathematically as $\mathsf{Adv}^{zk}_{\mathcal{A}}(\lambda)$, which can be mathematically expressed as

$$\mathsf{Adv}^{zk}_{\mathcal{A}}(\lambda) = \left|\Pr\begin{bmatrix} \mathsf{KeyGen}(1^\lambda,\mathsf{C}) \to \mathsf{crs} \\ \mathcal{A}_1(\mathrm{g}) \to (\mathsf{s},\mathsf{w}) \\ \mathsf{Prove}(\mathsf{crs},\mathsf{s},\mathsf{w}) \to \pi \\ \mathcal{A}_2(\mathsf{crs},\mathsf{s},\mathsf{w},\pi) = 1 \end{bmatrix} - \Pr\begin{bmatrix} \mathsf{S}_1(1^\lambda,\mathsf{C}) \to \mathsf{crs} \\ \mathcal{A}_1(\mathrm{g}) \to (\mathsf{s},\mathsf{w}) \\ \mathsf{S}_2(\mathsf{crs},\mathsf{s}) \to \pi \\ \mathcal{A}_2(\mathsf{crs},\mathsf{s},\mathsf{w},\pi) = 1 \end{bmatrix}\right| \le \mathsf{negl}(\lambda). \tag{8.4}$$

8.5 Diffusion Model-Based Resource Allocation

To achieve a more strategic resource allocation, we first formulate an optimization problem considering the interactions of semantic extraction, AIGC inference, and rendering modules and propose a diffusion model-based resource allocation mechanism to obtain near-optimal strategies, as shown in Figure 8.5.

8.5.1 Joint Optimization Problem Formulation

Between physical users and MSPs, there are three key modules: the semantic, AIGC, and rendering modules. The focus of optimization should be on the computation and communication aspects within these modules and functions.

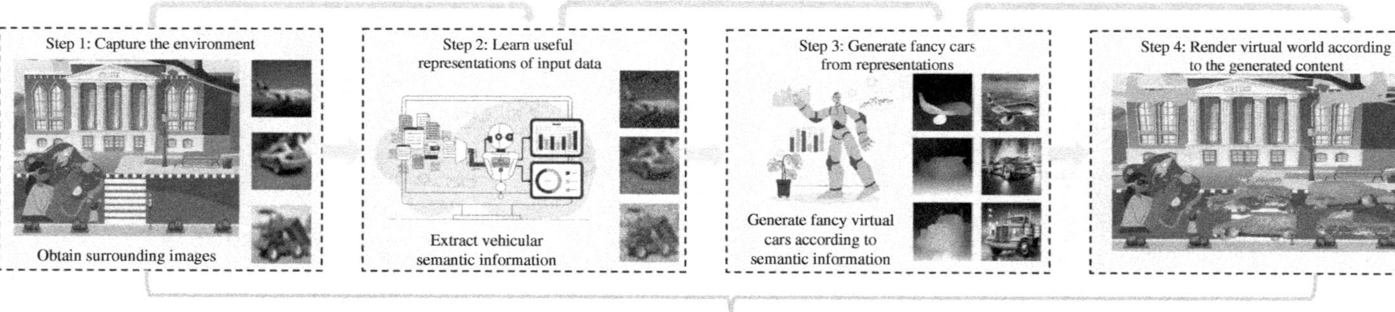

Figure 8.5 Strategic resource allocation mechanism. Source: [Lin et al., 2023c]/Cornell University/CC BY 4.0.

8.5.1.1 Semantic Extraction

Users at the edge employ a semantic encoder within the semantic framework to derive pertinent semantic information from the raw data collected by edge devices. The primary intent of this extraction is to extract meaningful content and reduce the number of bits sent across the network. The computational duration needed for this extraction is quantified by the metric $T_s^{\text{comp}} = \frac{Z_s}{C_s}$, where Z_s is the computational demand and C_s signifies the available computational bandwidth [Liu et al., 2019]. For improved fidelity in the transmitted data, this semantic output is forwarded to a module dedicated to AIGC, deployed by the MSPs. The latency for this data transmission is captured by the term T_s^{comm}, which is a function of the semantic rate R_s^a and the data volume D_s being transmitted. The rate R_s^a is computed considering several factors: the approximated semantic entropy H_s^a, the communication bandwidth W_s^a accessible between edge users and MSPs, and the mean count of symbols transferred, denoted as K_s^a [Yan et al., 2022]. Mathematically, the rate is given as $R_s^a = \frac{H_s^a}{K_s^a/W_s^a}$. Therefore, the transmission time T_s^{comm} from the semantic framework to the AIGC module can be precisely calculated as $T_s^{\text{comm}} = \frac{D_s}{R_s^a}$.

8.5.1.2 AIGC Inference

Once the MSPs receive the semantic information from the physical users, they utilize AIGC modules to perform inference tasks. These tasks are designed not only to generate digital content tailored to the user but also to enhance the quality of the original data. The time required for this AIGC-based computational processing is denoted by T_a^{comp}. This time is a function of the computational demands for the inference, represented as Z_a, and the available computational resources C_a. Thus, the relationship can be mathematically captured as $T_a^{\text{comp}} = \frac{Z_a}{C_a}$. To enhance end-user engagement and provide a rich interactive experience, the digital content that is newly created by the AIGC module is then transferred to a rendering module. This module, which is also deployed by the MSPs, creates a visually appealing graphical representation of the content. The latency associated with this transfer is represented by T_a^{comm}, and it depends on both the rate at which the data can be transferred, R_a^r, as well as the volume of the data, D_a. This relationship is summarized by the equation $T_a^{\text{comm}} = \frac{D_a}{R_a^r}$. Various factors, such as the available communication bandwidth W_a^r, channel gain g_a^r, the power level at which data is transmitted p_a^r, and the additive Gaussian noise σ_a^2, contribute to determining this rate R_a^r. Based on these parameters, the rate can be computed as $R_a^r = W_a^r \log\left(1 + \frac{g_a^r p_a^r}{\sigma_a^2}\right)$ [Xu et al., 2023a].

8.5.1.3 Graphics Rendering

The MSPs utilize the digital content generated through AIGC inference for graphics rendering, aimed at enhancing the user experience for physical users. The time required for this computation is denoted by T_r^{comp}, which is a function of both the available computational resources C_r and the resources required Z_r for graphics rendering within the MSPs. Mathematically, this is expressed as $T_r^{\text{comp}} = \frac{Z_r}{C_r}$. After the rendering process is complete in the virtual world, the results are fed back to the physical users to update their experiences on edge devices. The time required for this communication, T_r^{comm}, is determined by the transmission rate R_r^s between the MSPs and physical users and the data volume D_r of the feedback results used for rendering on the edge devices. The transmission rate R_r^s is influenced by the available bandwidth W_r^s, channel gain g_r^s, transmit power p_r^s, and additive Gaussian noise σ_r^2 [Xu et al., 2023a]. This rate is calculated as $R_r^s = W_r^s \log\left(1 + \frac{g_r^s p_r^s}{\sigma_r^2}\right)$.

8.5.1.4 Formulation

Considering the constraints of limited bandwidth $W_{\max}$ and acceptable transmission time $T_{\max}$ between physical users and MSPs, we can formulate the utility of the semantic, AIGC, and rendering modules as follows, in line with prior work [Liu et al., 2019; Yan et al., 2022]. The objective is to maximize the sum of the logarithms of the rates R_s^a, R_a^r, and R_r^s. By summing the log rates from different parts of the framework, we can effectively look to balance and maximize efficiency across the semantic, AIGC, and rendering modules to achieve system-wide optimization.

$$\begin{aligned} \max_{W_s^a, W_a^r, W_r^s} \quad & \log(1 + R_s^a) + \log(1 + R_a^r) + \log(1 + R_r^s) \\ \text{s.t.} \quad & T_s^{\text{comm}} + T_s^{\text{comp}} + T_a^{\text{comp}} + T_a^{\text{comm}} + T_r^{\text{comm}} + T_r^{\text{comp}} \le T_{\max} \\ & W_s^a + W_a^r + W_r^s \le W_{\max}. \end{aligned} \tag{8.5}$$

8.5.2 Strategic Joint Resource Allocation

Given the intricate nature of the utility function, which is influenced by rates that are themselves complex functions of multiple variables such as bandwidth, power, and noise [Zheng et al., 2023a], a straightforward optimization approach is not feasible. Inspired by existing research [Wang et al., 2022b], we combine diffusion models with deep reinforcement learning (DRL) to effectively navigate the complexities inherent in both the objective function and the associated constraints, thereby yielding a near-optimal strategy for resource allocation across the semantic, AIGC, and rendering modules.

8.5.2.1 Diffusion Model

Drawing on principles from nonequilibrium thermodynamics, diffusion models construct a Markov chain that incrementally adds and removes noises from samples over designated sampling steps, in both the forward and reverse processes, to facilitate the discovery of near-optimal solutions.

In the forward process, noise z_t is incrementally added to randomly selected samples x_t from a given data distribution. This noise is sampled from a standard normal distribution $\mathcal{N}(0, 1)$ at each selected step. The process occurs over a predefined number of steps T, and a predefined variance α_t is used to adjust the step sizes, where t is the specific step. The forward process can be expressed as

$$x_t = \sqrt{\alpha_t} x_{t-1} + \sqrt{1 - \alpha_t} z_t. \tag{8.6}$$

When the predefined variance α_t decreases, the impact of the added noise z_t increases, causing the samples to gradually lose the characteristics of their original distribution and become more like noise. According to (8.6), the forward process can be derived starting from x_0 as follows:

$$x_t = \sqrt{\alpha_t \alpha_{t-1} \cdots \alpha_2 \alpha_1} x_0 + \sqrt{1 - \alpha_t \alpha_{t-1} \cdots \alpha_2 \alpha_1}\, z. \tag{8.7}$$

Here z denotes the aggregated noise, which follows a new normal distribution $\mathcal{N}(0, \sigma_1^2 + \cdots + \sigma_t^2 + \cdots + \sigma_T^2)$. This aggregated noise is composed of individual noise terms $z_1, \ldots, z_t, \ldots, z_T$, each sampled from Gaussian distributions. The noised data x_t is fed into the diffusion models for training. The resulting output $\text{model}(x_t, t)$ is then used to compute the loss between the added noise and this output to train a better model.

In the reverse process, noise z is extracted from x_t through random sampling from the given distribution $\mathcal{N}(0, 1)$. The aim is to predict the noise as it evolves from step T back to step 0.

This process can be mathematically represented as follows:

$$x_{t-1} = \frac{1}{\sqrt{\alpha_t}}\left(x_t - \frac{1-\alpha_t}{\sqrt{1-\overline{\alpha}_t}}\,\mathrm{model}(x_t, t)\right) + \sqrt{\beta_t}z. \tag{8.8}$$

The forward and reverse processes are iteratively performed over a given number of steps T. This allows for both the sampling and removal of noise from the data, as well as the prediction of the noise.

8.5.2.2 Resource Allocation

Diffusion models are integrated into DRL to model resource allocation as a Markov decision process (MDP). This MDP is characterized by several components: state spaces $\mathcal{S}$, which include factors that adjust according to action spaces $\mathcal{A}$ like rates across the semantic, AIGC, and rendering modules; action spaces, which relate to the bandwidth allocated across these three modules; the environment, which could include factors like computational resources; the reward function, defined as the utility in the given formulation; and the initial state distribution, which specifies the initial conditions for the factors in the state spaces. The objective of the resource allocation mechanism based on diffusion models is to identify a policy $\pi(a^0|s \in \mathcal{S})$ that maximizes cumulative rewards, thereby optimizing bandwidth allocation across the semantic, AIGC, and rendering modules.

Step 1: Define state spaces $\mathcal{S}$. The state spaces capture the various conditions within the framework. These spaces include parameters that are essential for calculating transmission rates, computational resources, and data volumes across the semantic, AIGC, and rendering modules. Mathematically, the state spaces at step t can be represented as $S^t = [H_s^a, \sigma_a, \sigma_r, g_a^r, p_a^r, g_r^s, p_r^s, K_s^a, C_r, C_a]$. These parameters are randomly sampled from given distributions at each step.

Step 2: Define action spaces $\mathcal{A}$. The action spaces focus on the available bandwidth resources for the semantic, AIGC, and rendering modules. Mathematically, the action spaces at step t can be represented as $\mathcal{A}^t = [W_s^a, W_a^r, W_r^s]$.

Step 3: Define the reward function $\mathcal{R} = [\mathcal{S}, \mathcal{A}]$. The reward function is set as the utility of the modules, provided that the utility constraints are met. If the constraints are not satisfied, the reward function is set to zero as a punitive measure, guiding the diffusion model to search for a better strategy. Based on the defined state and action spaces, the diffusion models aim to learn the distribution of parameters and establish a mapping between states and actions. The objective is to output a near-optimal policy $\pi(a^0|s \in \mathcal{S})$ that maximizes cumulative rewards, thereby achieving strategic resource allocation over multiple steps.

Step 4: Utilize the diffusion model to identify a strategic policy for optimal resource allocation. To start the training phase, the diffusion models should be provided with various hyperparameters such as the number of diffusion steps, the size of the data batch, and the amplitude of exploration noise. The diffusion model architecture is set up with a pair of critic networks and their respective target networks, each having unique weight configurations. During each training epoch, a random Gaussian distribution c^T is generated to explore resource allocation policies with a multistep loop, where the algorithm first assesses the current environmental state. The actions are then influenced by Gaussian noise to determine the subsequent action by applying a reverse diffusion process to denoise the present action $p(a^i|a^{i+1}, s)$ and add some exploratory noise. Upon executing the action, a reward is calculated based on a utility function, and the environmental data are saved in a replay buffer. For model optimization, a random subset of records is pulled from this buffer to update the critic networks through loss and policy gradient calculations. The target networks are updated accordingly as the final step.

Step 5: Derive a nearly optimal approach for resource allocation. The trained diffusion models are utilized to produce an effective near-optimal policy $\pi(a^0|s \in S)$ with the given environmental states. The near-optimal strategy for resource allocation is then formulated by applying the reverse process of diffusion models to predict Gaussian noise.

8.6 Simulation Results

In this section, we first conduct experiments to demonstrate the performance of the ZKP-based verification mechanism to achieve the authenticity of semantic information before transmitting it to MSPs. Then we simulate the diffusion model-based resource allocation mechanism to achieve efficient utilization of communication resources.

8.6.1 Authenticity

Experiments were conducted on the Revisited Paris image benchmark [Radenović et al., 2018] to evaluate the verification mechanism between physical edge users and MSPs. Utilizing pretrained networks from ImageNet [Russakovsky et al., 2015] and AlexNet [Krizhevsky et al., 2017], we carried out the verification simulations over 100 iterations. For the parameters not explicitly mentioned, we relied on the configurations presented in Tolias et al. [2019].

Figure 8.6 illustrates the performance loss of semantic attacks, employing various loss functions [Global, Hist, Tensor]. As the number of iterations progresses, it becomes apparent that the Tensor loss function converges at a slower rate compared to Global and Hist. As depicted in Figure 8.7, we analyze the performance without the verification mechanism using three loss functions: Global, Hist, and Tensor. Figure 8.7a,b highlight that the Global and Hist functions stabilize around the 40th iteration, demonstrating a more rapid convergence compared to the Tensor function.

As shown in Figure 8.8, we present a comparison of semantic similarities with or without the verification mechanism. Figure 8.9 illustrates the verification mechanism in terms of semantic similarity between the adversarial semantic information versus either authentic or carrier information utilizing Global, Hist, and Tensor loss functions.

When combining Figure 8.7 with Figure 8.9, it is evident that the adversarial semantic information exhibits distinct semantic similarities with the verification mechanism. Specifically, as

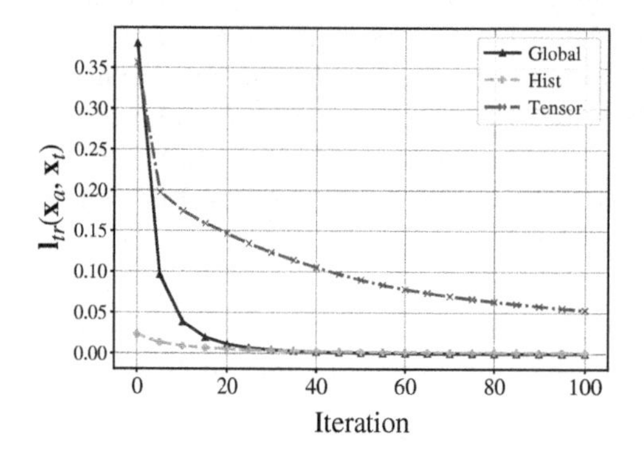

Figure 8.6 Performance loss of attacks. Source: [Lin et al., 2023a]/IEEE/CC-BY 4.0.

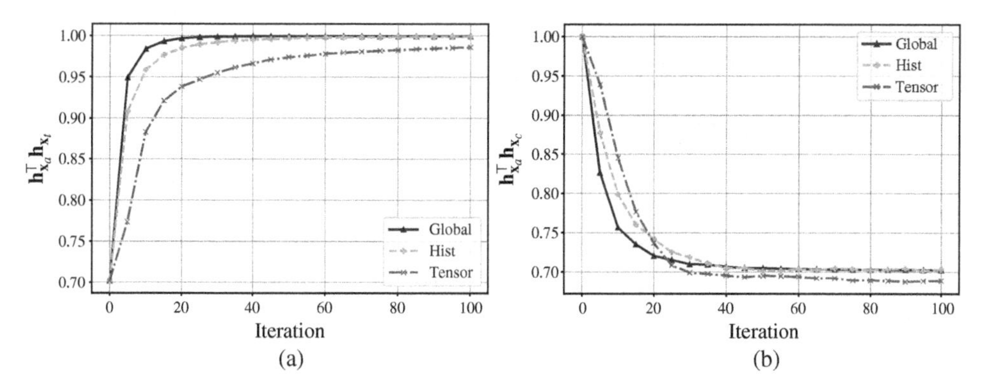

Figure 8.7 Semantic similarity without the verification mechanism. (a) $\mathbf{x_a}$ and $\mathbf{x_t}$ and (b) $\mathbf{x_a}$ and $\mathbf{x_c}$ [Lin et al., 2023a]/IEEE/CC-BY 4.0.

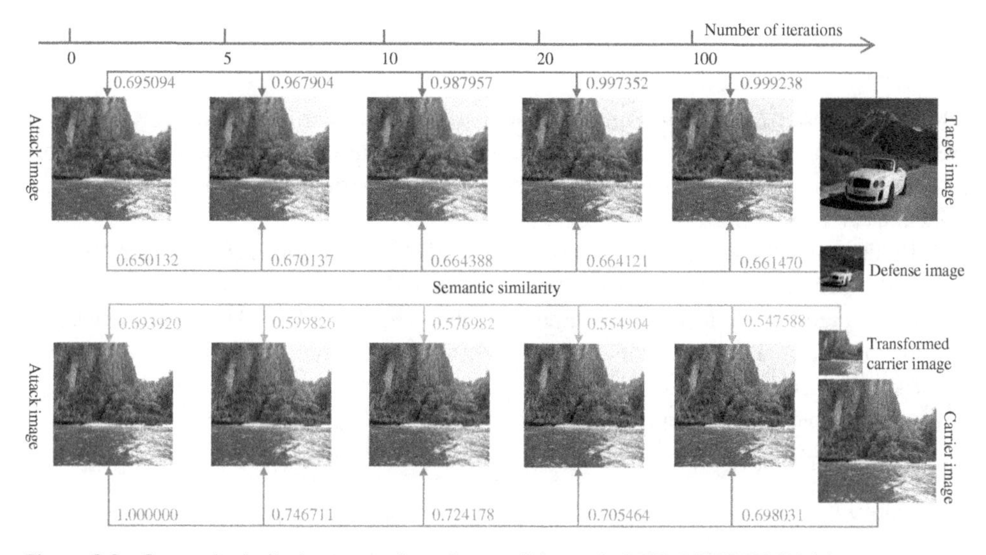

Figure 8.8 Semantic similarity comparison. Source: [Lin et al., 2023a]/IEEE/CC BY 4.0.

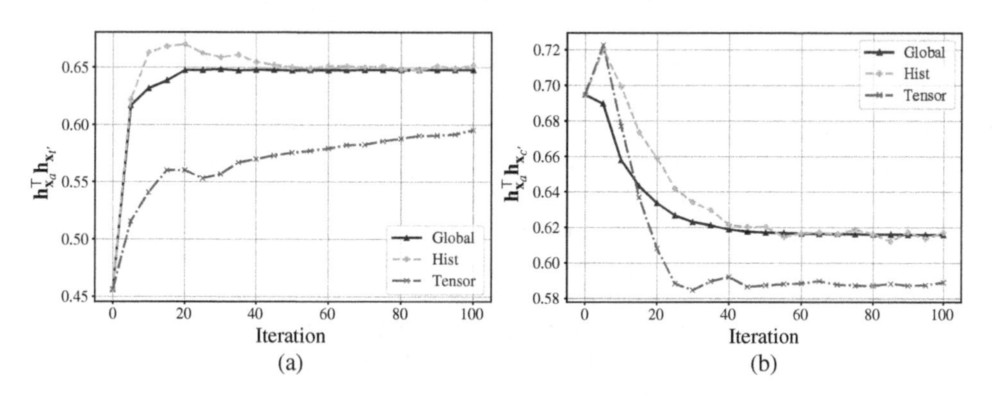

Figure 8.9 Semantic similarity performance with the verification mechanism. (a) The adversarial versus the authentic and (b) the adversarial versus the carrier. Source: [Lin et al., 2023a]/IEEE/CC-BY 4.0.

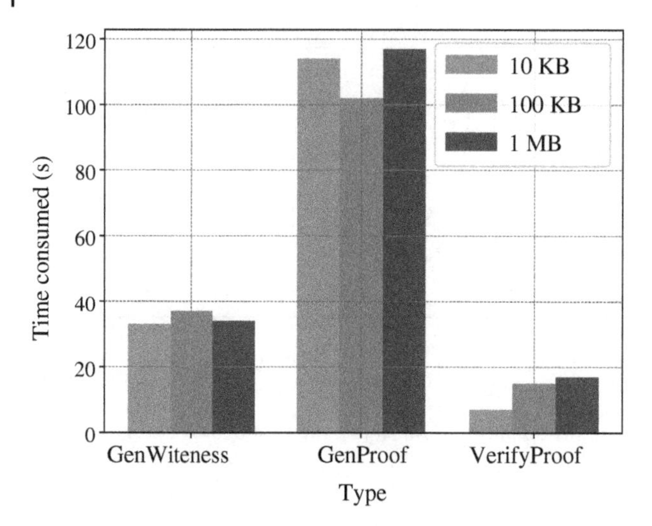

Figure 8.10 ZKP computation time. Source: [Lin et al., 2023a]/IEEE/CC-BY 4.0.

observed in Figure 8.7a compared to Figure 8.9a, the semantic similarity can exhibit a reduction of up to 35% between adversarial and authentic semantic information. Similarly, the comparison between Figures 8.7b and 8.9b indicates a potential 10% decrease in semantic similarity between adversarial and carrier semantic information.

Figure 8.10 illustrates the computational overhead of ZKP for operations such as GenWitness, GenProof, and VerifyProof, across varying data sizes of 10 KB, 100 KB, and 1 MB. A notable observation from Figure 8.10b is that the computational overhead of ZKP is largely influenced by the GenProof operation. The sizes of the semantic information, on the other hand, have minimal impact on the proposed scheme. Additionally, the computational demand for verifying proofs is less intensive compared to the generation of proofs.

8.6.2 Efficiency

The proposed diffusion model-based resource allocation mechanism (Diffusion) is compared with the proximal policy optimization (PPO) algorithm [Schulman et al., 2017] and the random algorithm (Separate) to find nearly optimal bandwidth distribution strategies among the semantic, AIGC, and rendering modules. PPO is an on-policy, model-free actor-critic algorithm that leverages a clipped surrogate objective to enhance both learning stability and efficiency. Separate is a nonintegrated baseline that allocates resources among the modules in a random fashion. The learning rates for Diffusion and PPO are set at 3e-7 and 3e-6, respectively. Unless stated otherwise, it is assumed that both methods function under the same set of parameters and environmental conditions. Other hyperparameters are given in Lin et al. [2023c]. The training iterations are configured to last for 3,000 epochs, featuring a buffer capacity of 1,000,000, exploration noise set at 0.01, 10 steps per epoch, and 100 iterations for each data collection phase.

Figure 8.11 presents a reward analysis that compares Diffusion with PPO and Separate. Throughout the training phase, it is evident that the Diffusion yields substantially higher rewards than either PPO or Separate. Figure 8.12 offers a side-by-side comparison of utility values calculated through near-optimal actions for different network states. Specifically, we consider four scenarios: PPO_1, PPO_2, Diffusion_1, and Diffusion_2, each generated under distinct network states for evaluation.

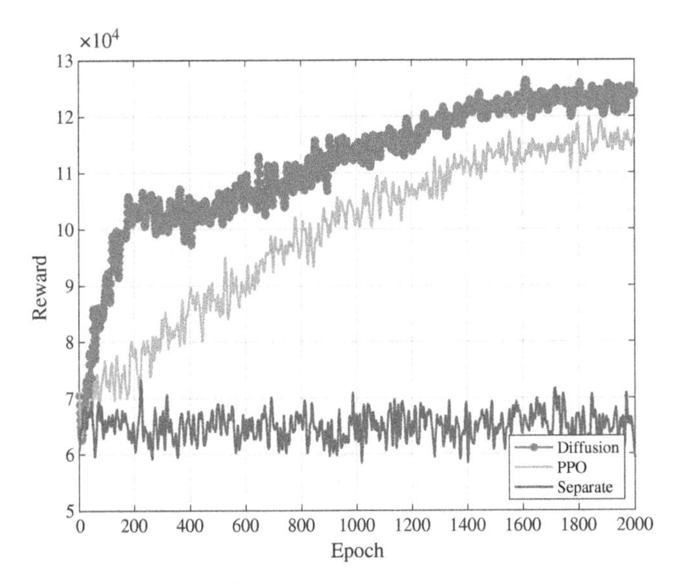

Figure 8.11 Training curves of the joint resource allocation. Source: [Lin et al., 2023c]/arXivLabs/CC-BY 4.0.

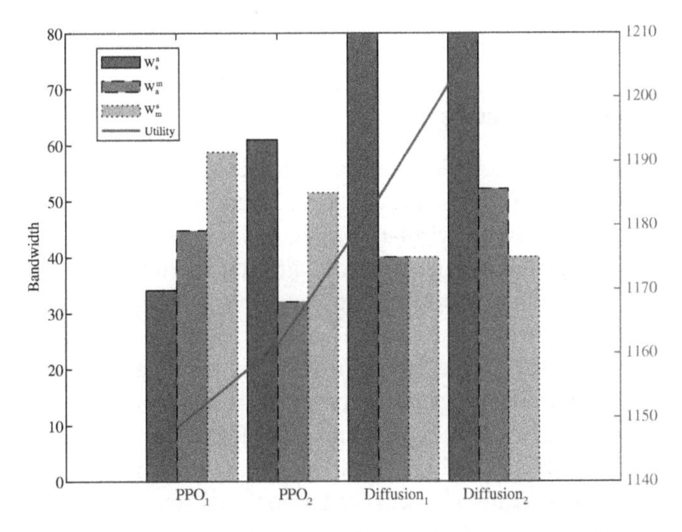

Figure 8.12 Generated utility of diffusion compared with PPO. Source: [Lin et al., 2023c]/arXivLabs/CC-BY 4.0.

The Diffusion-based strategies, Diffusion_1 and Diffusion_2, outperform PPO_1 and PPO_2 generated by the PPO method. The primary reason lies in the adaptability of the diffusion-based resource allocation approach.

8.7 Future Directions

To provide a clearer perspective on future directions, we use the classic application of the metaverse, e.g., virtual transportation [Ng et al., 2022], as a representative example in the subsequent points.

8.7.1 Incentives for Semantic Information

As AI technologies continue to advance, the generation of semantic information from AI-derived data introduces new vulnerabilities that could compromise the integrity of the metaverse. For example, malicious actors could deploy AI algorithms to fabricate counterfeit environmental scenes and then proceed to extract semantic information from these manipulated images. Such actions could corrupt vital sectors like virtual transportation systems, create misinformation, and degrade the overall reliability of the metaverse. Given these challenges, the importance of obtaining genuine semantic information from physical users becomes increasingly critical. To encourage the sharing of authentic semantic information, incentives become a significant aspect that needs further exploration. The emerging economic models within the metaverse [Huang et al., 2022; Lin et al., 2023d], offer a foundation for developing such incentives. These models leverage NFTs, auction mechanisms, and game theory-based strategy optimization to encourage user participation. However, the design and implementation of these incentive mechanisms are not trivial and require an understanding of human behavior, market dynamics, and the technical intricacies involved in data validation and resource allocation. Future research should focus on developing robust and adaptable incentive mechanisms that not only encourage the sharing of authentic semantic data but also account for potential loopholes or exploits that may arise. For instance, it is necessary to differentiate between AI-generated and human-generated semantic information effectively and ensure that the incentive mechanisms are not susceptible to collusion or other forms of strategic manipulation.

8.7.2 Privacy Preserving for SemCom in the Metaverse

While the integration of SemCom and the metaverse can facilitate seamless user experiences, it also presents privacy concerns. Given that semantic information is directly derived from the raw data of physical users, transmitting this information to render virtual worlds could expose user privacy. For instance, as noted in Ng et al. [2022], key segmentation from surrounding transportation images is extracted as semantic information. If such information is relayed to MSPs, they might deduce the real-world locations of the physical users along with other users. Federated learning [McMahan et al., 2017] offers a promising method for privacy preservation, as it shares model parameters rather than raw data. In this scenario, a semantic encoder can be employed to extract model tensors from raw data similar to the process in federated learning. This prevents MSPs and other users from inferring real-world locations. Additionally, in a blockchain-driven metaverse where model tensors are accessible and immutable to all participants, physical users who want to exit the virtual world permanently can utilize federated unlearning [Liu et al., 2021] to thoroughly erase a user's contributed model tensors, ensuring that other participants can no longer access them.

8.7.3 Multitask Semantic Extraction Across Virtual Worlds

The complexity of the metaverse demands diverse kinds of semantic information for its various virtual worlds. In many instances, a single virtual environment may require multiple types of semantic extraction tasks to be performed concurrently. For example, while physical users may be tasked with capturing semantic information to represent the condition of traffic lights, they may simultaneously need to collect data reflecting the level of traffic congestion. The ability to integrate and correlate these different sets of semantic data effectively is essential for enhancing the depth and breadth of services such as a virtual transportation system. Leveraging multitask semantic extraction not only improves the richness of the environment but also optimizes the allocation of computational and communication resources. Domain adaptation techniques in the semantic encoder and decoder could serve as a robust solution for efficiently transmitting these diverse, task-specific features [Zhang et al., 2022a]. Such multitask approaches also raise exciting avenues for future research. Questions arise concerning how to build intelligent systems that can autonomously determine which sets of semantic information should be prioritized or how to dynamically adapt to the ever-changing needs of various tasks in real time.

8.7.4 Lightweight Semantic Communication Before Transmission

The emergence of edge devices with limited computational resources poses a significant challenge for the deployment of complex semantic encoders. These devices often need to perform multiple tasks, including data collection, encoding, and transmission, all within their constrained resource envelope. As such, the development of lightweight SemCom strategies becomes increasingly important. The objective is to simplify the computational requirements for encoding semantic information without sacrificing the quality or integrity of the transmitted data. A potential avenue for achieving this balance is to employ finite-bit constellations, allowing for efficient semantic encoding and decoding even on less powerful edge devices [Xie and Qin, 2020]. The move toward lightweight solutions opens new research directions, including how to optimize these reduced-complexity algorithms for various types of semantic information and how to ensure that they remain robust against potential errors or attacks. Additionally, questions surrounding the trade-offs between computational efficiency and data fidelity in such lightweight models become increasingly pertinent.

8.7.5 Challenges in Practical Deployment

The operational implementation of SemCom within the metaverse necessitates a complex coordination of computational and communicative resources across diverse layers and multiple service providers. This undertaking is made particularly challenging due to limitations such as restricted access to the fundamental infrastructure, varying deployment environments, and specialized requirements for different use-cases. Key issues that arise in this context include the need for seamless interoperability among disparate systems, optimization of resource allocation to meet both computational and communication needs, and effective performance management to ensure reliable service delivery. These challenges are accentuated by the dynamic nature of the metaverse, where changes in user behavior, technology adoption rates, and emerging functionalities can rapidly alter system requirements. Navigating this labyrinth of constraints and variables requires a multidisciplinary approach that combines insights from network architecture, distributed computing, and information theory. Further research in these areas is essential for overcoming these practical deployment challenges and ensuring that SemCom can be reliably and effectively integrated into the ever-evolving landscape of the metaverse.

8.8 Conclusion

In this chapter, we first introduce a unified framework for SemCom in the metaverse with semantic extraction, AIGC inference, and rendering modules, emphasizing its significant advantages and transformative potential for user interactions. We then explore the motivation for integrating SemCom in the metaverse and pinpoint the current research gaps in these areas. To address authenticity concerns, we first point out a training-based semantic attack mechanism and detail a semantic verification mechanism based on ZKP, ensuring the integrity of semantic information before its transmission. In addition, we propose a diffusion model-based approach for resource allocation, aiming to dynamically enhance utility across the various SemCom and metaverse modules. Our simulations provide the performance of the authenticity and efficiency of the proposed mechanisms. Finally, we discuss the future directions of SemCom in metaverse, including incentives for semantic information, privacy-preserving for SemCom in the metaverse, multitask semantic extraction across virtual worlds, lightweight SemCom before transmission, and challenges in practical deployment.

References

Kenneth R Castleman. *Digital Image Processing*. Prentice Hall Press, 1996.

Hongyang Du, Jiacheng Wang, Dusit Niyato, Jiawen Kang, Zehui Xiong, Mohsen Guizani, and Dong In Kim. Rethinking wireless communication security in semantic Internet of Things. *IEEE Wireless Communications*, 30(3):36–43, 2023a.

Hongyang Du, Jiacheng Wang, Dusit Niyato, Jiawen Kang, Zehui Xiong, and Dong In Kim. AI-generated incentive mechanism and full-duplex semantic communications for information sharing. *IEEE Journal on Selected Areas in Communications*, 41(9):2981–2997, 2023b.

Hongyang Du, Jiacheng Wang, Dusit Niyato, Jiawen Kang, Zehui Xiong, Junshan Zhang, and Xuemin Shen. Semantic communications for wireless sensing: RIS-aided encoding and self-supervised decoding. *IEEE Journal on Selected Areas in Communications*, 41(8):2547–2562, 2023c.

Hongyang Du, Ruichen Zhang, Yinqiu Liu, Jiacheng Wang, Yijing Lin, Zonghang Li, Dusit Niyato, Jiawen Kang, Zehui Xiong, Shuguang Cui, et al. Beyond deep reinforcement learning: A tutorial on generative diffusion models in network optimization. *arXiv preprint arXiv:2308.05384*, 2023d.

Yongkai Fan, Binyuan Xu, Linlin Zhang, Jinbao Song, Albert Zomaya, and Kuan-Ching Li. Validating the integrity of convolutional neural network predictions based on zero-knowledge proof. *Information Sciences*, 625:125–140, 2023.

Hisham S Galal and Amr M Youssef. Aegis: Privacy-preserving market for non-fungible tokens. *IEEE Transactions on Network Science and Engineering*, 10(1):92–102, 2022.

Tianxiao Han, Jiancheng Tang, Qianqian Yang, Yiping Duan, Zhaoyang Zhang, and Zhiguo Shi. Generative model based highly efficient semantic communication approach for image transmission. In *ICASSP 2023-2023 IEEE International Conference on Acoustics, Speech and Signal Processing (ICASSP)*, pages 1–5. IEEE, 2023.

Qiyu Hu, Guangyi Zhang, Zhijin Qin, Yunlong Cai, Guanding Yu, and Geoffrey Ye Li. Robust semantic communications with masked VQ-VAE enabled codebook. *arXiv preprint arXiv:2206.04011*, 2022.

Qiyu Hu, Guangyi Zhang, Zhijin Qin, Yunlong Cai, Guanding Yu, and Geoffrey Ye Li. Robust semantic communications with masked VQ-VAE enabled codebook. *IEEE Transactions on Wireless Communications*, 22(12):8707–8722, 2023.

Huawei Huang, Qinnan Zhang, Taotao Li, Qinglin Yang, Zhaokang Yin, Junhao Wu, Zehui Xiong, Jianming Zhu, Jiajing Wu, and Zibin Zheng. Economic systems in metaverse: Basics, state of the art, and challenges. *arXiv preprint arXiv:2212.05803*, 2022.

Iden3. zkSnark circuit compiler. https://github.com/iden3/circom, 2022.

Lotfi Ismail, Dusit Niyato, Sumei Sun, Dong In Kim, Melike Erol-Kantarci, and Chunyan Miao. Semantic information market for the Metaverse: An auction based approach. In *2022 IEEE Future Networks World Forum (FNWF)*, pages 628–633. IEEE, 2022.

Alex Krizhevsky, Ilya Sutskever, and Geoffrey E Hinton. ImageNet classification with deep convolutional neural networks. *Communications of the ACM*, 60(6):84–90, 2017.

Zi Qin Liew, Hongyang Du, Wei Yang Bryan Lim, Zehui Xiong, Dusit Niyato, and Han Yu. Economics of semantic communication in metaverse: An auction approach. In *2023 IEEE 20th Consumer Communications & Networking Conference (CCNC)*, pages 398–403. IEEE, 2023.

Yijing Lin, Hongyang Du, Dusit Niyato, Jiangtian Nie, Jiayi Zhang, Yanyu Cheng, and Zhaohui Yang. Blockchain-aided secure semantic communication for AI-generated content in metaverse. *IEEE Open Journal of the Computer Society*, 4:72–83, 2023a.

Yijing Lin, Zhipeng Gao, Hongyang Du, Dusit Niyato, Jiawen Kang, Ruilong Deng, and Xuemin Sherman Shen. A unified blockchain-semantic framework for wireless edge intelligence enabled web 3.0. *IEEE Wireless Communications*, 31(2):126–133, 2023b.

Yijing Lin, Zhipeng Gao, Hongyang Du, Dusit Niyato, Jiawen Kang, Abbas Jamalipour, and Xuemin Sherman Shen. A unified framework for integrating semantic communication and AI-generated content in metaverse. *IEEE Network*, 38(4):174–181, 2023c. doi: 10.1109/MNET.2023.3321539.

Yijing Lin, Zhipeng Gao, Yaofeng Tu, Hongyang Du, Dusit Niyato, Jiawen Kang, and Hui Yang. A blockchain-based semantic exchange framework for web 3.0 toward participatory economy. *IEEE Communications Magazine*, 61(8):94–100, 2023d.

Yi Liu, Huimin Yu, Shengli Xie, and Yan Zhang. Deep reinforcement learning for offloading and resource allocation in vehicle edge computing and networks. *IEEE Transactions on Vehicular Technology*, 68(11):11158–11168, 2019.

Gaoyang Liu, Xiaoqiang Ma, Yang Yang, Chen Wang, and Jiangchuan Liu. FedEraser: Enabling efficient client-level data removal from federated learning models. In *2021 IEEE/ACM 29th International Symposium on Quality of Service (IWQOS)*, pages 1–10. IEEE, 2021.

Maheshi U Lokumarambage, Vishnu Sai Sankeerth Gowrisetty, Hossein Rezaei, Thushan Sivalingam, Nandana Rajatheva, and Anil Fernando. Wireless end-to-end image transmission system using semantic communications. *IEEE Access*, 11:37149–37163, 2023.

Xinlai Luo, Zhiyong Chen, Meixia Tao, and Feng Yang. Encrypted semantic communication using adversarial training for privacy preserving. *arXiv preprint arXiv:2209.09008*, 2022a.

Xinlai Luo, Benshun Yin, Zhiyong Chen, Bin Xia, and Jiangzhou Wang. Autoencoder-based semantic communication systems with relay channels. In *2022 IEEE International Conference on Communications Workshops (ICC Workshops)*, pages 711–716. IEEE, 2022b.

Xinlai Luo, Zhiyong Chen, Meixia Tao, and Feng Yang. Encrypted semantic communication using adversarial training for privacy preserving. *IEEE Communications Letters*, 27(6):1486–1490, 2023.

Brendan McMahan, Eider Moore, Daniel Ramage, Seth Hampson, and Blaise Aguera y Arcas. Communication-efficient learning of deep networks from decentralized data. In *Artificial Intelligence and Statistics*, pages 1273–1282. PMLR, 2017.

Wei Chong Ng, Hongyang Du, Wei Yang Bryan Lim, Zehui Xiong, Dusit Niyato, and Chunyan Miao. Stochastic resource allocation for semantic communication-aided virtual transportation networks in the metaverse. *arXiv preprint arXiv:2208.14661*, 2022.

Filip Radenović, Ahmet Iscen, Giorgos Tolias, Yannis Avrithis, and Ondřej Chum. Revisiting Oxford and Paris: Large-scale image retrieval benchmarking. In *Proceedings of the IEEE Conference on Computer Vision and Pattern Recognition*, pages 5706–5715, 2018.

Olga Russakovsky, Jia Deng, Hao Su, Jonathan Krause, Sanjeev Satheesh, Sean Ma, Zhiheng Huang, Andrej Karpathy, Aditya Khosla, Michael Bernstein, et al. ImageNet large scale visual recognition challenge. *International Journal of Computer Vision*, 115(3):211–252, 2015.

John Schulman, Filip Wolski, Prafulla Dhariwal, Alec Radford, and Oleg Klimov. Proximal policy optimization algorithms. *arXiv preprint arXiv:1707.06347*, 2017.

Rui Song, Shang Gao, Yubo Song, and Bin Xiao. A traceable and privacy-preserving data exchange scheme based on non-fungible token and zero-knowledge. In *2022 IEEE 42nd International Conference on Distributed Computing Systems (ICDCS)*, pages 224–234. IEEE, 2022.

Neal Stephenson. *Snow Crash: A Novel*. Spectra, 2003.

Jiawei Su, Zhixin Liu, Yuan-ai Xie, Kai Ma, Hongyang Du, Jiawen Kang, and Dusit Niyato. Semantic communication-based dynamic resource allocation in D2D vehicular networks. *IEEE Transactions on Vehicular Technology*, 72(8):10784–10796, 2023.

Giorgos Tolias, Filip Radenovic, and Ondrej Chum. Targeted mismatch adversarial attack: Query with a flower to retrieve the tower. In *Proceedings of the IEEE/CVF International Conference on Computer Vision*, pages 5037–5046, 2019.

Zhiguo Wan, Yan Zhou, and Kui Ren. zk-AuthFeed: Protecting Data Feed to Smart Contracts with Authenticated Zero Knowledge Proof. *IEEE Transactions on Dependable and Secure Computing*, 20(2):1335–1347, 2022.

Yining Wang, Mingzhe Chen, Walid Saad, Tao Luo, Shuguang Cui, and H Vincent Poor. Performance optimization for semantic communications: An attention-based learning approach. In *2021 IEEE Global Communications Conference (GLOBECOM)*, pages 1–6. IEEE, 2021.

Weilun Wang, Jianmin Bao, Wengang Zhou, Dongdong Chen, Dong Chen, Lu Yuan, and Houqiang Li. Semantic image synthesis via diffusion models. *arXiv preprint arXiv:2207.00050*, 2022a.

Zhendong Wang, Jonathan J Hunt, and Mingyuan Zhou. Diffusion policies as an expressive policy class for offline reinforcement learning. *arXiv preprint arXiv:2208.06193*, 2022b.

Jiacheng Wang, Hongyang Du, Zengshan Tian, Dusit Niyato, Jiawen Kang, and Xuemin Shen. Semantic-aware sensing information transmission for metaverse: A contest theoretic approach. *IEEE Transactions on Wireless Communications*, 22(8):5214–5228, 2023.

Zhenzi Weng and Zhijin Qin. Semantic communication systems for speech transmission . *IEEE Journal on Selected Areas in Communications*, 39(8):2434–2444, 2021.

Huiqiang Xie and Zhijin Qin. A lite distributed semantic communication system for Internet of Things. *IEEE Journal on Selected Areas in Communications*, 39(1):142–153, 2020.

Minrui Xu, Dusit Niyato, Junlong Chen, Hongliang Zhang, Jiawen Kang, Zehui Xiong, Shiwen Mao, and Zhu Han. Generative AI-empowered simulation for autonomous driving in vehicular mixed reality metaverses. *arXiv preprint arXiv:2302.08418*, 2023a.

Wei Xu, Zhaohui Yang, Derrick Wing Kwan Ng, Marco Levorato, Yonina C Eldar, and Mérouane Debbah. Edge learning for B5G networks with distributed signal processing: Semantic communication, edge computing, and wireless sensing. *IEEE Journal of Selected Topics in Signal Processing*, 17(1):9–39, 2023b.

Lei Yan, Zhijin Qin, Rui Zhang, Yongzhao Li, and Geoffrey Ye Li. QoE-aware resource allocation for semantic communication networks. In *GLOBECOM 2022-2022 IEEE Global Communications Conference*, pages 3272–3277. IEEE, 2022.

Wanting Yang, Hongyang Du, Zi Qin Liew, Wei Yang Bryan Lim, Zehui Xiong, Dusit Niyato, Xuefen Chi, Xuemin Sherman Shen, and Chunyan Miao. Semantic communications for future internet: Fundamentals, applications, and challenges. *IEEE Communications Surveys & Tutorials*, 25(1):213–250, 2022.

Guangyi Zhang, Qiyu Hu, Zhijin Qin, Yunlong Cai, and Guanding Yu. A unified multi-task semantic communication system with domain adaptation. In *GLOBECOM 2022-2022 IEEE Global Communications Conference*, pages 3971–3976. IEEE, 2022a.

Hongwei Zhang, Shuo Shao, Meixia Tao, Xiaoyan Bi, and Khaled B Letaief. Deep learning-enabled semantic communication systems with task-unaware transmitter and dynamic data. *IEEE Journal on Selected Areas in Communications*, 41(1):170–185, 2022b.

Bowen Zhang, Zhijin Qin, and Geoffrey Ye Li. Semantic communications with variable-length coding for extended reality. *arXiv preprint arXiv:2302.08645*, 2023a.

Haijun Zhang, Hongyu Wang, Yabo Li, Keping Long, and Arumugam Nallanathan. DRL-driven dynamic resource allocation for task-oriented semantic communication. *IEEE Transactions on Communications*, 71(7):3992–4004, 2023b.

Jiakang Zheng, Jiayi Zhang, Emil Björnson, and Bo Ai. Impact of channel aging on cell-free massive MIMO over spatially correlated channels. *IEEE Transactions on Wireless Communications*, 20(10):6451–6466, 2021.

Jiakang Zheng, Jiayi Zhang, Emil Björnson, Zhetao Li, and Bo Ai. Cell-free massive MIMO-OFDM for high-speed train communications. *IEEE Journal on Selected Areas in Communications*, 40(10):2823–2839, 2022.

Jiakang Zheng, Jiayi Zhang, Julian Cheng, Victor C M Leung, Derrick Wing Kwan Ng, and Bo Ai. Asynchronous cell-free massive MIMO with rate-splitting. *IEEE Journal on Selected Areas in Communications*, 41(5):1366–1382, 2023a.

Jiakang Zheng, Jiayi Zhang, Hongyang Du, Dusit Niyato, Bo Ai, Mérouane Debbah, and Khaled B Letaief. Mobile cell-free massive MIMO: Challenges, solutions, and future directions. *arXiv preprint arXiv:2302.02566*, 2023b.

Qingyang Zhou, Rongpeng Li, Zhifeng Zhao, Chenghui Peng, and Honggang Zhang. Semantic communication with adaptive universal transformer. *IEEE Wireless Communications Letters*, 11(3):453–457, 2021.

9

Large Language Model-Assisted Semantic Communication Systems

Shuaishuai Guo[1,2], Yanhu Wang[1,2], Biqian Feng[3], and Chenyuan Feng[4]

[1] School of Control Science and Engineering, Shandong University, Jinan, China
[2] Shandong Key Laboratory of Wireless Communication Technologies, Shandong University, Jinan, China
[3] Department of Electronic Engineering, Shanghai Jiao Tong University, China
[4] Eurecom, France

9.1 Introduction

Semantic communications prioritize the efficient transmission of semantic information while minimizing semantic-level loss, emphasizing this over bit-level errors in the transmission process [Qin et al., 2021]. There are three critical application domains that urgently necessitate semantic communications. The first involves the rapid transmission of substantial data volumes within short timeframes, such as in augmented reality/virtual reality (AR/VR), intelligent transportation/factory systems, and healthcare [Gündüz et al., 2023]. The second pertains to human–machine communications, as humans naturally communicate effectively at a semantic level and seek to engage machines at the same level of understanding. The third application domain encompasses communications in challenging environments characterized by severe channel attenuation such as deep space [Ha et al., 2022] and underwater communications [Karasalo et al., 2013].

Currently, there are two primary approaches to implementing semantic communications. The first method follows the conventional source–channel separation paradigm, wherein transmitters heavily compress data before transmitting semantic information across the channel. This approach involves source compression and subsequent decompression to mitigate semantic loss. By introducing semantic encoding and decoding, it seamlessly integrates with existing communication systems employing block designs [Niu et al., 2022]. However, this method isolates the source and channel components, potentially falling short of an optimal joint design. The second method involves implementing semantic communications through joint source–channel coding (JSCC) designs, leveraging the capabilities of deep learning (DL) [Farsad et al., 2018; Bourtsoulatze et al., 2019; Kurka and Gündüz, 2020; Xie et al., 2021]. For instance, the study [Farsad et al., 2018] have utilized bidirectional long short-term memory networks (BLSTM) as JSCC for text transmission, effectively minimizing word error rates. Moreover, Xie et al. [2021] proposed a DL-based method named "Deep-JSCC" aimed at minimizing semantic-level loss. For further insights into recent advancements in semantic communications, readers are encouraged to refer to Lan et al. [2021], Luo et al. [2022], and Gündüz et al. [2023].

While semantic communication schemes based on the JSCC architecture demonstrate remarkable performance, particularly in low signal-to-noise ratio (SNR) regimes, several issues warrant further exploration. Primarily, many prevalent semantic communication schemes focus primarily

Wireless Semantic Communications: Concepts, Principles, and Challenges, First Edition.
Edited by Yao Sun, Lan Zhang, Dusit Niyato, and Muhammad Ali Imran.

on extracting semantic information at the model level. Despite utilizing sophisticated structures and transmission schemes, their objective learning functions at the bit level—such as mean squared error (MSE) or cross-entropy (CE) loss—inevitably create a semantic gap. Besides, traditionally guided by Shannon's source-channel separation theorem [Shannon, 1948], existing communication systems rely on separate source and channel coding designs with layered structures. In contrast, many semantic communication schemes within the JSCC framework directly encode raw data into continuous-valued feature vectors for transmission. Unfortunately, this approach conflicts with established communication systems, necessitating significant updates across devices and systems, thereby incurring substantial costs. Recent efforts [Fu et al., 2023; Hu et al., 2023; Xie et al., 2023] have proposed leveraging vector quantization to bridge this compatibility gap. This strategy involves the joint maintenance of a codebook by both the transmitter and receiver. The transmitter maps the encoded continuous-valued feature vector to a corresponding codeword within the codebook and transmits the index of this codeword. Upon receiving the index, the receiver retrieves the associated codeword from the codebook to reconstruct the original feature vector. However, despite these advancements, the universality of semantic-level JSCC techniques remains constrained by the limitations of DL technology. Specifically, the need for redesigning and retraining the transceiver persists when either the receiver's task or encountered channels undergo changes. Moreover, the cost of training becomes prohibitive, particularly when dealing with large network models within the transceiver architecture.

To address the aforementioned challenges, we propose a novel signal shaping method for semantic communications (SSSC) that aimed at minimizing semantic loss. Additionally, we delve into the realm of semantic importance-aware communications (SIAC) by introducing a cross-layer manager into existing communication systems. Drawing inspiration from human-to-human semantic communications, where emphasis on "important" words occurs through louder speech or repeated usage, our focus lies in quantifying semantic importance and designing SIAC systems accordingly. There are several key issues with these schemes: How can we accurately measure semantic loss between different messages? How can we identify these "important" elements akin to human communication? Developing mathematical models capable of measuring semantic loss or quantifying semantic importance proves challenging due to its contextual nature, dependence on background knowledge, and numerous other contributing factors. This intricacy poses a significant hurdle in the development of such systems.

In this chapter, we leverage pretrained Large language models (LLMs) to enhance semantic communications, introducing two distinct schemes: SSSC and SIAC. In SSSC, we propose a signal shaping technique aimed at minimizing semantic loss within semantic communication systems featuring a limited number of message candidates. Specifically, we formulate a signal set optimization problem to minimize the semantic loss measured by the pretrained bidirectional encoder representations from transformers (BERT) model. The formulated problem is then transformed into a vector optimization subject to a power constraint and solved by an efficient projected gradient descent method. The SIAC scheme consists of two methods. The first method involves employing pre-trained generative LLMs, exemplified by ChatGPT-SIAC. This approach supports reliable cloud chat robot (CCR)-to-human semantic communications, where ChatGPT generates the messages for transmission. ChatGPT, in this application, emphasizes the essential words while creating the communication content. However, it is important to note that ChatGPT-SIAC requires access to the closed-source ChatGPT[1] and faces challenges due to the generative LLM's

1 As of March 1, 2023, the license for using ChatGPT is subject to the OpenAI Platform License v1.0. Under the license, one is granted access to ChatGPT and is permitted to use it to interact with the model.

inherent randomness. To address these challenges and expand the scope of SIAC applications, we introduce another method utilizing pretrained discriminative LLMs. Taking the open-source BERT[2] model as an example, we present the BERT-SIAC transmission scheme. By assessing the semantic importance of frames provided by ChatGPT or BERT, the transmitter can implement SIAC strategies, enhancing the reliability of transmitting crucial frames. Our experimental results demonstrate that both ChatGPT-SIAC and BERT-SIAC significantly outperform existing equal-priority communication strategies in terms of semantic loss.

The rest of this chapter is organized as follows. First, we introduce the SSSC system using pretrained LLMs in Section 9.2. We then present SIAC scheme using pretrained LLMs in Section 9.3. The future direction of using LLMs for semantic correction is pointed out in Section 9.4. Section 9.5 concludes the chapter.

9.2 SSSC Using Pretrained LLMs

9.2.1 System Model

In this chapter, we initially examine a message semantic communication system depicted in Figure 9.1. Within this framework, a message $\mathbf{m}_i$ chosen from a small message set $\mathcal{M}$ of size M is mapped to the ith signal vector $\mathbf{x}_i \in \mathbb{C}^N$, where N represents the number of channel uses. Signals carrying messages may be wrongly detected when passing through the noisy channel. Traditional communication systems commonly use metrics like bit/symbol/message error rates for system design, neglecting the semantic loss inherent in various message error detections. Until now, an efficient method to measure semantic loss has been lacking. Thanks to the advance of DL and its applications in nature language processing (NLP), some pretrained LLMs provide an efficient way to quantify the semantic similarity between two different messages. Leveraging the pretrained BERT model from Devlin et al. [2018], we define the semantic loss $A(\mathbf{m}_i, \mathbf{m}_j)$ between messages $\mathbf{m}_i$ and $\mathbf{m}_j$ as follows:

$$A(\mathbf{m}_i, \mathbf{m}_j) = 1 - \phi(\mathbf{m}_i, \mathbf{m}_j), \tag{9.1}$$

where $\phi(\mathbf{m}_i, \mathbf{m}_j) = \frac{B_\psi(\mathbf{m}_i)^T B_\psi(\mathbf{m}_j)}{||B_\psi(\mathbf{m}_i)||\cdot||B_\psi(\mathbf{m}_j)||}$ represents the semantic similarity of two messages $\mathbf{m}_i$ and $\mathbf{m}_j$. $B_\psi(\cdot)$ stands for the pretrained BERT model. In Figure 9.2, we show a table listing the semantic similarity of four messages. It can be seen that the value of $\phi(\mathbf{m}_i, \mathbf{m}_j)$ can well reflect the semantic similarity between $\mathbf{m}_i$ and $\mathbf{m}_j$.

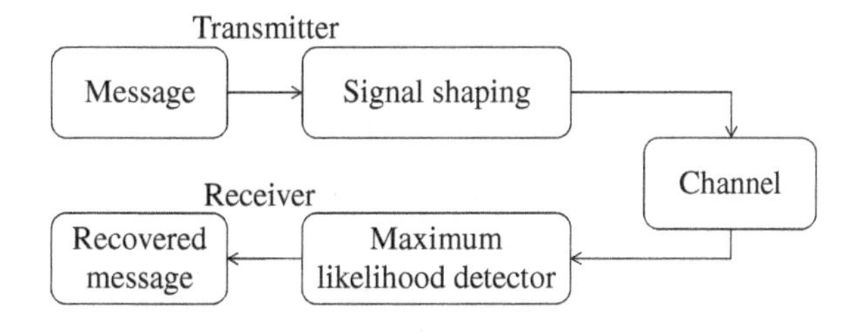

Figure 9.1 Demonstration of signal-shaping semantic communication systems, which target to minimize the error rate, in reducing the semantic loss.

2 BERT is released under the Apache License 2.0, which is a permissive open-source license. The Apache License 2.0 allows one to use, modify, and distribute BERT for both commercial and noncommercial purposes. It grants you the freedom to use the software without restrictions, subject to certain conditions outlined in the license.

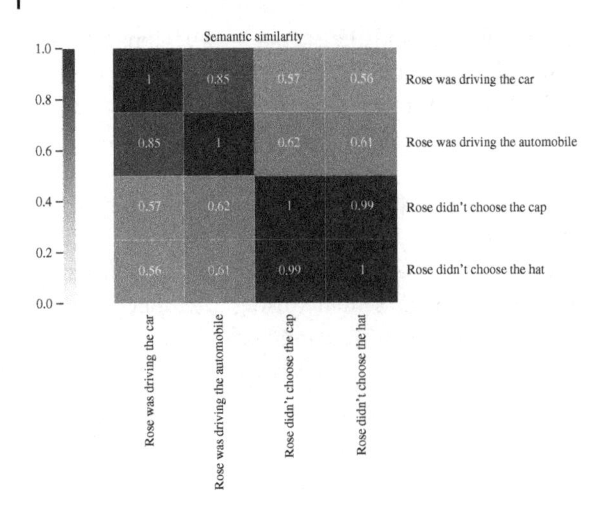

Figure 9.2 An example to show the semantic similarity computed by the pretrained BERT model.

Without loss of generality, the signal vectors are assumed to meet an average normalized power constraint, i.e., $\mathbf{E}(||\mathbf{x}_i||_2^2) \leq 1$. Given all messages being transmitted with an equal probability $\frac{1}{M}$, the power constraint can be formulated as

$$\frac{1}{M}\sum_{i=1}^{M}\mathbf{x}_i^H\mathbf{x}_i \leq 1. \tag{9.2}$$

The channel considered in this chapter is the additive white Gaussian noise (AWGN) channel that adds noise to achieve a given SNR γ. Let $\mathcal{X}$ represent the set of all legitimate transmitted signal vectors corresponding to M messages, i.e., $\mathcal{X} = \{\mathbf{x}_1, \mathbf{x}_2, \mathbf{x}_3, \dots, \mathbf{x}_M\}$. At the receiver, the maximum likelihood (ML) detection can be performed by

$$\hat{\mathbf{x}}_i = \arg\max_{\mathbf{x}_i \in \mathcal{X}} \mathbf{p}(\mathbf{y}|\mathbf{x}_i). \tag{9.3}$$

As $\mathbf{p}(\mathbf{y}|\mathbf{x}_i) \propto \exp(-||\mathbf{y}-\mathbf{x}_i||_2^2)$, the ML detector can be further expressed as

$$\hat{\mathbf{x}}_i = \arg\max_{\mathbf{x}_i \in \mathcal{X}} ||\mathbf{y}-\mathbf{x}_i||_2^2. \tag{9.4}$$

9.2.2 Problem Formalization

According to Tse and Viswanath [2005], the probability of pairwise error detection for any $\mathbf{x}_i \neq \mathbf{x}_j$ using the ML detector can be expressed as

$$P(\mathbf{x}_i, \mathbf{x}_j) = Q\left(\sqrt{\frac{\gamma||\mathbf{x}_i - \mathbf{x}_j||_2^2}{2}}\right), \tag{9.5}$$

where $Q(x) = \frac{1}{\sqrt{2\pi}}\int_x^{\infty}\exp\left(-\frac{u^2}{2}\right)du$. The corresponding semantic loss caused by the wrong detection between $\mathbf{x}_i$ and $\mathbf{x}_j$ can be expressed by

$$SL(\mathbf{m}_i, \mathbf{m}_j) = A(i,j)Q\left(\sqrt{\frac{\gamma||\mathbf{x}_i - \mathbf{x}_j||_2^2}{2}}\right). \tag{9.6}$$

For the transmission of all messages, a union upper bound on the average semantic loss can be expressed as

$$\overline{SL}(\mathcal{X}) = \frac{1}{M}\sum_{i=1}^{M}\sum_{j=1,j\neq i}^{M} SL(\mathbf{m}_i, \mathbf{m}_j). \tag{9.7}$$

Substituting Eq. (9.6) into Eq. (9.7), we can compute the upper bound for the semantic loss as

$$\overline{SL}(\mathcal{X}) = \frac{1}{M}\sum_{i=1}^{M}\sum_{j=1,j\neq i}^{M} A(i,j)Q\left(\sqrt{\frac{||\gamma\mathbf{x}_i - \mathbf{x}_j||_2^2}{2}}\right). \tag{9.8}$$

Therefore, the optimization problem of signal shaping to reduce the semantic loss can be formulated as:

$$\begin{aligned} (\mathbf{P1}):\ \text{Given} &: \gamma, A(i,j), \forall i,j \\ \text{Find} &: \mathcal{X} = \{\mathbf{x}_1, \mathbf{x}_2, \ldots, \mathbf{x}_M\} \\ \text{Minimize} &: \overline{SL}(\mathcal{X}) \\ \text{Subject to} &: \frac{1}{M}\sum_{i=1}^{M}\mathbf{x}_i^H\mathbf{x}_i \leq 1. \end{aligned} \tag{9.9}$$

9.2.3 Problem Transformation

Problem (**P1**) is a set optimization problem, which is difficult to solve. To make it tractable, we first transform it as a vector optimization problem and then propose an efficient projected gradient descent algorithm to deal with it.

First, we rewrite the power of the Euclidean distance between signal vectors as

$$||\mathbf{x}_i - \mathbf{x}_j||_2^2 = ||\mathbf{G}\mathbf{D}_\mathbf{z}(\mathbf{e}_i - \mathbf{e}_j)||_2^2, \tag{9.10}$$

where

$$\begin{aligned} \mathbf{G} &= \overbrace{[\mathbf{I}_N, \mathbf{I}_N, \ldots, \mathbf{I}_N]}^{M} \in \mathbb{C}^{N\times MN}, \\ \mathbf{D}_\mathbf{z} &= \text{diag}(\mathbf{z}) \in \mathbb{C}^{MN\times MN}, \\ \mathbf{z} &= [\mathbf{x}_1^T, \mathbf{x}_2^T, \ldots, \mathbf{x}_M^T]^T \in \mathbb{C}^{MN\times 1}, \\ \mathbf{e}_i &= \mathbf{g}_i \otimes \mathbf{x}_i \in \mathbb{C}^{MN\times 1},\ \mathbf{e}_j = \mathbf{g}_j \otimes \mathbf{x}_j \in \mathbb{C}^{MN\times 1}, \end{aligned}$$

and $\mathbf{g}_i \in \mathbb{C}^{M\times 1}$, $\mathbf{g}_i \in \mathbb{C}^{M\times 1}$ represent the ith and jth one-hot vectors with all zeros except a one at the ith position and jth position, respectively. Expanding the expression, we have

$$\begin{aligned} ||\mathbf{G}\mathbf{D}_\mathbf{z}(\mathbf{e}_i - \mathbf{e}_j)||_2^2 &= (\mathbf{e}_i - \mathbf{e}_j)^H\mathbf{D}_\mathbf{z}^H\mathbf{G}^H\mathbf{G}\mathbf{D}_\mathbf{z}(\mathbf{e}_i - \mathbf{e}_j) \\ &= \text{trace}\left(\mathbf{D}_\mathbf{z}^H\mathbf{R}_\mathbf{G}\mathbf{D}_\mathbf{z}\Delta\mathbf{E}_{i,j}\right), \end{aligned} \tag{9.11}$$

where $\mathbf{R}_\mathbf{G} = \mathbf{G}^H\mathbf{G}$ and $\mathbf{E}_{ij} = (\mathbf{e}_i - \mathbf{e}_j)(\mathbf{e}_i - \mathbf{e}_j)^H$.

Summarizing the aforementioned and based on the rule $\text{trace}(\mathbf{D}_\mathbf{u}\mathbf{A}\mathbf{D}_\mathbf{v}\mathbf{B}^T) = \mathbf{u}^H(\mathbf{A} \odot \mathbf{B})\mathbf{v}$ [Zhang, 2017], we have

$$||\mathbf{x}_i - \mathbf{x}_j||_2^2 = \mathbf{z}^H\mathbf{W}_{ij}\mathbf{z}, \tag{9.12}$$

where $\mathbf{W}_{ij} = \mathbf{R}_\mathbf{G} \odot \Delta\mathbf{E}_{ij}^T$.

With such transformation, the upper bound for the semantic loss can be rewritten as

$$\overline{SL}(\mathbf{z}) = \frac{1}{M}\sum_{i=1}^{M}\sum_{j=1,j\neq i}^{M} A(i,j)Q\left(\sqrt{\frac{\gamma\mathbf{z}^H\mathbf{W}_{ij}\mathbf{z}}{2}}\right), \tag{9.13}$$

and the power constraint can be rewritten as

$$\mathbf{z}^H\mathbf{z} \leq M. \tag{9.14}$$

Therefore, the set optimization problem can be reformulated as a vector optimization problem:

$$\begin{aligned} (\mathbf{P2}):\ \text{Given}\ &: \gamma, \mathbf{W}_{ij}, A(i,j), \forall i,j \\ \text{Find}\ &: \mathbf{z} \\ \text{Minimize}\ &: \overline{SL}(\mathbf{z}) \\ \text{Subject to}\ &: \mathbf{z}^H\mathbf{z} \leq M. \end{aligned} \tag{9.15}$$

9.2.4 Projected Gradient Descent Optimization

Problem (**P2**) is a non-convex problem, and the optimal solution to the problem (**P2**) is not unique.[3] To solve the problem (**P2**), we formulate the Lagrangian function as

$$L(\mathbf{z}, \lambda) = \overline{SL}(\mathbf{z}) + \lambda(\mathbf{z}^H\mathbf{z} - M). \tag{9.16}$$

According to the Karush–Kuhn–Tucker (KKT) conditions, the optimal solutions to the problem (**P2**) should satisfy

$$\begin{cases} \nabla_{\mathbf{z}} L(\mathbf{z}, \lambda) = 0, \\ \lambda(\mathbf{z}^H\mathbf{z} - M) = 0, \\ \lambda \geq 0. \end{cases} \tag{9.17}$$

Because $L(\mathbf{z}, \lambda)$ monotonically decreases with the power, thus it is minimized when the power constraint is met with strict equality, i.e., $\lambda(\mathbf{z}^H\mathbf{z} - M) = 0$. The first condition in Eq. (9.17) can be expressed as

$$\nabla_{\mathbf{z}} L(\mathbf{z}, \lambda) = \left[\Omega(\mathbf{z}) + 2\mu\mathbf{I}_{MN}\right]\mathbf{z} = 0, \tag{9.18}$$

where

$$\Omega(\mathbf{z}) = -\frac{1}{M}\sum_{i=1}^{M}\sum_{j=1,j\neq i}^{M}\sqrt{\frac{\gamma A(i,j)^2}{4\pi\mathbf{z}^H\mathbf{W}_{ij}\mathbf{z}}} \cdot e^{-\frac{\gamma\mathbf{z}^H\mathbf{W}_{ij}\mathbf{z}}{4}} \cdot \mathbf{W}_{ij}. \tag{9.19}$$

Clearly, the closed-form solution to Eq. (9.18) is difficult to obtain. In this chapter, we resort to a projected gradient descent method to find a good solution. In detail, we first compute the gradient descent direction in the kth iteration as

$$\mathbf{g}_k = -\Omega(\mathbf{z}_k)\mathbf{z}_k. \tag{9.20}$$

To ensure the power of the solution remains unchanged, we perform a projection by

$$\mathbf{g}_k^{\perp} = \mathbf{g}_k - \frac{\mathbf{z}_k^H\mathbf{g}_k\mathbf{z}_k}{||\mathbf{z}_k||_2^2}, \tag{9.21}$$

---

3 It is obvious that a same phase rotation on all signal vectors will not change their mutual Euclidean distances. Thus, a phase rotation on the optimal solution is still an optimal solution to achieve the minimum semantic loss.

such that $\mathbf{z}_k^H \mathbf{g}_k^{\perp} = 0$. After projection, we update the solution to be

$$\mathbf{z}_{k+1} = \cos\theta \cdot \mathbf{z}_k + \sin\theta \cdot \sqrt{M} \frac{\mathbf{g}_k^{\perp}}{\left\| \mathbf{g}_k^{\perp} \right\|_2}, \tag{9.22}$$

where $\theta \in [0, \frac{\pi}{2}]$ can be obtained by solving

$$\theta = \arg \min_{\theta \in [0, \frac{\pi}{2}]} \overline{SL}(\mathbf{z}_{k+1}). \tag{9.23}$$

By updating the solution until the stop criterion $\frac{\|\mathbf{g}_k^{\perp}\|_2}{\|\mathbf{g}_k\|_2} \leq \epsilon$ is met, we can obtain a good solution. For clarity, we summarize the projected gradient descent algorithm in Algorithm 9.1.

Algorithm 9.1 Projected gradient descent algorithm for minimizing semantic loss in semantic communications

Initialize $k = 1$, ϵ, and $\mathbf{z}_1$ with $\mathbf{z}_1^H \mathbf{z}_1 = M$.
repeat
  Compute the gradient descent direction by Eq. (9.20).
  Perform projection by Eq. (9.21).
  Search θ by solving Eq. (9.23).
  Update $\mathbf{z}_{k+1}$ by Eq. (9.22).
  $k \leftarrow k + 1$
until $\frac{\|\mathbf{g}_k^{\perp}\|_2}{\|\mathbf{g}_k\|_2} \leq \epsilon$
Output $\mathbf{z} = \mathbf{z}_k$.

9.2.5 Convergence and Computational Complexity Analysis

The convergence of the algorithm can be attributed to the lower bound of the semantic loss, as described in the following proposition.

Proposition 9.1 Algorithm 9.1 always guarantees $\overline{SL}(\mathbf{z}_{k+1}) \leq \overline{SL}(\mathbf{z}_k)$.

Proof: For a sufficient small θ with $\theta \to 0$, we can derive the first-order Taylor expansion of $\overline{SL}(\mathbf{z}_{k+1})$ based on Eq. (9.22) as

$$\overline{SL}(\mathbf{z}_{k+1}) \approx \overline{SL}(\mathbf{z}_k) - \mathbf{g}_k^H \cdot \sqrt{M} \frac{\mathbf{g}_k^{\perp}}{\left\| \mathbf{g}_k^{\perp} \right\|_2} \cdot \theta. \tag{9.24}$$

Since $\mathbf{g}_k^H \mathbf{g}_k^{\perp} = (1 - \cos^2\alpha) || \mathbf{g}_k ||_2^2 \geq 0$, where $\alpha = \arccos \frac{<\mathbf{g}_k, \mathbf{z}_k>}{||\mathbf{g}_k||_2 ||\mathbf{z}_k||_2}$, we have

$$\overline{SL}(\mathbf{z}_{k+1}) \leq \overline{SL}(\mathbf{z}_k). \tag{9.25}$$

The primary computational complexity of Algorithm 9.1 arises from computing the gradient descent direction using Eqs. (9.19) and (9.20). Its complexity can be assessed as $\mathcal{O}(N_{iter} M^4 N^2)$, where N_{iter} denotes the number of iterations executed by Algorithm 9.1. Despite the high computational complexity, especially for larger M, conducting signal shaping offline mitigates this issue. Therefore, the computational overhead is manageable and does not pose a significant challenge.

9.2.6 Experimental Results

In this section, we initially examine the convergence of the proposed algorithm and its sensitivity to the randomly generated initial solution $\mathbf{z}_1$ through simulations. For the simulations, we set $M = 16$, $N = 1, \gamma = 10$ dB, and $\epsilon = 10^{-2}$. Algorithm 9.1 is executed 10 times with 10 distinct initial solutions. The outcomes are depicted in Figure 9.3, revealing that Algorithm 9.1 achieves rapid convergence while displaying sensitivity to the choice of the initial solution. To address this sensitivity, we optimize the signal shaping using 50 randomly generated initial solutions. Subsequently, we select the solution that minimizes the semantic loss for further simulations.

Next, we present the optimized signal designs intended for message semantic communication systems featuring $M = 4, 8, 16$ candidate messages. For these designs, we maintain other parameters at $N = 1$, $\epsilon = 10^{-2}$, and $\gamma = 10$ dB. The signal designs are depicted in Figure 9.4, showcasing irregular constellations distinct from traditional ones. This divergence arises from considering the semantic meanings of individual signal points. Points with higher semantic similarities are positioned closer together, allowing more space for additional signal constellation points.

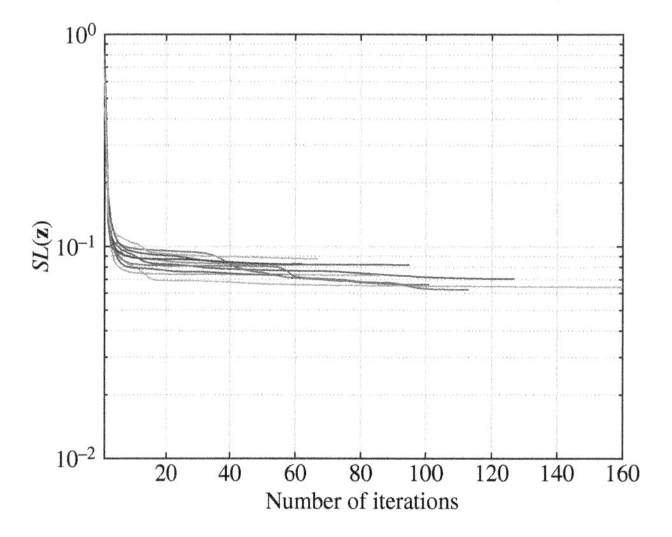

Figure 9.3 Convergence property of Algorithm 9.1 with 10 different randomly generated solutions.

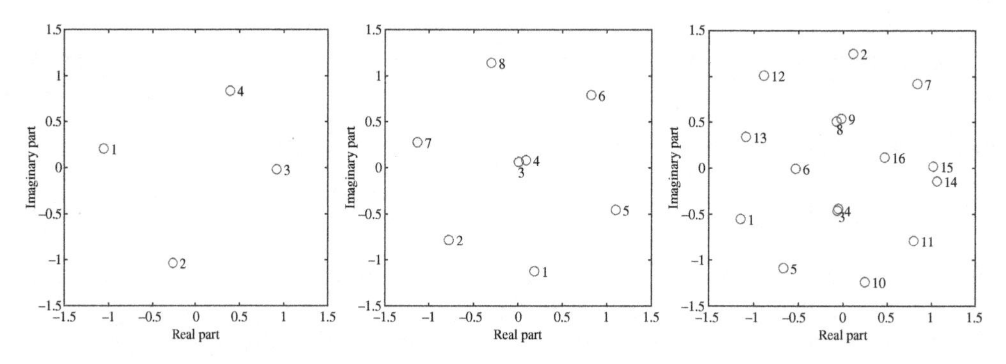

Figure 9.4 Signal constellation designs for semantic communication systems with $M = 4, 8,$ and 16 messages and $N = 1$ at $\gamma = 10$ dB.

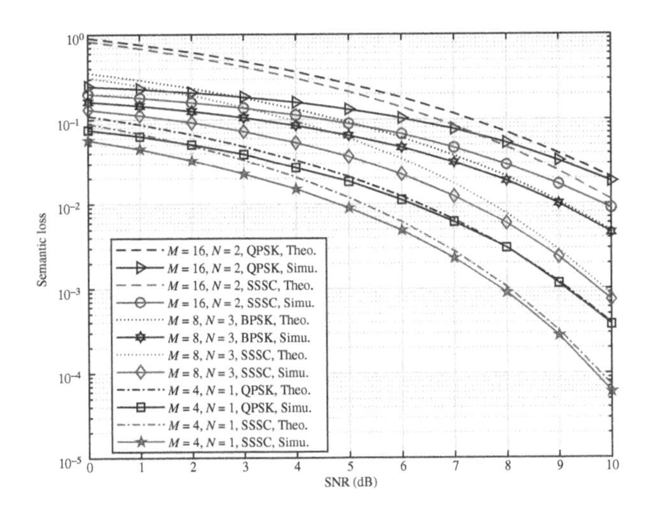

Figure 9.5 Semantic loss of different signal-shaping methods.

Furthermore, to demonstrate the superiority of the proposed SSSC method, we conduct a comparative analysis against signal designs capable of achieving the minimum message error rate, such as binary phase shift keying (BPSK) and quadrature phase shift keying (QPSK). The simulation results, presented in Figure 9.5, highlight that considering the semantic meaning of messages yields substantial performance advantages in reducing semantic loss within semantic communication systems. Specifically, in scenarios where $M = 4, N = 1$, SSSC outperforms QPSK by 1 dB at a semantic loss of 10^{-3}. For setups involving $M = 8$ and $N = 3$, SSSC demonstrates an approximately 0.8 dB improvement over BPSK at a semantic loss of 10^{-2}. Moreover, in configurations with $M = 16, N = 2$, SSSC surpasses QPSK by 1.3 dB at a semantic loss of 2×10^{-2}. Additionally, we provide numerical results for the theoretical upper bound of semantic loss. These results indicate that the theoretical upper bound tightly correlates with the high SNR regime.

9.3 SIAC Using Pretrained LLMs

9.3.1 System Model

In this part, our focus centers on exploring a communication system designed to ensure the reliable transmission of semantic meaning within data frames, despite limitations in communication and network resources between the transmitter and receiver, as depicted in Figure 9.6. This system holds potential applications in various domains, such as CCR-to-human communications or agent-to-agent semantic communications. At the transmitter end, we consider a batch of messages ready for transmission. In scenarios like CCR-to-human communications, these messages might be generated by generative LLMs, such as ChatGPT, or selected from a knowledge base system. Before transmission, these messages are segmented into a total of S frames. Conventional data-oriented communication systems often overlook the semantic meaning of the transmitted data. Consequently, our aim in this chapter is to minimize semantic loss during data transmission. To achieve this, we prioritize quantifying the importance of each frame and implement semantic importance-aware priority-based communications for individual frames. Our proposed solution involves a cross-layer design where a pretrained LLM is integrated within or connected by the cross-layer manager, as illustrated in Figure 9.7. This manager interfaces with all layers, and by

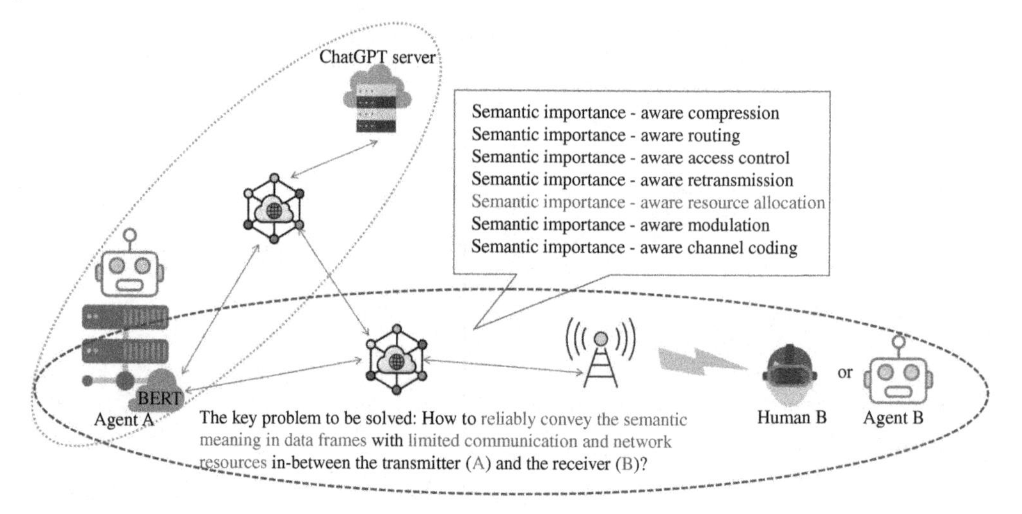

Figure 9.6 Demonstration of SIAC systems, which target to reliably transmit the semantic meaning in data frames.

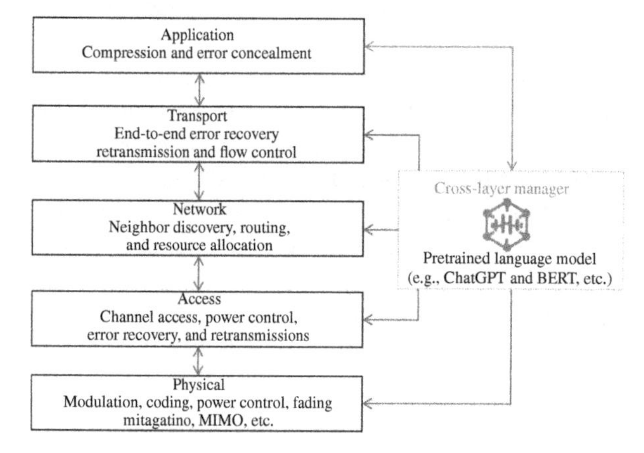

Figure 9.7 SIAC cross-layer structure with pretrained LLMs (e.g., ChatGPT and BERT) embedded in/connected by the cross-layer manager.

collaborating with the application layer, the pretrained LLM assesses and outputs the semantic importance of frames. This frame importance information is passed to the physical, access, network, transport, and application layers, allowing for priority-based parameter configurations across one or multiple layers. To facilitate this process, we present a range of priority-based communication technologies that could be adopted:

- *Physical layer*: Importance-aware modulation, coding, power control, fading mitigation, and multiple-input multiple-output (MIMO) setups;
- *Access layer*: Importance-aware channel access, power control, error recovery, and retransmissions;
- *Network layer*: Importance-aware routing and resource allocation;
- *Transport layer*: Importance-aware error recovery, retransmissions, and flow control;
- *Applications layer*: Importance-aware compression and error concealment.

The crux of implementing SIAC lies in identifying the semantic importance of a frame. Nonetheless, developing a mathematical model to precisely quantify this semantic importance proves challenging due to its contextual dependency, reliance on background knowledge, and numerous other contributing factors. In Sections 9.3.2 and 9.3.3, we will elucidate our methodologies for quantifying semantic importance utilizing pretrained LLMs.

9.3.2 Semantic Importance Quantification Using Generative LLMs like ChatGPT

In CCR-to-human communications, the most advanced and widely utilized chat robots presently rely on ChatGPT for generating content. Given that ChatGPT generates communication content, leveraging its capabilities to recognize the semantic importance within that content is a straightforward approach. To measure the semantic importance of frames, we define frame importance based on the number of words they encompass, as exemplified in Table 9.1. The greater the number of important words within a frame, the higher its overall importance. With this definition in mind, ChatGPT can prioritize emphasizing the crucial words while generating communication content. Quantifying the semantic importance of a frame becomes achievable by tallying the count of these pivotal words within it.

However, this method encounters two primary challenges. The first challenge pertains to the necessity of accessing the closed-source ChatGPT, limiting its applicability. The second challenge arises from ChatGPT's nature as a generative model, which inherently incorporates randomness. This variability is evident in the diverse outputs of ChatGPT, as demonstrated in the second output in Table 9.1. Consequently, ChatGPT may recognize different critical words for the same content, presenting a potential inconsistency. Since word importance recognition occurs prior to transmission and can be performed frequently, the impact of this randomness can be mitigated. One approach to address this issue involves minimizing randomness by aggregating multiple results through techniques like majority voting or arithmetic averaging, as illustrated in Table 9.1.

Table 9.1 Important ("1") and non-important ("0") words in "It is an important step towards equal rights for all passengers."

Frame Data	ChatGPT Output 1	ChatGPT Output 2	ChatGPT Output 3	Majority Vote	Arithmetic Average
It	0	0	0	0	0
is	0	0	0	0	0
an	0	0	0	0	0
important	1	1	1	1	1
step	1	1	1	1	1
towards	0	0	0	0	0
equal	1	1	1	1	1
rights	1	1	1	1	1
for	0	0	0	0	0
all	1	0	1	1	0.67
passengers	1	1	1	1	1
Sum	6	5	6	6	5.67

9.3.3 Semantic Importance Quantification Using Discriminative LLMs like BERT

To enhance the versatility of SIAC's application scenarios and circumvent the unpredictability inherent in a generative LLM, we introduce an importance indicator for words/frames utilizing pretrained discriminative LLMs such as BERT. Our approach involves defining word/frame importance based on the semantic loss resulting from the absence of a specific word/frame. A higher loss signifies greater importance of the word/frame in conveying semantic meaning. We calculate this semantic loss using the formula defined in Eq. (9.1).

To illustrate this concept, we present an example in Table 9.2. In the sentence "It is an important step toward equal rights for all passengers," words such as "It," "is," and "for" are considered less crucial. However, if words like "step," "rights," and "passengers" are omitted, it could significantly alter the sentence's intended meaning. In Table 9.2, it is evident that "step," "rights," and "passengers," exhibit greater semantic loss, indicating their higher importance in conveying meaning.

9.3.4 Overhead Analysis

Incorporating pretrained LLMs to assess frame importance before transmission unavoidably results in increased costs and delays. Within this section, we categorize application scenarios into three distinct groups and analyze the incremental costs and delays for each separately. To facilitate clear understanding of this classification, we employ examples involving the use of a CCR (Agent **A**) aiding a visually impaired individual (Human **B**) in completing daily-life tasks.

9.3.4.1 Communication Content Is Generated by Generative LLMs like ChatGPT

An illustrative scenario involves Agent **A** assisting visually impaired individual **B** by providing answers using ChatGPT. In these scenarios, **A** simply adjusts the input prompts for ChatGPT to highlight important words concurrently. Our empirical studies in Section 9.3.6 demonstrate that this adjustment incurs only a minimal increase in delay compared to generating communication content without emphasizing the important words.

9.3.4.2 Communication Content Has Already Been Pre-generated by Other Sources

A typical scenario involves Agent **A** reading a book or newspaper to assist the visually impaired individual, **B**. In such cases, the communication content can be entirely preprocessed offline, i.e., before transmission from **A** to **B**. During online transmission, the transmitter only needs to prioritize communication based on the importance of each frame.

Table 9.2 Semantic loss of each word in the sentence "It is an important step towards equal rights for all passengers."

Frame Data	Semantic Loss	Frame Data	Semantic Loss
It	0.0274	equal	0.0330
is	0.0163	rights	0.0705
an	0.0574	for	0.0167
important	0.0164	all	0.0084
step	0.0544	passengers	0.0665
towards	0.0754	—	—

9.3.4.3 Communication Content Is Generated by Other Sources in Real-Time

A typical scenario involves Agent **A** offering real-time voice navigation services for the visually impaired individual, **B**. In such applications, it is crucial to consider the delay and computational cost associated with processing the content. Let t_p denote the incremental delay in processing a frame. This delay can fluctuate due to network latency, processing request time, and the response duration of the language models. The response time of these models is further influenced by factors like input text complexity, available computational resources, and server load. Original LLMs boast an extensive number of parameters (e.g., BERT with 340 million parameters, and ChatGPT based on generative pre-trained transformer [GPT]-3.5 with 175 billion parameters), demanding substantial computational resources for inference and potentially causing significant delays. However, despite these costs and delays, employing such models remains invaluable in certain applications, ensuring reliable semantic information delivery. For instance, providing navigation information to the visually impaired **B** at a semantic level could prevent serious hazards. To mitigate these costs and delays, one approach is to use a self-hosted version of ChatGPT/BERT or similar LLMs. This strategy ensures that inference costs and delays depend solely on specific hardware and software configurations. Additionally, researchers have developed smaller, optimized versions of LLMs like DistilBERT and MobileBERT, with fewer parameters. These models incur lower inference costs and can be utilized in applications with more constrained computational resources such as mobile devices or embedded systems.

9.3.5 Semantic Importance-aware Power Allocation

To mitigate semantic loss, we can craft semantic importance-aware priority-based communication strategies within various layers: physical, access, network, transport, or application. Numerous communication techniques stand open for redesigning, offering diverse avenues for implementation. In this section, we focus on illustrating semantic importance-aware power allocation as a demonstrative case.

It is essential to note that resource allocation traverses multiple layers within the protocol stack, such as the physical and access layers, as depicted in Figure 9.7. Specifically, in this chapter, our emphasis lies on semantic importance-aware power allocation within the physical layer. Assuming that all frames undergo transmission through a Rayleigh fading channel, the outage probability of the ith frame can be mathematically expressed as

$$P_i^{out} = P\left(\frac{p_i|h|^2}{\sigma^2} < \gamma_{th}\right) = P\left(|h|^2 < \frac{\gamma_{th}\sigma^2}{p_i}\right), \tag{9.26}$$

where p_i signifies the power allocated for transmitting the ith frame, h represents the channel coefficient that adheres to a zero-mean unit-variance complex Gaussian distribution. Additionally, σ^2 denotes the noise variance, and γ_{th} stands for the SNR threshold. Let $g = |h|^2$ denote the channel gain. Given that h follows a zero-mean unit-variance complex Gaussian distribution, g follows an exponential distribution with a mean of 1. The cumulative distribution function (CDF) of g can be expressed as

$$F_g(g) = 1 - \exp(-g). \tag{9.27}$$

Let $g_i = \frac{\gamma_{th}\sigma^2}{p_i}$. Then, the outage probability of the ith frame can be expressed as

$$P_i^{out} = F_g(g_i) = 1 - \exp\left(-\frac{\gamma_{th}\sigma^2}{p_i}\right). \tag{9.28}$$

Hence, the optimization of power allocation to minimize the expected important word errors/semantic loss can be expressed as

$$\min_{\sum_{i=1}^{N} p_i = P_{total}} \sum_{i=1}^{N} w_i P_i^{out}, \tag{9.29}$$

where P_{total} is the total power of all frames. In ChatGPT-SIAC, w_i represents the count of important words contained within the ith frame. Meanwhile, within BERT-SIAC, w_i signifies the semantic importance of the ith frame, a metric defined by the BERT model. The problem outlined in Eq. (9.29) typifies a manifold optimization challenge, and for its resolution, we have employed the Manopt toolbox [Boumal et al., 2014]. To facilitate reproducibility and accessibility, we have made the implementation code available on GitHub[4].

9.3.6 Experimental Results

In this section, we demonstrate our experimental results. In our experiments, we assess the effectiveness of SIAC using the European Parliament proceedings dataset [Koehn, 2005], encompassing approximately 2 million sentences and 53 million words. To prepare the dataset, we conduct preprocessing and create 100 batches, each comprising 100 words. During the experiments, we handle one batch at a time. We organize the data by packaging every five words into a frame, resulting in a total of 20 frames intended for transmission in each batch. To ascertain the significance of each frame, we employ both ChatGPT and BERT for quantification. For transparency and access, all data and detailed experiment information can be found on our GitHub repository.

In the conducted experiments, we compared three distinct schemes: ChatGPT-SIAC, BERT-SIAC, and equal-priority transmission. Our experimental setup involved fixing σ^2 at

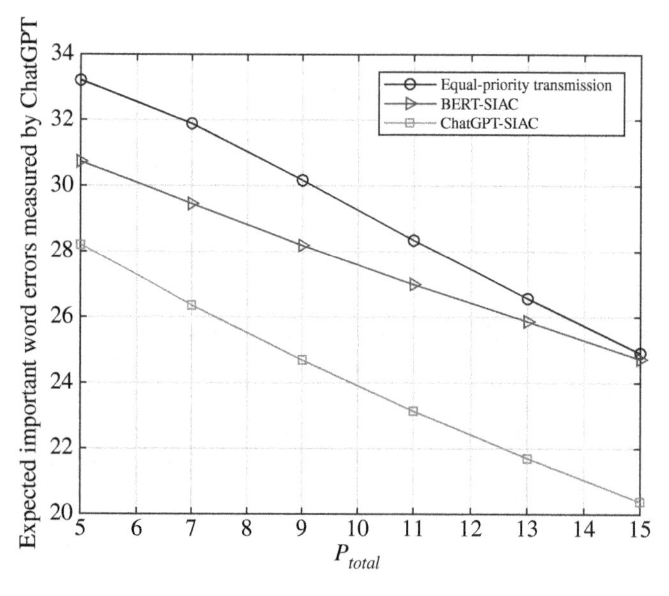

Figure 9.8 Expected important words errors versus the total power, where the word importance is indicated by ChatGPT.

4 https://github.com/SSG2019.

1 W and γ_{th} at 0 dB, while varying the total power P_{total} within the range of 5–15 W. To allocate power among the 20 frames within a batch, we solved the optimization problem outlined in Eq. (9.29). Our comparison of these schemes focused on evaluating two performance metrics affected by frame outages. The first metric, termed "expected important word errors," is derived based on whether a word is deemed important or not, determined by ChatGPT's output. Figure 9.8 depicts the experimental results, revealing ChatGPT-SIAC's optimal performance, as it aims to minimize these expected important word errors. BERT-SIAC also outperforms equal-priority transmission, particularly evident when the transmission power budget is limited. However, the performance gain diminishes with increasing total power. The second metric, termed "expected semantic loss" and defined in Eq. (9.1) by the BERT model utilizing semantic similarity, is showcased in Figure 9.9. Here, BERT-SIAC emerges as optimal due to its power allocation strategy directly linked to minimizing the expected semantic loss. Additionally, ChatGPT-SIAC exhibits lower semantic loss compared to equal-priority transmission. These comparative analyses under both performance metrics validate the superiority of the proposed SIAC schemes, affirming their advantages in mitigating semantic loss and minimizing important word errors in the transmission process.

Additionally, we conducted an empirical study to examine the average word generation rate of ChatGPT under varying prompts. During this study, we presented ChatGPT with two types of prompts: an original prompt and a copied prompt modified by appending the instruction, "Please highlight the important words to the semantic meanings in every sentence using boldface fonts." Subsequently, we tracked and recorded the word generation speed within the chat dialogue for both scenarios. The findings of this study are visually depicted in Figure 9.10. The experiment results indicate a slight decrease in word generation speed when simultaneously highlighting the important words, suggesting a marginal increase in delay.

Moreover, we recorded the frame importance quantification time during our experiments, and the results are depicted in Figure 9.11. The findings indicate that ChatGPT swiftly outputs the frame importance, benefitting from its operation on robust servers. Conversely, a self-hosted BERT model operating on an Intel(R) i7-11700@2.5 GHz and NVIDIA GeForce GTX 3090 requires

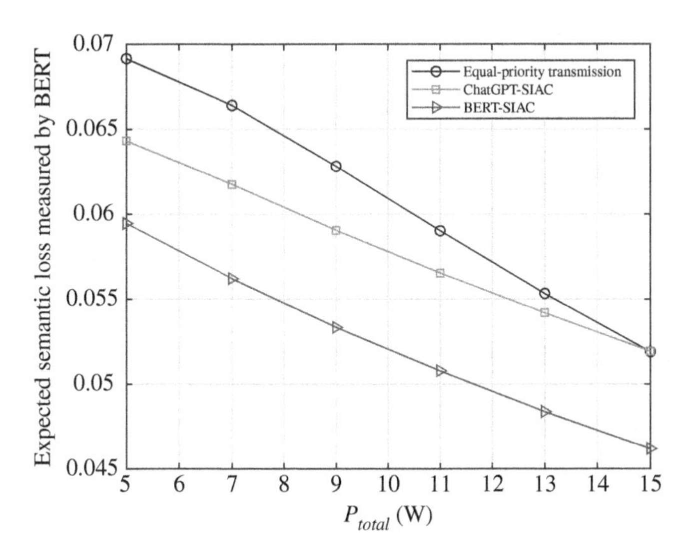

Figure 9.9 Expected semantic loss versus the total power, where the semantic loss is measured by BERT using Eq. (9.1).

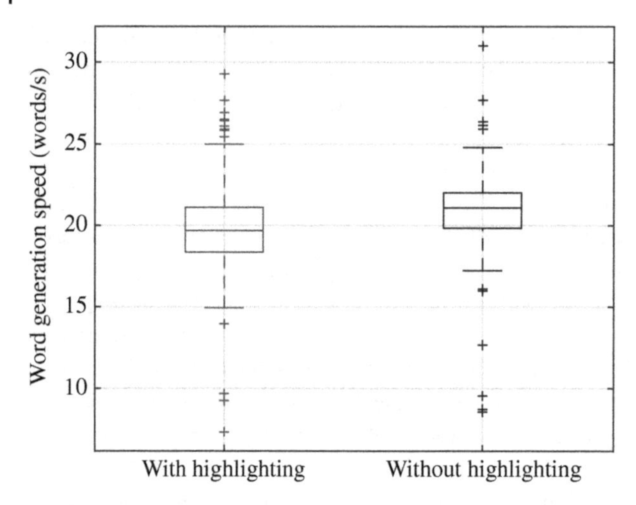

Figure 9.10 Word generation speed of ChatGPT with and without highlighting the important words.

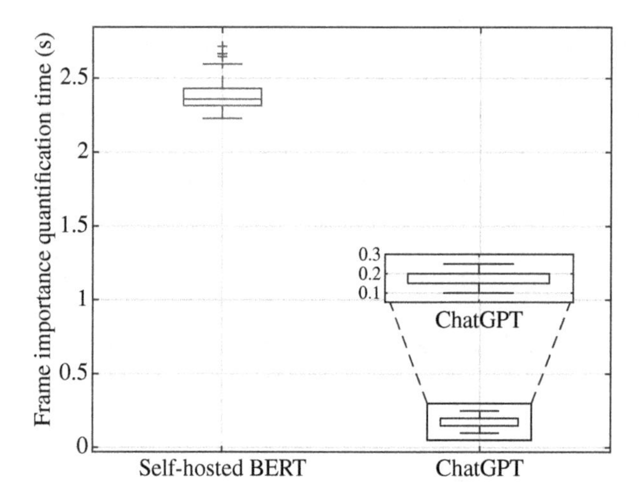

Figure 9.11 Frame importance quantification time of self-hosted BERT and ChatGPT.

relatively more time, operating at a pace of approximately 25 frames per minute. This pace might be acceptable for applications tolerant of delays; however, for time-sensitive applications, one might consider substituting the BERT model with a simplified version to mitigate inference delays.

9.4 Future Direction of Using LLMs: Semantic Correction

In Section 9.3, we quantified the semantic importance and investigated SIACs to reliably convey the semantics with limited communication and network resources. However, in the face of random wireless channel fading and random noise, we still cannot guarantee the complete reliable transmission of information. When the semantic importance-aware signals reach the receiver, they are detected as bits and corrected in the presence of correctable errors. Those frames that cannot be corrected will be erased. Erased or lost frames affect the receiver's interpretation and decision-making of the information. To address this issue, we can use LLMs for semantic correction.

9.4.1 ChatGPT-powered Semantic Correction

ChatGPT has been developed by training extensively on textual data, enhancing its ability to understand language contexts, grammar rules, and the use of common vocabulary. Drawing upon the knowledge gained during its training, ChatGPT can effectively employ contextual information and linguistic patterns to predict the most probable words or content within a given context. This capability enables us to utilize ChatGPT for completing erased or missing parts in messages. Specifically, we substitute the missing parts of the message with [MASK]. Following this, we input them into ChatGPT with a prompt such as "Fill the position of [MASK] in the text." Table 9.3 provides an example. It can be seen that ChatGPT is able to correct most missing frame data. However, employing ChatGPT for semantic correction exhibits some randomness, as evident from rows 1 and 7.

9.4.2 BERT-Powered Semantic Correction

One of BERT's pretraining tasks is MLM, which stands for masked language model. During the training process, a portion of the input text is randomly masked (replaced with special tokens), and the model is required to predict the masked words based on the surrounding context. This allows us to utilize BERT for filling in missing frames in messages. Specifically, for a given text, the missing or erased parts can be replaced with special tokens (such as [MASK]). The processed text is then passed to the BERT model for encoding. The encoded text is fed through the BERT model, enabling the model to learn word vector representations for each position based on contextual information. For the masked parts, the model predicts potential words or phrases to fill in those positions. BERT typically generates a probability distribution for each word in the vocabulary, and the content to

Table 9.3 Utilizing ChatGPT for semantic correction of missing word in sentence "It is an important step towards equal rights for all passengers."

Lost Frame Data	1 ChatGPT Semantic Correction	2 ChatGPT Semantic Correction	3 ChatGPT Semantic Correction
It	It	*This*	*Inclusion*
is	*was*	*was*	*was*
an	an	an	an
important	important	important	important
step	step	step	step
towards	*toward*	towards	towards
equal	*protecting*	*ensuring*	equal
rights	rights	rights	rights
for	for	for	for
all	all	all	*airplane*
passengers	*people*	*people*	*people*

Table 9.4 Utilizing BERT for semantic correction of missing word in sentence "It is an important step towards equal rights for all passengers."

Lost Frame Data	BERT Semantic Correction	Lost Frame Data	BERT Semantic Correction
It	it	equal	equal
is	is	rights	*pay*
an	an	for	for
important	important	all	all
step	step	passengers	.
towards	towards	—	—

fill the missing parts is chosen based on the model's predicted probability distribution. The word or phrase with the highest probability is selected to fill in the missing parts.

To demonstrate the effectiveness of this idea, an example is listed in Table 9.4. Using BERT can correct most words; however, it fails to correct when crucial words like "rights" or "passengers" are missing. So, it is important to note that while LLMs have shown great potential in various natural language understanding and generation tasks, applying them in this context might require adaptation and optimization to effectively handle the specifics of error correction in transmitted data. Moreover, computational constraints and latency should also be considered when implementing such a system in real-time communication scenarios.

9.5 Conclusion

This chapter explored the utilization of LLMs in assisting semantic communications, presenting two innovative schemes: SSSC and SIAC. In SSSC, leveraging the pretrained BERT model, we defined semantic loss and formulated a signal set optimization problem. We introduced an efficient projected gradient descent algorithm to address this problem and compared our proposed signal design with existing approaches, achieving minimum message error rates. Simulation results substantiated the considerable gains offered by SSSC in reducing semantic loss. Moving to SIAC, we proposed a cross-layer design integrating a pretrained LLM within the cross-layer manager to quantify the semantic importance of data frames. Utilizing this quantified semantic importance, we explored semantic importance-aware power allocation. Unlike conventional deep JSCC-based semantic communication schemes, SIAC seamlessly integrates into current communication systems solely by introducing a cross-layer manager. Our experimental findings highlight that the proposed SIAC scheme achieves significantly lower semantic loss compared to existing equal-priority communications.

Acronyms

AR	augmented reality
AWGN	additive white Gaussian noise
BERT	bidirectional encoder representations from transformers

BLSTM	bidirectional long and short term memory network
BPSK	binary phase shift keying
ChatGPT	Chat generative pre-trained transformer
CCR	cloud chat robot
CDF	cumulative distribution function
CE	cross entropy
DL	deep learning
GPT	generative pre-trained transformer
JSCC	joint source-channel coding
KKT	Karush–Kuhn–Tucke
LLM	large language model
MSE	mean squared error
MIMO	multiple-input multiple-output
ML	maximum likelihood
NLP	natural language processing
QPSK	quadrature phase shift keying
SIAC	semantic importance-aware communications
SSSC	signal shaping semantic communications
SNR	signal-to-noise ratio
VR	virtual reality

References

N. Boumal, B. Mishra, P. Absil, and R. Sepulchre. Manopt, a Matlab toolbox for optimization on manifolds. *The Journal of Machine Learning Research*, 15(1):1455–1459, 2014.

E. Bourtsoulatze, D. B. Kurka, and D. Gündüz. Deep joint source-channel coding for wireless image transmission. In *ICASSP 2019 - 2019 IEEE International Conference on Acoustics, Speech and Signal Processing (ICASSP)*, pages 4774–4778, 2019. doi: 10.1109/ICASSP.2019.8683463.

J. Devlin, M. Chang, K. Lee, and K. Toutanova. BERT: Pre-training of deep bidirectional transformers for language understanding. *arXiv preprint,* arXiv:1810.04805, 2018.

N. Farsad, M. Rao, and A. Goldsmith. Deep learning for joint source-channel coding of text. In *2018 IEEE International Conference on Acoustics, Speech and Signal Processing (ICASSP)*, pages 2326–2330, 2018. doi: 10.1109/ICASSP.2018.8461983.

Q. Fu, H. Xie, Z. Qin, G. Slabaugh, and X. Tao. Vector quantized semantic communication system. *IEEE Wireless Communications Letters*, 12(6):982–986, 2023. doi: 10.1109/LWC.2023.3255221.

D. Gündüz, Z. Qin, I. E. Aguerri, H. S. Dhillon, Z. Yang, A. Yener, K. K. Wong, and C.-B. Chae. Beyond transmitting bits: Context, semantics, and task-oriented communications. *IEEE Journal on Selected Areas in Communications*, 41(1):5–41, 2023. doi: 10.1109/JSAC.2022.3223408.

T. Ha, D. Lee, J. Oh, Y. Jeon, C. Lee, and S. Cho. DTN-based multi-link bundle protocol architecture for deep space communications. In *2022 13th International Conference on Information and Communication Technology Convergence (ICTC)*, pages 1164–1166, 2022. doi: 10.1109/ICTC55196.2022.9952925.

Q. Hu, G. Zhang, Z. Qin, Y. Cai, G. Yu, and G. Y. Li. Robust semantic communications with masked VQ-VAE enabled codebook. *IEEE Transactions on Wireless Communications*, 22(12):8707–8722, 2023. doi: 10.1109/TWC.2023.3265201.

I. Karasalo, T. Öberg, B. Nilsson, and S. Ivansson. A single-carrier turbo-coded system for underwater communications. *IEEE Journal of Oceanic Engineering*, 38(4):666–677, 2013. doi: 10.1109/JOE.2013.2278892.

Philipp Koehn. Europarl: A parallel corpus for statistical machine translation. In *Proceedings of Machine Translation Summit X: Papers*, pages 79–86, 2005.

D. B. Kurka and D. Gündüz. Deep joint source-channel coding of images with feedback. In *ICASSP 2020 - 2020 IEEE International Conference on Acoustics, Speech and Signal Processing (ICASSP)*, pages 5235–5239, 2020. doi: 10.1109/ICASSP40776.2020.9054216.

Q. Lan, D. Wen, Z. Zhang, Q. Zeng, X. Chen, P. Popovski, and K. Huang. What is semantic communication? A view on conveying meaning in the era of machine intelligence. *Journal of Communications and Information Networks*, 6(4):336–371, 2021. doi: 10.23919/JCIN.2021.9663101.

X. Luo, H. Chen, and Q. Guo. Semantic communications: Overview, open issues, and future research directions. *IEEE Wireless Communications*, 29(1):210–219, 2022. doi: 10.1109/MWC.101.2100269.

Kai Niu, Jincheng Dai, Shengshi Yao, Sixian Wang, Zhongwei Si, Xiaoqi Qin, and Ping Zhang. A paradigm shift toward semantic communications. *IEEE Communications Magazine*, 60(11):113–119, 2022. doi: 10.1109/MCOM.001.2200099.

Z. Qin, X. Tao, J. Lu, W. Tong, and G. Ye Li. Semantic communications: Principles and challenges. *arXiv preprint*, arXiv:2201.01389, 2021.

C. E. Shannon. A mathematical theory of communication. *The Bell System Technical Journal*, 27(3):379–423, 1948. doi: 10.1002/j.1538-7305.1948.tb01338.x.

D. Tse and P. Viswanath. *Fundamentals of Wireless Communication*. Cambridge University Press, 2005.

H. Xie, Z. Qin, G. Y. Li, and B. Juang. Deep learning enabled semantic communication systems. *IEEE Transactions on Signal Processing*, 69:2663–2675, 2021. doi: 10.1109/TSP.2021.3071210.

S. Xie, S. Ma, M. Ding, Y. Shi, M. Tang, and Y. Wu. Robust information bottleneck for task-oriented communication with digital modulation. *IEEE Journal on Selected Areas in Communications*, 41(8):2577–2591, 2023. doi: 10.1109/JSAC.2023.3288252.

X. Zhang. *Matrix Analysis and Applications*. Cambridge University Press, 2017.

10

RIS-Enhanced Semantic Communication

Bohao Wang, Ruopeng Xu, Zhaohui Yang, and Chongwen Huang

College of Information Science and Electronic Engineering, Zhejiang University, Hangzhou, Xihu Region, China

10.1 RIS-Empowered Communications

Reconfigurable intelligent surfaces (RISs) represent a cutting-edge innovation in wireless communications, leveraging advances in material science and electromagnetic wave manipulation to achieve the goal of programming the environment.

10.1.1 What Is RIS?

RISs are essentially artificially structured surfaces composed of numerous tiny elements, each capable of individually controlling electromagnetic waves. The genesis of RIS technology is deeply rooted in the early explorations of metamaterials—artificial electromagnetic media engineered at a subwavelength scale [Di Renzo et al., 2020]. Initially, the primary applications of metamaterials were focused on revolutionary concepts such as cloaking, which involves rendering objects invisible or undetectable, and subwavelength imaging [Engheta and Ziolkowski, 2006]. However, the evolving demands for more efficient and higher-capacity wireless communication systems led researchers to pivot their focus. This shift in research, highlighted in Cui et al. [2014], explored how metamaterials could be leveraged to improve signal transmission and reception. Such explorations have significantly contributed to the development of the RIS concept, a topic further elaborated in Huang et al. [2019].

10.1.2 RIS-Assisted Communication System

RIS-assisted communication systems are introducing transformative features and benefits to the realm of wireless communication, significantly altering its landscape with several key attributes.

First and foremost, RIS technology enables dynamic manipulation of electromagnetic wave properties such as phase, amplitude, and polarization, thereby altering the wavefront of reflected, transmitted, refracted, or scattered waves [Di Renzo et al., 2020]. This level of programmability renders the wireless communication environment both adaptable and controllable, representing a substantial leap forward in communication system design.

Furthermore, the design of most RIS units is oriented toward near-passivity. This means they typically do not rely on active power sources for signal processing, leading to markedly lower

Wireless Semantic Communications: Concepts, Principles, and Challenges, First Edition.
Edited by Yao Sun, Lan Zhang, Dusit Niyato, and Muhammad Ali Imran.

energy consumption, as elaborated in Liu et al. [2021]. This aspect of RIS, combined with its relatively lower hardware complexity and cost, positions it favorably against more traditional active antenna systems, such as phased arrays. RIS can thus be envisaged as an economical, adaptive thin composite material—similar to a sheet of wallpaper—capable of being applied to various surfaces like walls, buildings, and ceilings.

Moreover, the ability to individually tailor the behavior of each element within an RIS provides an unprecedented degree of customization in wavefront shaping, as interpreted in Di Renzo et al. [2020]. Such versatility allows for the optimization of the RIS based on specific network requirements and varying environmental conditions, offering a tailored solution to diverse communication challenges.

The advent of RIS-assisted communication systems has opened the door to a plethora of application scenarios, each addressing distinct challenges in modern wireless networks. The scope of these applications is vast, encompassing enhancements in signal coverage and quality [Basharat et al., 2021], significant improvements in spectral and energy efficiency [Huang et al., 2019], as well as amplifying network capacity and diminishing interference [Abrardo et al., 2021]. Focusing on the aspect of energy efficiency, as detailed in Huang et al. [2019], the study develops a realistic RIS power consumption model to formulate the energy efficiency (EE) maximization problem to optimize the RIS phase shifts and the downlink transmit powers under maximum power and minimum quality of service (QoS) constraints as follows:

$$\max_{\mathbf{\Phi},\mathbf{P}} \frac{\sum_{k=1}^{K} \log_2\left(1 + p_k\sigma^{-2}\right)}{\xi \sum_{k=1}^{K} p_k + P_{BS} + KP_{UE} + NP_n(b)}, \tag{10.1a}$$

$$\text{s.t.} \log_2\left(1 + p_k\sigma^{-2}\right) \geq R_{min,k}\ \forall k = 1, 2, \ldots, K, \tag{10.1b}$$

$$\text{tr}((\mathbf{H}_2\mathbf{\Phi}\mathbf{H}_1)^{+}\mathbf{P}(\mathbf{H}_2\mathbf{\Phi}\mathbf{H}_1)^{+H}) \leq P_{max}, \tag{10.1c}$$

$$|\phi_n| = 1\ \forall n = 1, 2, \ldots, N, \tag{10.1d}$$

where $R_{min,k}$ denotes the individual QoS constraint of the kth user. Also, constraint (10.1c) ensures that the base station (BS) transmit power is kept below the maximum feasible threshold P_{max}, while constraint (10.1d) accounts for the fact that each RIS reflecting element can only provide a phase shift, without amplifying the incoming signal.

To tackle this non-convex optimization problem, the authors propose gradient- and sequential fractional programming (SFP)-based approaches. The achievable EE performance is illustrated in Figure 10.1. We can see that the proposed algorithms for the RIS-based system case significantly outperform the derived algorithm for the relay-assisted one. The EE of the RIS-based system is 300% larger than that of the one based on the AF relay when $P_{max} \geq 32$ dBm. This is a direct consequence of the fact that the former system exhibits a much lower energy consumption compared to the latter one.

10.2 Beamforming Design for RISs Enhanced Semantic Communications

10.2.1 System Model

A single-cell network with multiple distributed RISs and one multi-antenna BS serving several single-antenna users is shown in Figure 10.2. The number of RISs and users is denoted by L and

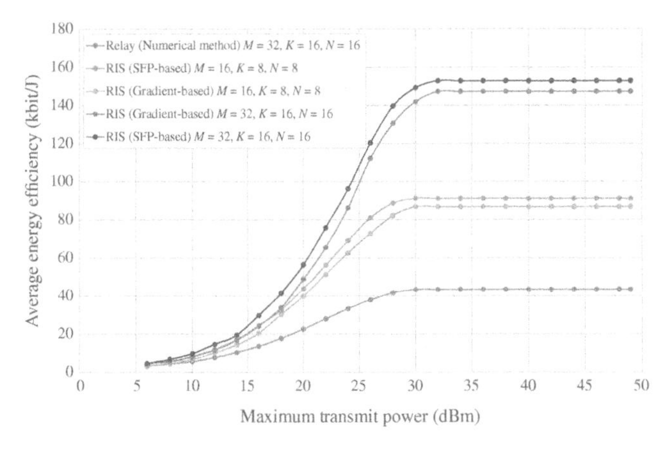

Figure 10.1 Average EE using either RIS or AF relay versus P_{max} for $R_{min} = 0$ bps/Hz and: (a) $M = 32$, $K = 16$, $N = 16$; and (b) $M = 16$, $K = 8$, $N = 8$. Adapted from [Huang et al., 2019].

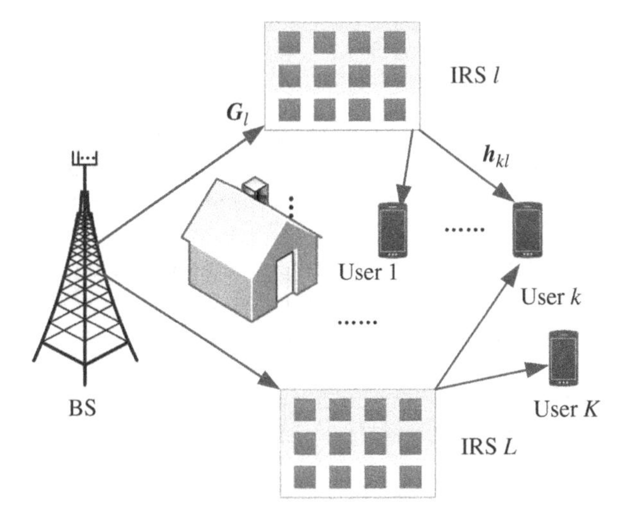

Figure 10.2 A wireless communication system with distributed RISs, one BS, and multiple users.

K, respectively. The number of antennas at the BS is denoted by N. Without loss of generality, the number of elements for each RIS is the same, i.e., N_0. The BS is equipped with uniform linear array (ULA), while RISs utilize uniform planer array (UPA). Then we can get the normalized array response vector of the BS and RISs to infer the channel gain, $\boldsymbol{G}_l$, from the BS to RIS l and $\boldsymbol{h}_{kl}$ between RIS l and user k. What is more, we let $x_{kl} \in \{0, 1\}$ denote the user association between user k and RIS l. Assume that the direct signal between the BS and all users is blocked due to obstacles and high buildings. Considering the signals reflected from multiple RISs, the received signal at user k, similar to Li et al. [2020], is as follows:

$$y_k = \sum_{l=1}^{L} x_{kl} \boldsymbol{h}_{kl}^H \boldsymbol{\Theta}_l \boldsymbol{G}_l \boldsymbol{s} + n_k, \tag{10.2}$$

where $\boldsymbol{\Theta}_l = \text{diag}(e^{j\theta_{l1}}, \dots, e^{j\theta_{lN_0}}) \in \mathbb{C}^{N_0 \times N_0}$ with $\theta_{ln} \in [0, 2\pi]$, and $l \in \mathcal{L} = \{1, \dots, L\}$, and $n \in \mathcal{N}_0 = \{1, \dots, N_0\}$, where $\text{diag}(e^{j\theta_{l1}}, \dots, e^{j\theta_{lN_0}})$ is a diagonal matrix with $e^{j\theta_{l1}}, \dots, e^{j\theta_{lN_0}}$ being its diagonal elements, and $n_k \sim \mathcal{CN}(0, \sigma^2)$ is the additive white Gaussian noise.

Based on (10.2), the achievable rate of user k can be given by

$$r_k = B \log_2 \left(1 + \frac{p_k \left| \sum_{l=1}^{L} x_{kl} \boldsymbol{h}_{kl}^H \boldsymbol{\Theta}_l \boldsymbol{G}_l \boldsymbol{w}_k \right|^2}{\sum_{i=1, i \neq k}^{K} p_i \left| \sum_{l=1}^{L} \boldsymbol{h}_{kl}^H \boldsymbol{\Theta}_l \boldsymbol{G}_l \boldsymbol{w}_i \right|^2 + \sigma^2} \right), \tag{10.3}$$

where B is the bandwidth of the system, p_k is the allocated power of user k, and $\boldsymbol{w}_k$ is the beamforming vector of user k with satisfying $||\boldsymbol{w}_k|| = 1$.

Beamforming is one of the most important technologies in the communication system because it allows for the optimization of wireless communication systems by directing the transmission of signals toward the intended receiver. However, the joint beamforming design of multi-RIS systems is much more complicated due to the existence of a large amount of RIS elements and cooperative communications between different RISs [You et al., 2020]. Combining the advantages of both distributed RISs and semantic communication, there remain works investigating a joint communication and computation framework for distributed RISs-assited semantic communication.

10.2.2 Problem Formulation

The framework integrates semantic communication, which involves joint source and channel coding, wireless resource allocation, and learning resource allocation. This is envisioned to be a key technology for realizing the future metaverse through providing seamless and reliable transmission.

First, the compute-then-transmit protocol is adopted for semantic communication systems, as shown in Figure 10.3, which involves the extraction of semantic information at the transmitter. The algorithm takes into account the computation resources required for this process, ensuring efficient allocation.

Then, the framework formulates the joint communication and computation problem as an optimization problem, with the goal of maximizing the sum rate of the system under total transmit

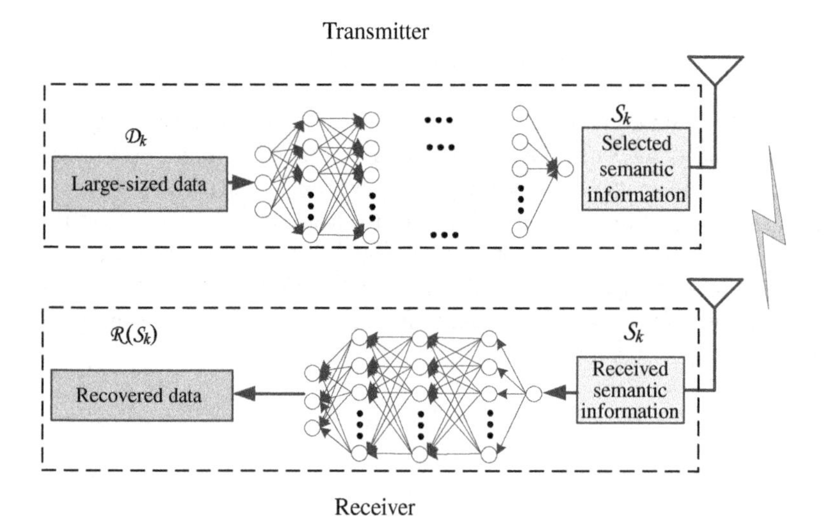

Figure 10.3 The compute-then-transmit protocol for semantic communication systems.

power, phase shift, RIS-user association, and semantic compression ratio constraints as follows:

$$\max_{z} \sum_{k=1}^{K} B\frac{1-a_k z_k^{-b_k}}{z_k} r_k, \tag{10.4a}$$

$$\text{s.t.} \quad \sum_{k=1}^{K} x_{kl} \leq K_0, \quad \forall l, \tag{10.4b}$$

$$\sum_{l=1}^{L} x_{kl} \leq L_0, \quad \forall k, \tag{10.4c}$$

$$\sum_{k=1}^{K} p_k \leq P, \tag{10.4d}$$

$$\|\boldsymbol{w}_k\| = 1, \quad \forall k, \tag{10.4e}$$

$$0 < z_k \leq 1, \quad \forall k, \tag{10.4f}$$

$$p_k \geq 0, x_{kl} \in \{0,1\}, \theta_{ln} \in [0, 2\pi], \quad \forall l, k, n, \tag{10.4g}$$

where $\theta = [\theta_{11}, \dots, \theta_{1N_0}, \dots, \theta_{LN_0}]^T$, $\boldsymbol{w} = [\boldsymbol{w}_1; \dots; \boldsymbol{w}_K]$, $\boldsymbol{x} = [x_{11}, \dots, x_{KL}]^T$, $\boldsymbol{z} = [z_1, \dots, z_K]^T$, K_0 is the maximum number of associated users for each RIS, L_0 is the maximum number of associated RISs for each user, and P is the maximum transmit power of the BS.

10.2.3 Algorithm Design

To solve the formulated optimization problem, the first step is to solve the RIS-user association problem, which involves associating each user with one or more RISs to optimize the communication and computation resources. The chapter proposes a many-to-many matching algorithm to solve this problem, which sorts the channel gains of each user reflected by each RIS in a descending order and obtains a stable matching with low complexity. The next step is to optimize the semantic compression ratio, which involves joint source and channel coding to extract the semantic information. The paper formulates this problem as an optimization problem and obtains the optimal semantic compression ratio in closed form, subject to certain constraints. The final step is to optimize the phase shift of the RISs to improve the signal-to-noise ratio (SNR) and increase the range and capacity of the wireless communication system. A tensor-based beamforming approach is proposed to optimize the beamforming for multiple RISs serving one user, contributing to the overall optimization of the system.

Let us take one RIS l serving multiple users as an example. The set of users associated with RIS l is denoted by $\mathcal{U}_l$. In the time division scheme, the whole RIS l is occupied by all served users simultaneously on the same bandwidth. Mathematically, the joint transmit and passive beamforming optimization problem can be formulated as

$$\max_{\boldsymbol{\theta}_l, \boldsymbol{w}_k} \quad |\boldsymbol{h}_{kl}^H \text{diag}(\boldsymbol{\theta}_l) \boldsymbol{G}_l \boldsymbol{w}_k|, \tag{10.5}$$

$$\text{s.t.} \quad \boldsymbol{h}_{il}^H \boldsymbol{\Theta}_l \boldsymbol{G}_l \boldsymbol{w}_k = 0, \quad \forall i \neq k, i \in \mathcal{U}_l, \tag{10.5a}$$

$$\theta_{ln} \in [0, 2\pi], \quad \forall n = 1, \dots, N_0, \tag{10.5b}$$

$$\|\boldsymbol{w}_k\| = 1, \tag{10.5c}$$

where $\boldsymbol{\theta}_l = [\theta_{l1}, \dots, \theta_{lN_0}]$.

Problem (10.5) can be solved in two steps, i.e., phase shift optimization in the first step and transmit beamforming in the second step. In the first step, we optimize the phase shift $\boldsymbol{\theta}_l$. The objective function in (10.5) can be written as

$$\begin{aligned}
&\boldsymbol{h}_{kl}^H \mathrm{diag}(\boldsymbol{\theta}_l)\boldsymbol{G}_l \\
&\quad = \boldsymbol{a}_I^H(\varphi_{lk}^a, \varphi_{lk}^e)\mathrm{diag}(\boldsymbol{\theta}_l)\boldsymbol{a}_I(\varphi_l^a, \varphi_l^e)\boldsymbol{a}_B^H(\varphi_l) \\
&\quad = \boldsymbol{a}_I^H(\varphi_{lk}^a, \varphi_{lk}^e) \odot \boldsymbol{a}_I^T(\varphi_l^a, \varphi_l^e)\boldsymbol{\theta}_l\boldsymbol{a}_B^H(\varphi_l) \\
&\quad = \boldsymbol{a}^H\left(\frac{2\pi d_I}{\lambda}(\sin(\varphi_{lk}^e)\cos(\varphi_{lk}^a) - \sin(\varphi_l^e)\cos(\varphi_l^a)), M_1\right) \\
&\qquad \otimes \boldsymbol{a}^H\left(\frac{2\pi d_I}{\lambda}(\cos(\varphi_{lk}^e) - \cos(\varphi_l^e)), M_2\right)\boldsymbol{\theta}_l\boldsymbol{a}_B^H(\varphi_l).
\end{aligned} \tag{10.6}$$

Then, the data rate of user k is

$$\frac{1}{|\mathcal{U}_l|}B\log_2\left(1 + \frac{p_k\left|\boldsymbol{h}_{kl}^H\boldsymbol{\Theta}_l\boldsymbol{G}_l\boldsymbol{w}_k\right|^2}{\sigma^2}\right). \tag{10.7}$$

10.2.4 Simulation Results

The work considers two kinds of distributed RISs: distributed RISs scheme 1, with a large distance between every two RISs, and distributed RISs scheme 2, with a small distance between every two RISs. Figures 10.4 and 10.5 show the sum rate versus sum transmit power under two different user distributions. It can be observed that both distributed RIS schemes yield a higher sum rate compared to the centralized RIS. For user distribution 1, it can be found that distributed RISs with a small distance between every two RISs can achieve higher performance than the distributed RISs with a large distance between every two RISs.

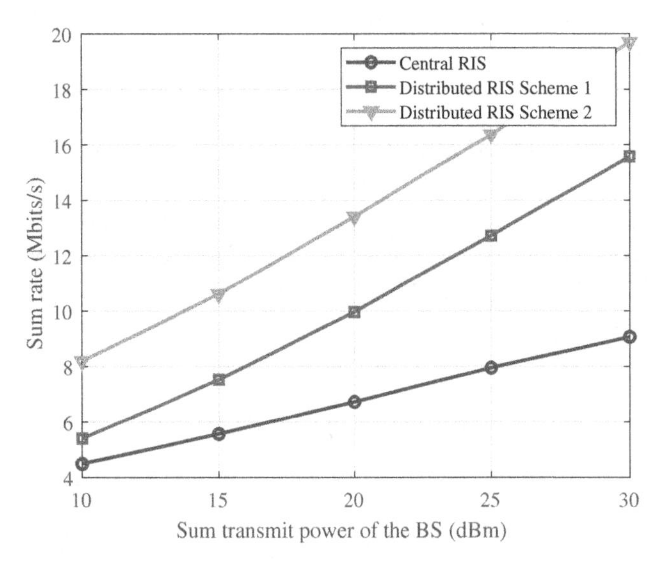

Figure 10.4 Sum rate versus sum transmit power under user distribution 1.

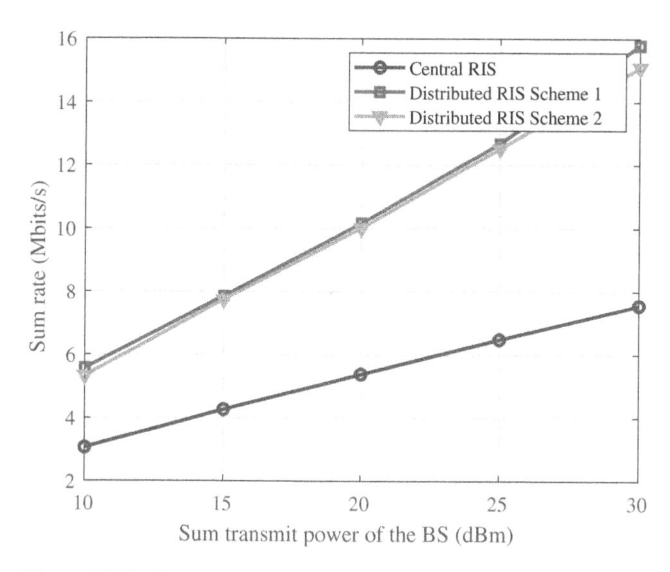

Figure 10.5 Sum rate versus sum transmit power under user distribution 2.

10.3 Privacy Protection in RIS-Assisted Semantic Communication System

The integration of RIS into semantic communication systems enhances signal propagation and efficiency but poses unique privacy challenges due to the potential exposure of semantic information, as explored in Wang et al. [2023], Qin et al. [2023], and Zhao et al. [2022].

Wang et al. [2023] address the problem of vulnerability of semantic information during transmission in RIS-assisted systems. The approach involves using simultaneous transmitting and reflecting (STAR)-RIS to enhance signal transmission between a BS and a user (Bob) while simultaneously creating interference specifically for eavesdroppers. This is achieved by designing the reflection-coefficient vector (RCV) of STAR-RIS to generate either task-level or SNR-level interference at Eve, effectively preventing Eve from inferring task-related information and fooling Eve into incorrect task results. The system aims to separate the desired semantic signal for Bob and the interference for Eve without affecting the communication quality between the BS and Bob. Simulation results demonstrate the effectiveness of this approach in protecting semantic communication privacy. As depicted in Figure 10.6, all methods exhibit stronger interference to Eve as the compressed rate decreases. This observation suggests that when the transmitted information is highly compressed, it becomes more susceptible to attacks. It is also important to note that the task success rate at Bob also decreases with the compressed rate. This is because the compression process eliminates less important content for the task, thereby reducing the redundancy and weakening the capacity for error correction of the transmitted information itself.

Zhao et al. [2022] improve secret key generation for enhancing privacy by introducing a physical-layer key generation (PKG) scheme named "semantic secret keys" (SKey). Figure 10.7 illustrates how to generate the semantic keys at the transmitter for the ith data. The proposed physical-layer semantic encryption scheme consists of two key steps: semantic encryption and semantic obfuscation. In the semantic encryption step, SKeys are generated to encrypt semantic data by computing bilingual evaluation understudy (BLEU) scores and the corresponding weights in 1-gram, 2-gram, 3-gram, and 4-gram settings. The SKey is then used to encrypt

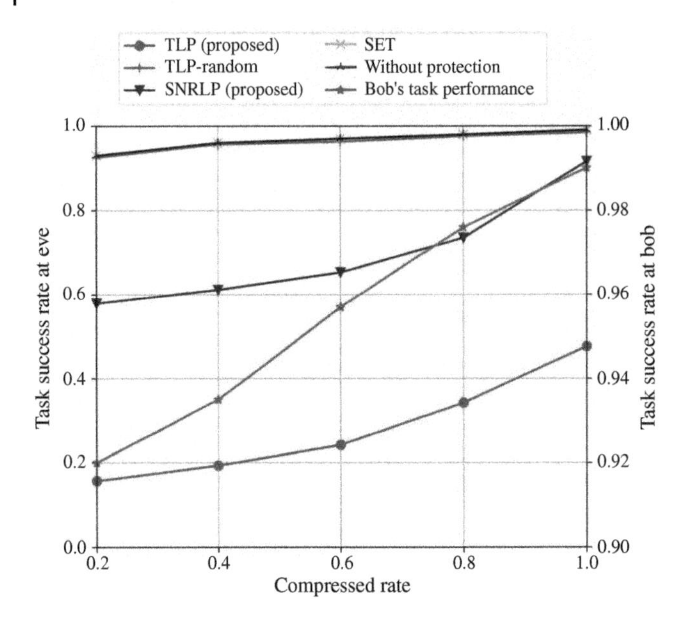

Figure 10.6 Task success rate versus compressed rate. Source: [Wang et al., 2023]/arXivLabs/CC-BY 4.0.

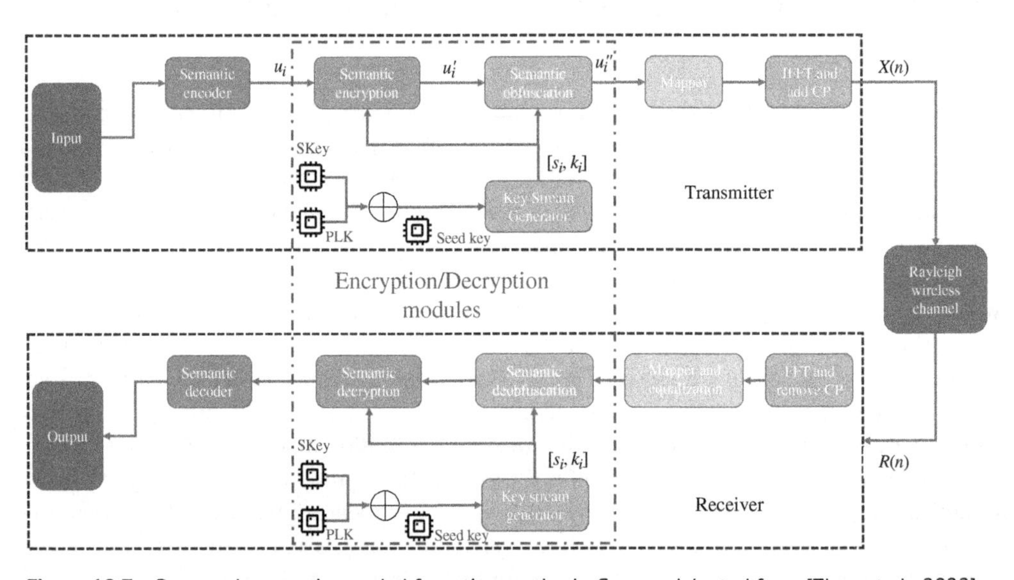

Figure 10.7 Proposed encryption and obfuscation methods. Source: Adapted from [Zhao et al., 2022].

the semantic representation of the data. In the semantic obfuscation step, a subcarrier-level obfuscation technique is introduced to further secure semantic communications by inserting dummy data between the encrypted data. The obfuscated data are then transmitted using orthogonal frequency-division multiplexing (OFDM) to obtain a complex data point $X(n)$, which is sent to the wireless channel. The received signal $R(n)$ is then demodulated, semantically deobfuscated, decrypted, and interpreted by the receiver. This scheme enhances the security of deep learning-based semantic communications by exploiting the randomness inherent in semantic drifts and RIS-assisted channels. SKey extracts random features from the semantic communication system to form a randomly varying switch sequence for the RIS-assisted channel.

This approach significantly improves the secret key generation rate and ensures better security for semantic communication systems.

10.4 AI for RIS-Assisted Semantic Communications

Thanks to the advancements in artificial intelligence (AI), the field of semantic communication has gained significant attention recently [Hu et al., 2022; Jiang et al., 2023; Shi et al., 2023; Weng and Qin, 2021; Xie et al., 2021]. Inspired by recent advancements in RIS-based on-the-air DNN, there remains some work exploring the potential of enabling semantic communication on RISs [Chen et al., 2023; Du et al., 2023b].

10.4.1 RIS-Based On-the-Air DDNN Semantic Communications

Chen et al. [2023] aim to motivate a paradigm shift from the mainstream research on hardware-based semantic communications toward RIS-based on-the-air diffractional deep neural network (DDNN) semantic communications. We can see in Figure 10.8 that semantic processing and communication are typically handled through dedicated hardware devices in the conventional hardware-based semantic communications. These systems focus on physical-layer optimizations but are limited in their ability to dynamically adapt to changing communication environments or to understand and prioritize data based on its semantic content. However, in the proposed RIS-based on-the-air semantic communications, by incorporating AI, particularly the proposed DDNN approach, RIS-based systems can not only optimize signal transmission at the physical

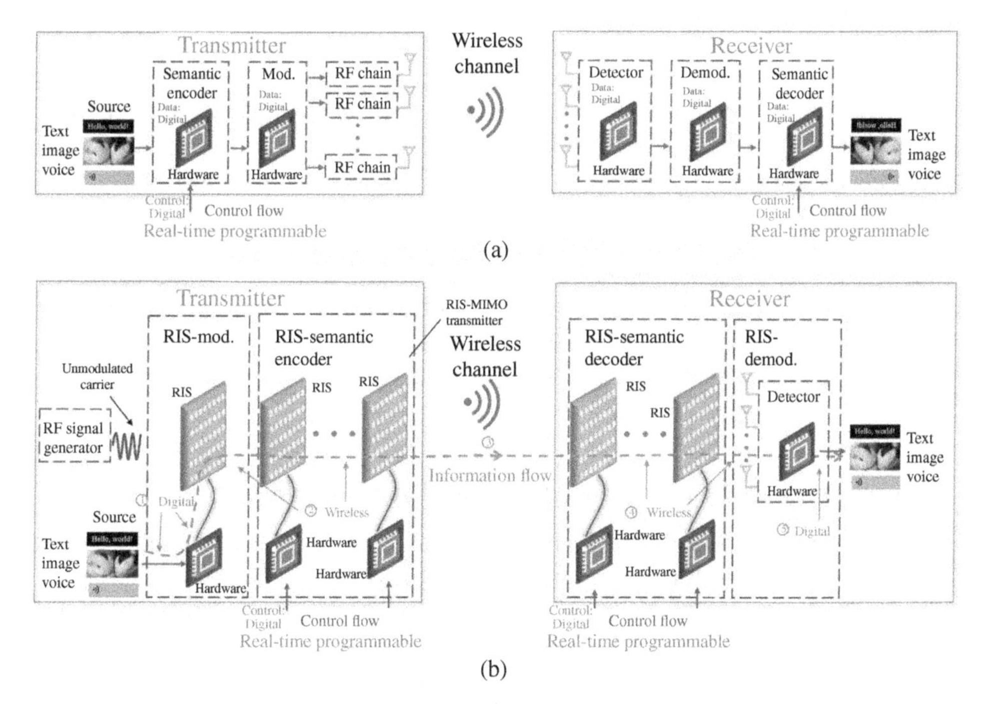

Figure 10.8 Comparison between traditional semantic communications and RIS-based on-the-air semantic communications. (a) Conventional hardware-based semantic communications and (b) RIS-based on-the-air semantic communications. Source: [Chen et al., 2023]/arXivLabs/CC-BY 4.0.

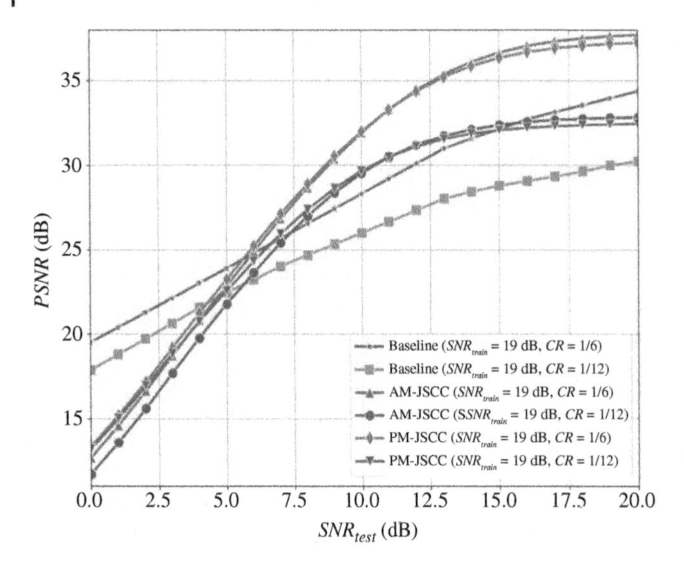

Figure 10.9 PSNR of the proposed scheme on CIFAR-10 test images with respect to SNR over an additive white gaussian noise (AWGN) channel. Source: [Chen et al., 2023]/arXivLabs/CC-BY 4.0.

level but also prioritize data based on its semantic significance. This leads to more intelligent and context-aware communication, adapting to the environment and the meaning of the transmitted data.

The authors compare the peak signal-to-noise ratio (PSNR) performances of the proposed scheme and the baseline scheme, as shown in Figure 10.9. It is evident that in most cases, the proposed AM-JSCC and PM-JSCC schemes outperform the baseline scheme. The primary reason is that the parameters in the proposed neural network are in the form of complex values. Despite requiring approximately four times as many floating-point operations, complex-valued convolutional neural networks (CvCNN) surpasses CNN in terms of image reconstruction. Moreover, since the proposed scheme operates directly on the wireless signal, there is no additional computation cost associated with CvCNN. As a result, the proposed scheme achieves the same performance improvement as conventional CvCNN without incurring any extra computational overhead. The baseline scheme demonstrates improved robustness only when the channel conditions are very poor, such as when the signal-to-noise ratio (SNR) is below 3 dB. Similar performance can be observed for the amplitude-modulated joint source-channel coding (AM-JSCC) and phase-modulated JSCC (PM-JSCC) schemes, as they only differ in the modulation of the input signal.

10.4.2 Encoding and Decoding of RIS-aided Semantic Communications

Based on the challenges of high data volume and limited bandwidth in wireless communications, Du et al. [2023b] present a novel design of RIS hardware for encoding several signal spectrums into one MetaSpectrum. The authors use a semantic hash sampling method for efficient encoding and a self-supervised learning method for decoding MetaSpectrums.

As shown in Figure 10.10, the "inverse" means that the processing of source messages is no longer to extract semantic information but to combine multiple source messages into one hyper-source message for transmission or storage. Subsequently, through decoding, the semantic information of

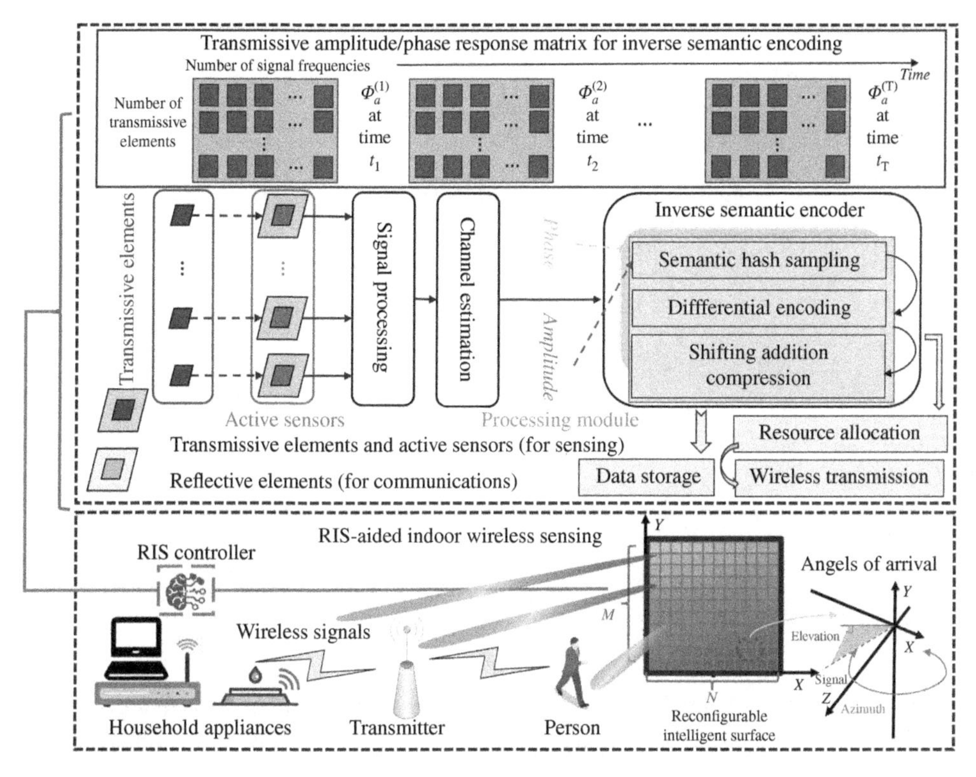

Figure 10.10 The framework of the proposed inverse semantic-aware wireless sensing system. Source: Adapted from [Du et al., 2023b].

the hyper-source message, i.e., source messages, can be obtained to support multiple different tasks. The key focus is on reducing data volume while maintaining the effectiveness of wireless sensing tasks. This is achieved through an innovative RIS-aided encoding method and a self-supervised decoding technique.

Unlike most RIS research works that consider only the phase response matrix of RIS, the work considers L-shaped RIS to make full use of the amplitude response matrix to help the system design for wireless sensing. The overall channel frequency response (CFR) obtained by the L-shaped transmissive element array on the RIS can be expressed as a CFR matrix as follows:

$$\boldsymbol{H}_{xoy}^{(t)} = \left[\boldsymbol{H}_x^{(t)}\ \boldsymbol{H}_0^{(t)}\ \boldsymbol{H}_y^{(t)}\right] = \left[\underbrace{\begin{matrix} h_{f_1}^{[1,0]} & \cdots & h_{f_1}^{[M,0]} \\ \vdots & \ddots & \vdots \\ h_{f_K}^{[1,0]} & \cdots & h_{f_K}^{[M,0]} \end{matrix}}_{\boldsymbol{H}_x} \ \underbrace{\begin{matrix} h_{f_1}^{[0,0]} \\ \vdots \\ h_{f_K}^{[0,0]} \end{matrix}}_{\boldsymbol{H}_0} \ \underbrace{\begin{matrix} h_{f_1}^{[0,1]} & \cdots & h_{f_1}^{[0,N]} \\ \vdots & \ddots & \vdots \\ h_{f_K}^{[0,1]} & \cdots & h_{f_K}^{[0,N]} \end{matrix}}_{\boldsymbol{H}_y}\right].$$

Then, one can observe that every element in the CFR matrix is a complex number, which denotes the amplitude and phase of the CFR. Taking $H_x^{(t)}$ as an example, it can be further decomposed into the amplitude and phase spectrums as

$$\boldsymbol{H}_X^{(t)} \to \left\{\boldsymbol{H}_{X_a}^{(t)}, \boldsymbol{H}_{X_p}^{(t)}\right\} = \left\{\underbrace{\begin{bmatrix} \left\|h_{f_1}^{[1,0]}\right\| & \cdots & \left\|h_{f_1}^{[M,0]}\right\| \\ \vdots & \ddots & \vdots \\ \left\|h_{f_K}^{[1,0]}\right\| & \cdots & \left\|h_{f_K}^{[M,0]}\right\| \end{bmatrix}}_{amplitude\ matrix}, \underbrace{\begin{bmatrix} \angle h_{f_1}^{[1,0]} & \cdots & \angle h_{f_1}^{[M,0]} \\ \vdots & \ddots & \vdots \\ \angle h_{f_K}^{[0,1]} & \cdots & \angle h_{f_K}^{[M,0]} \end{bmatrix}}_{phase\ matrix}\right\},$$

Additionally, zero compensation processing is performed on the amplitude spectrums X_A and phase spectrums X_P. Then, the MetaSpectrums of amplitude Z_A and phase Z_P are derived as the encoding results as follows:

$$Z_A = \sum_{i=1}^{T} X_A\{i\}, \tag{10.8}$$

$$Z_P = \sum_{i=1}^{T} X_P\{i\}. \tag{10.9}$$

For the inverse semantic-aware decoding, the objective is to decode x_A and x_P from z_A and z_P, which are the vectorized Z_A and Z_P, respectively. The decoding objective function can be formulated as

$$\min_{\boldsymbol{x}_A, \boldsymbol{x}_P} \alpha_1 \left\|\boldsymbol{z}_A - \boldsymbol{\Phi}\boldsymbol{x}_A\right\|^2 + \alpha_2 \left\|\boldsymbol{z}_P - \boldsymbol{\Phi}\boldsymbol{x}_P\right\|^2, \tag{10.10}$$

Figure 10.11 depicts the average PSNR values of 10 experiments versus the number of outer loop decoding iterations, with or without the codebook $\phi_A^{(i)}$. If the codebook is available, the PSNR values of both the amplitude and phase spectrums are increasing as the number of iterations increases and gradually reach a plateau after about 10 iterations. Although minor fluctuations occur at higher iteration steps because of the dynamic nature of the decoding process, the overall trend demonstrates the effectiveness of the self-supervised decoding scheme in recovering the original signal spectrums. However, if no codebook is available or the codebook is wrong, the PSNR values decrease as

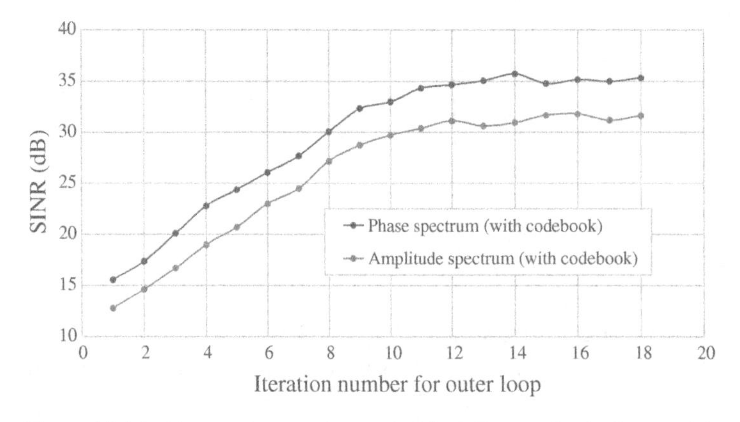

Figure 10.11 Proposed encryption and obfuscation methods. Source: Adapted from [Du et al., 2023b].

the number of iterations increases. The reason is that the parameters of the two decoding network, i.e., θ_A and θ_P, are learned according to a wrong objective function.

10.5 Conclusion

RIS technology is emerging as a cornerstone in the evolution of 5.5G and 6G communication systems, primarily for its unparalleled proficiency in precisely controlling electromagnetic wave propagation. Its application in enhancing signal propagation and manipulation within the realm of semantic communication represents a cutting-edge and innovative development. This chapter delves into the forefront of advancements in RIS for semantic communication, focusing on three critical domains: First, it explores the optimization of beamforming in RIS-aided systems, aiming to enhance communications in intricate digital environments such as the metaverse. Second, it investigates the application of physical-layer strategies in RIS systems to bolster privacy protection within semantic communication frameworks. Finally, it examines the integration of deep learning techniques for the advanced interpreting, prioritizing, encoding, and decoding of semantic transmission in wireless communications. These thematic areas collectively underscore the transformative potential of RIS in reshaping communication paradigms. They highlight the emphasis on efficiency, security, and intelligent data processing, paving the way for groundbreaking advancements in semantic communication through the use of RIS technology.

The integration of RIS into semantic communication systems presents several challenges. First, implementing RIS in a semantic communication framework requires advanced techniques to control these surfaces for encoding semantic information. This involves complex signal processing and wavefront shaping, which can be challenging to implement efficiently. Furthermore, semantic communication involves understanding and processing the meaning or intent of the transmitted data, which is fundamentally different from traditional data transmission methods. Integrating RIS, which primarily deals with physical-layer modifications, with the higher-level semantic processing layers adds complexity to the system design. Another critical challenge lies in ensuring the energy efficiency and scalability of RIS-enabled semantic communication systems. As the system complexity increases with the integration of RIS and semantic processing, maintaining energy efficiency becomes more difficult, especially in large-scale deployments. Moreover, the real-world performance of RIS in semantic communication is susceptible to a variety of environmental factors, including physical obstacles, weather conditions, and user mobility. Ensuring consistent and reliable performance under these varying conditions presents a formidable challenge for RIS-aided semantic communication systems, necessitating robust and adaptable solutions.

Acronyms

RIS	reconfigurable intelligent surface
BS	base station
ULA	uniform linear array
UPA	uniform planer array
NOMA	non-orthogonal multiple access
ZF	zero forcing
STAR	simultaneous transmitting and reflecting
OFDM	orthogonal frequency-division multiplexing

SNR	signal-to-noise ratio
PSNR	peak signal-to-noise ratio
QoS	quality of service
AI	artificial intelligence
DDNN	diffractional deep neural network
CvCNN	complex-valued convolutional neural networks

References

A. Abrardo, D. Dardari, M. Di Renzo, and X. Qian. MIMO interference channels assisted by reconfigurable intelligent surfaces: Mutual coupling aware sum-rate optimization based on a mutual impedance channel model. *IEEE Wireless Communications Letters*, 10(12):2624–2628, Aug. 2021.

S. Basharat, S. A. Hassan, H. Pervaiz, A. Mahmood, Z. Ding, and M. Gidlund. Reconfigurable intelligent surfaces: Potentials, applications, and challenges for 6G wireless networks. *IEEE Wireless Communications Letters*, 28(6):184–191, Sept. 2021.

S. Chen, Y. Hui, Y. Qin, Y. Yuan, W. Meng, X. Luo, and H. Chen. RIS-based on-the-air semantic communications–a diffractional deep neural network approach. *arXiv preprint arXiv:2312.00535*, Dec. 2023.

T. Cui, M. Qi, X. Wan, J. Zhao, and Q. Cheng. Coding metamaterials, digital metamaterials and programmable metamaterials. *Light: Science & Applications*, 3(10):e218, Oct. 2014.

M. Di Renzo, A. Zappone, M. Debbah, M.-S. Alouini, C. Yuen, J. de Rosny, and S. Tretyakov. Smart radio environments empowered by reconfigurable intelligent surfaces: How it works, state of research, and the road ahead. *IEEE Journal on Selected Areas in Communications*, 38(11):2450–2525, Jul. 2020. doi: 10.1109/JSAC.2020.3007211.

H. Du, J. Wang, D. Niyato, J. Kang, Z. Xiong, J. Zhang, and X. Shen. Lightweight wireless sensing through RIS and inverse semantic communications. In *Proceedings of Wireless Communications and Networking Conference (WCNC)*. IEEE, Mar. 2023a.

H. Du, J. Wang, D. Niyato, J. Kang, Z. Xiong, J. Zhang, and X. Shen. Semantic communications for wireless sensing: RIS-aided encoding and self-supervised decoding. *IEEE Journal on Selected Areas in Communications*, 41(8):2547–2562, Jun. 2023b.

N. Engheta and R. Ziolkowski. *Metamaterials: Physics and Engineering Explorations*. John Wiley & Sons, 2006.

J. Hu, H. Zhang, K. Bian, Z. Han, H. Vincent Poor, and L. Song. MetaSketch: Wireless semantic segmentation by reconfigurable intelligent surfaces. *IEEE Transactions on Wireless Communications*, 21(8):5916–5929, Jan. 2022.

C. Huang, A. Zappone, G. C. Alexandropoulos, M. Debbah, and C. Yuen. Reconfigurable intelligent surfaces for energy efficiency in wireless communication. *IEEE Transactions on Wireless Communications*, 18(8):4157–4170, Aug. 2019.

P. Jiang, C. Wen, S. Jin, and G. Li. RIS-enhanced semantic communications adaptive to user requirements. *arXiv preprint arXiv:2307.16100*, Aug. 2023.

Z. Li, M. Hua, Q. Wang, and Q. Song. Weighted sum-rate maximization for multi-IRS aided cooperative transmission. *IEEE Wireless Communications Letters*, 9(10):1620–1624, Jun. 2020.

Y. Liu, X. Liu, X. Mu, T. Hou, J. Xu, M. Di Renzo, and N. Al-Dhahir. Reconfigurable intelligent surfaces: Principles and opportunities. *IEEE Communications Surveys & Tutorials*, 23(3):1546–1577, May 2021.

Q. Qin, Y. Rong, G. Nan, S. Wu, X. Zhang, Q. Cui, and X. Tao. Securing semantic communications with physical-layer semantic encryption and obfuscation. *arXiv preprint arXiv:2304.10147*, Apr. 2023.

J. Shi, T. Chan, H. Pan, and T. Lok. Reconfigurable intelligent surface assisted semantic communication systems. *arXiv preprint arXiv:2306.09650*, Jun. 2023.

Y. Wang, W. Yang, P. Guan, Y. Zhao, and Z. Xiong. Star-RIS-assisted privacy protection in semantic communication system. *arXiv preprint arXiv:2306.12675*, Jun. 2023.

Z. Weng and Z. Qin. Semantic communication systems for speech transmission. *IEEE Journal on Selected Areas in Communications*, 39(8):2434–2444, Jun. 2021.

H. Xie, Z. Qin, G. Li, and B. Juang. Deep learning enabled semantic communication systems. *IEEE Transactions on Signal Processing*, 69:2663–2675, Apr. 2021.

C. You, B. Zheng, and R. Zhang. Wireless communication via double IRS: Channel estimation and passive beamforming designs. *IEEE Wireless Communications Letters*, 10(2):431–435, Oct. 2020.

R. Zhao, Q. Qin, N. Xu, G. Nan, Q. Cui, and X. Tao. SemKey: Boosting secret key generation for RIS-assisted semantic communication systems. In *Proceedings of Vehicular Technology Conference (VTC2022-Fall)*. IEEE, Sept. 2022.

Index

Wireless Semantic Communications: Concepts, Principles, and Challenges, First Edition.
Edited by Yao Sun, Lan Zhang, Dusit Niyato, and Muhammad Ali Imran.